AF441798

Handbook of Water Purity and Quality

Handbook of Water Purity and Quality

Satinder Ahuja
1061 Rutledge Court,
Calabash, NC, USA

AMSTERDAM • BOSTON • HEIDELBERG • LONDON
NEW YORK • OXFORD • PARIS • SAN DIEGO
SAN FRANCISCO • SINGAPORE • SYDNEY • TOKYO

Academic Press is an imprint of Elsevier

Academic Press is an imprint of Elsevier
360 Park Avenue South, New York, NY 10010–1710
30 Corporate Drive, Suite 400, Burlington, MA 01803, USA
525 B Street, Suite 1900, San Diego, CA 92101-4495, USA
32 Jamestown Road, London NW1 7BY, UK
Radarweg 29, PO Box 211, 1000 AE Amsterdam, The Netherlands

First edition 2009

Copyright © 2009 Elsevier Inc. All rights reserved.

No part of this publication may be reproduced, stored in a retrieval system or transmitted in any
form or by any means electronic, mechanical, photocopying, recording or otherwise without the
prior written permission of the publisher

Permissions may be sought directly from Elsevier's Science & Technology Rights
Department in Oxford, UK: phone (+44) (0) 1865 843830; fax (+44) (0) 1865 853333;
email: permissions@elsevier.com. Alternatively you can submit your request online by visiting
the Elsevier web site at http://elsevier.com/locate/permissions, and selecting Obtaining permission
to use Elsevier material

Notice
No responsibility is assumed by the publisher for any injury and/or damage to persons or property
as a matter of products liability, negligence or otherwise, or from any use or operation of any methods,
products, instructions or ideas contained in the material herein. Because of rapid advances in the
medical sciences, in particular, independent verification of diagnoses and drug dosages should be made

Library of Congress Cataloging-in-Publication Data
Handbook of water purity and quality / edited by Satinder Ahuja. – 1st ed.
 p. cm.
 Includes bibliographical references and index.
 ISBN 978-0-1237-4192-9 (alk. paper)
 1. Water–Pollution–Handbooks, manuals, etc.. 2. Water quality management–Handbooks,
manuals, etc. 3. Water quality–Measurement–Handbooks, manuals, etc.. 4. Water–Purification–
Handbooks, manuals, etc. I. Ahuja, Satinder, 1933-
 TD420.H36 2009
 628.1'61–dc22

2009015898

British Library Cataloguing in Publication Data
A catalogue record for this book is available from the British Library

ISBN: 978-0-12-374192-9

For information on all Academic Press publications
visit our web site at elsevierdirect.com

Printed and bound in Great Britain

09 10 10 9 8 7 6 5 4 3 2 1

Working together to grow
libraries in developing countries

www.elsevier.com | www.bookaid.org | www.sabre.org

ELSEVIER BOOK AID
International Sabre Foundation

Contents

3. Water Quality Issues in Eastern Africa
Shem O. Wandiga and Vincent O. Madadi

4. Effect of Human Land Development on Water Quality
Michael A. Mallin

8. Microbiological Threats to Water Quality
Lawrence B. Cahoon and Bongkeun Song

9. Monitoring Inorganic Compounds
Pamela Heckel and Tracy Dombek

10. Radionuclides in Surface Water and Groundwater
Kate M. Campbell

11. Volatile and Semivolatile Contaminants
Satinder Ahuja

12. Monitoring Disinfectants
Taha F. Marhaba

13. The Evolution of Analytical Technology and Its Impact on Water-Quality Studies for Selected Herbicides and Their Degradation Products in Water
Michael T. Meyer and Elisabeth A. Scribner

14. Monitoring of Pharmaceutical Residues in Sewage Effluents

Zulin Zhang, Darren P. Grover, and John L. Zhou

15. Monitoring for Terrorist-Related Contamination

Dan Kroll

16. Groundwater Arsenic Removal Technologies Based on Sorbents: Field Applications and Sustainability

Sad Ahamed, Abul Hussam, and Abul K.M. Munir

Water is the second most essential material for human survival. Life as we know it would not be possible without water. However, availability of pure water for human use is a major issue worldwide. The sad fact is that pollution of fresh water (drinking water) is a problem for about half of the world's population The United Nations estimates that 2.7 billion people will face a water shortage by 2025. It is well known that Earth is composed largely of water; however, fresh water comprises only 3% of the total water available to us. Of that, only 0.06% is easily accessible for human use. Over 80 countries in the world suffer from a water deficit, and an estimated 1.2 billion people drink unclean water. Each year there are about 250 million cases of water-related diseases, which result in roughly 5 million to 10 million deaths. A significant number of these deaths are caused by the ingestion of water contaminated with pathogenic bacteria, viruses, or parasites responsible for cholera, typhoid, schistosomiasis, dysentery, and other diarrheal diseases.

Water-related problems affect not only the less developed countries; the problems plague most advanced countries in the world as well, including the United States—which is facing a water crisis. Most experts agree that the U.S. water policy is in chaos. Decision making about allocation, infrastructure, repair, and pollution is spread across hundreds of federal, state, and local agencies. Over 700 different chemicals have been found in U.S. drinking water when it comes out of the tap. The U.S. Environmental Protection Agency (EPA) classifies 129 of these chemicals as being particularly dangerous, and it has set standards for approximately 90 contaminants in drinking water. An EPA report in 1996 noted that about one in ten community tap water systems (serving about one-seventh of the U.S. population) violated EPA's tap water treatment or contaminant standards, and 28% of tap water systems violated significant water-monitoring or reporting requirements. It has been estimated that 80 million to 100 million Americans drink tap water that contains very significant trihalomethane levels (over 40 ppb). Drinking bottled water does not assure the absence of contaminants.

This problem of water quality and purity becomes even more significant when you consider that our water is being constantly contaminated from the pollutants we add to the atmosphere from industries, fossil fuel used for automobiles and airplanes, and various other means that are recycled in the form of rain and snow that enter our water supply. In addition, improper wastewater disposal can lead to additional contamination. Furthermore, fertilizers, insecticides, pesticides, and herbicides that are used on our crops, golf courses, and lawns find their way into our water supplies. Improper disposal of hospital waste can lead to addition of disinfectants and other pharmaceuticals in water. The metabolites or unabsorbed pharmaceuticals find their way into water through urine and excreta.

We have known for some time now that water that we call potable may contain many trace and ultratrace contaminants. Despite our best attempts to purify water for drinking, it should be recognized that trace or ultratrace amounts (at or below parts-per-billion level) of about every substance present in untreated water is likely to be found in drinking water even with well-thought-out purification and reprocessing systems. To monitor contaminants in water, it is necessary to perform analyses at ultratrace levels. This was highlighted by me in the 1978 Metrochem meeting in a paper entitled "In Search of Femtogram": a femtogram is 10^{-15}g or 1 part per quadrillion. It was pointed out that it was essential to analyze very low quantities of various contaminants to fully understand the impact of various chemicals on our body. For example, it has been reported that dioxin (2,3,7,8-tetrachloro-dibenzodioxin) can cause abortion in monkeys at the 200 ppt (parts-per-trillion) level, and PCBs at 0.43 ppb level can weaken the backbones of trout.

This book provides a rich source of methods for analyzing water to assure its safety from natural and deliberate contaminants including those that are added because of carelessness of human endeavors. The first four chapters provide an overview of the subject and discuss major water-related issues in developing and developed countries. It should be noted that human development has great impact on water quality. Chapters 5–7 cover issues of sampling for water analysis, regulatory considerations, and forensics in water quality and purity investigations. Chapters 8–14 cover microbial as well as chemical contaminations from inorganic compounds, radionuclides, volatile and semivolatile compounds, disinfectants, herbicides, and pharmaceuticals including endocrine disruptors. Potential terrorist-related contamination and how it should be monitored is covered in Chapter 15.

The luxury of municipal water purification for human consumption is not available to a very large number of people in the world. This necessitates water purification by other means, as is exemplified by the horrendous problem of arsenic-contaminated groundwater. Chapter 2 delineates this worldwide problem that affects even advanced countries. To help solve the problem, a million-dollar Grainger prize was offered to an inventor for designing a simple filter that would economically remediate the arsenic contamination of water. Chapter 16 describes such a filter that can remove arsenic sufficiently to protect the health of a large number of people.

I would like to thank all the authors for their valuable contributions, which I am sure will be of great help to scientists, administrators, managers, and policy makers who are involved with water-related issues.

Satinder (Sut) Ahuja

Contributors

Numbers in parentheses indicate the pages on which the author's contributions begin.

Sad Ahamed (**381**), Center for Clean Water and Sustainable Technologies, Department of Chemistry and Biochemistry, George Mason University, Fairfax, VA 22030, USA

Satinder Ahuja (**1, 17, 237**), 1061 Rutledge Court, Calabash, NC, USA

Lawrence B. Cahoon (**129, 179**), Department of Biology and Marine Biology and Center for Marine Science, University of North Carolina Wilmington, Wilmington, NC 28403, USA

Kate M. Campbell (**213**), U.S. Geological Survey, 345 Middlefield Rd, MS 495, Menlo Park, CA 94025, USA

Dipankar Chakraborti (**93**), School of Environmental Studies, Jadavpur University, Kolkata 700 032, India

Robert H. Cutting (**129**), Department of Environmental Studies, University of North Carolina Wilmington, Wilmington, NC 28403, USA

Bhaskar Das (**93**), School of Environmental Studies, Jadavpur University, Kolkata 700 032, India

Tracy Dombek (**195**), NADP Program Office, Illinois State Water Survey, 2204 Griffith Drive, Champaign, IL 61820-7495, USA

Darren P. Grover (**317**), Department of Biology and Environmental Science, University of Sussex, Falmer, Brighton BN1 9QG, UK

Roy C. Haught (**145**), U.S. Environmental Protection Agency, National Risk Management Research Laboratory. Water Supply and Water Resources Division, Cincinnati, OH, USA

Pamela Heckel (**195**), College of Medicine, Department of Environmental Health, University of Cincinnati, Cincinnati, OH 45267-0056, USA

Abul Hussam (**381**), Center for Clean Water and Sustainable Technologies, Department of Chemistry and Biochemistry, George Mason University, Fairfax, VA 22030, USA

Dan Kroll (**345**), Hach Homeland Security Technologies, 5600 Lindbergh Drive, Loveland, CO 80539, USA

Vincent O. Madadi (**39**), Department of Chemistry, School of Physical Sciences, College of Biological and Physical Sciences, University of Nairobi, Nairobi, Kenya

Michael A. Mallin (**65**), Center for Marine Science, University of North Carolina Wilmington, Wilmington, NC 28409, USA

Taha F. Marhaba (**259**), Department of Civil and Environmental Engineering, New Jersey Institute of Technology, 323 MLK Blvd., Newark, NJ 07102, USA

Michael T. Meyer (**291**), US Geological Survey, Lawrence, KS, USA

Abul K.M. Munir (**381**), Manob Sakti Unnayan Kendro (MSUK), Kushtia, Bangladesh

Craig L. Patterson (**145**), U.S. Environmental Protection Agency, National Risk Management Research Laboratory. Water Supply and Water Resources Division, Cincinnati, OH, USA

Mohammad M. Rahman (93), School of Environmental Studies, Jadavpur University, Kolkata 700 032, India

Elisabeth A. Scribner (291), US Geological Survey, Lawrence, KS, USA

Bongkeun Song (179), Department of Biology and Marine Biology and Center for Marine Science, University of North Carolina Wilmington, Wilmington, NC 28403, USA

Shem O. Wandiga (39), Department of Chemistry, School of Physical Sciences, College of Biological and Physical Sciences, University of Nairobi, Nairobi, Kenya

Zulin Zhang (317), The Macaulay Institute, Craigiebuckler, Aberdeen AB15 8QH, UK

John L. Zhou (317), Department of Biology and Environmental Science, University of Sussex, Falmer, Brighton BN1 9QG, UK

Chapter 1

Overview

Satinder Ahuja

Ahuja Academy of Water Quality, UNCW, Calabash, NC, USA

INTRODUCTION

Water is the most essential material for human survival, after air. Without water, life as we know it would not be possible (Ahuja, 1980, 1986). Fortunately, air is purified adequately by nature with a minimum of help from us. However, today this is not the case with water. Our civilization has managed to pollute our surface water and even groundwater; this necessitates purification for drinking (see Chapter 2). The expression "clean as freshly driven snow" or "pure rainwater" is not true any more. In the past, rain was nature's way of providing freshwater; now, however, rain is usually contaminated with various pollutants that we put in the atmosphere. Here are some important facts about the availability, quality, and purity of our water supplies:

- Even though Earth is composed largely of water, freshwater comprises only 3% of the total water available to us. Of that, only 0.06% is easily accessible.

Handbook of Water Purity and Quality
Copyright © 2009 by Elsevier Inc. All rights of reproduction in any form reserved.

- Over 80 countries in the world suffer from a water deficit.
- Today an estimated 1.2 billion people drink unclean water.
- The United Nations estimates that 2.7 billion people will face a water shortage by 2025.
- Water-related diseases kill 5 million –10 million people, mostly children, around the world.

Even an advanced country like the United States is facing a water crisis. Most experts agree that the U.S. water policy is in chaos. Decision making about allocation, infrastructure, repair, and pollution is spread across hundreds of federal, state, and local agencies. Over 700 different chemicals have been found in U.S. drinking water when it comes out of the tap! The United States Environmental Protection Agency (EPA) classifies 129 of these chemicals as being particularly dangerous.

The sad fact is that pollution of freshwater (drinking water) is a problem for about half of the world's population. Each year there are about 250 million cases of water-related diseases, with roughly 5 million –10 million deaths. Diseases caused by the ingestion of water contaminated with pathogenic bacteria, viruses, or parasites include:

- Cholera
- Typhoid
- Schistosomiasis
- Dysentery and other diarrheal diseases

Drinking water, including bottled water, may reasonably be expected to contain at least small amounts of some contaminants, but the presence of contaminants does not necessarily indicate that water poses a health risk. The EPA sets standards for approximately 90 contaminants in drinking water. Those standards, along with each contaminant's likely source and its health effects, are available at www.epa.gov/safewater/mcl.html. An EPA report in 1996 noted that about one in ten community tap water systems (serving about one-seventh of the U.S. population) violated EPA's tap water treatment or contaminant standards, and 28% of tap water systems violated significant water-monitoring or -reporting requirements. In addition, the tap water of more than 32 million Americans exceeds 2 parts per billion (ppb) arsenic (the California Proposition 65 warning level, applicable to bottled water, is 5 ppb), and 80 million –100 million Americans drink tap water that contains very significant trihalomethane levels (over 40 ppb). Thus, while much tap water is supplied by systems that violate EPA standards or that serve water containing substantial levels of risky contaminants, the majority of the country's tap water apparently passes EPA standards. Therefore, while much tap water is indeed risky there is no assurance that bottled water is any safer than tap water. Laboratories tested most waters for about half of the drinking water contaminants regulated by FDA (to control costs). They found the following:

- Nearly one in four of the waters tested (23 of the 103 waters, or 22%) violated strict applicable state (California) limits for bottled water in at least one sample, most commonly for arsenic or certain cancer-causing man-made ("synthetic") organic compounds. Another three waters sold outside of California (3% of the national total) violated industry-recommended standards for synthetic organic compounds in at least one sample, but unlike in California, those industry standards were not enforceable in the states (Florida and Texas) in which they were sold.
- Nearly one in five tested waters (18 of the 103, or 17%) contained, in at least one sample, more bacteria than allowed under microbiological-purity "guidelines" (unenforceable sanitation guidelines based on heterotrophic plate count (HPC) bacteria levels in the water) adopted by some states, the industry, and the EU. The U.S. bottled water industry uses HPC guidelines, and there are European HPC standards applicable abroad to certain bottled waters, but there are no U.S. standards in light of strong bottler opposition to making such limits legally binding.

Water Pollution Worldwide

Earth is the "water planet." It is hard to comprehend why a planet with 71% of its surface covered by water would be facing a water shortage. As mentioned earlier, at least 80 countries already have water shortages that threaten health and economic activity. More than 1 billion people have no access to clean drinking water. And things are getting worse. The world population is growing at a fast pace. Farms, factories, and cities are using more water. Demand for water is doubling every 21 years—and faster in some areas. This suggests that we need to assure water quality and purity so that we do not face a dire water crisis in the near future.

Pollution of freshwater (drinking water) is a problem for about one-half of the world's population. Each year, there are about 250 million cases of water-related diseases, with roughly 5 million –10 million deaths. Each year, plastic waste in water and coastal areas kills up to:

- 100,000 marine mammals
- 1 million seabirds
- Immeasurable numbers of fish

The Pacific Ocean is the largest ocean realm on our planet, approximately the size of Africa—over 10 million square miles. There are large parts of the Pacific referred to as "plastic oceans," where enormous gyres are covered with plastic debris. The world's seas are beset by a variety of water pollution problems. See Table 1 for 10 of the worst areas.

TABLE 1 Major Bodies of Water/Areas with Serious Water Pollution Problems

Area	Micro-biological	Eutro-phication	Chemical	Suspended Solids	Solid Wastes	Thermal	Radio-nuclides	Spills
Gulf of Mexico	Severe Impact	Moderate Impact	Moderate Impact	Moderate Impact	Moderate Impact	None Known	None Known	Slight Impact
Caribbean Sea	Moderate Impact	Moderate Impact	Moderate Impact	Severe Impact	Moderate Impact	Slight Impact	Slight Impact	Severe Impact
Baltic Sea	Slight Impact	Severe Impact	Moderate Impact	Slight Impact	Slight Impact	None Known	Slight Impact	Moderate Impact
Aral Sea	Slight Impact	Severe Impact	Severe Impact	Severe Impact	Moderate Impact	Slight Impact	Slight Impact	Slight Impact
Yellow Sea	Moderate Impact	Severe Impact	Slight Impact	Slight Impact	Moderate Impact	Slight Impact	None Known	Moderate Impact
Bohal Sea	Moderate Impact	Severe Impact	Moderate Impact	Slight Impact	Moderate Impact	Slight Impact	None Known	Severe Impact
Congo Basin	Moderate Impact	Severe Impact	Moderate Impact	Moderate Impact	Severe Impact	None Known	None Known	Moderate Impact
Benguela Current	Moderate Impact	Moderate Impact	Severe Impact	Moderate Impact	Severe Impact	Slight Impact	Severe Impact	Severe Impact
Lake Victoria	Severe Impact	Severe Impact	Moderate Impact	Severe Impact	Slight Impact	None Known	None Known	None Known
Pacific Islands	Moderate Impact	Slight Impact	Moderate Impact	Moderate Impact	Severe Impact	Slight Impact	Severe Impact	Slight Impact

Source: Adapted form UNEP SEO Report, 2004–2005

To highlight the problems of worldwide water pollution, some of the water quality issues of various parts of the world are summarized in the following sections.

Developing Countries

The problems of water pollution in less developed countries in Africa, Asia, and Latin America are well known. Most travelers are advised not to drink local water. Microbial contamination of water is a major issue in these areas. In Africa, even the large bodies of water are polluted as is exemplified by Lake Victoria (see Chapter 3 for details). Asian rivers are considered by some to be the most polluted in the world. They have 3 times as many bacteria from human waste as the global average and 20 times more lead than rivers in industrialized countries. In 2004, water from half of the tested sections of China's seven major rivers was found to be undrinkable because of pollution. It has been reported that the Yangtze, China's longest river, is "cancerous" with pollution. The Yangtze rises in China's western mountains and passes through some of its most densely populated areas. Environmental experts fear pollution from untreated agricultural and industrial waste could turn the Yangtze into a "dead river" within 5 years. That would make it unable to sustain marine life or provide drinking water to the booming cities along its banks. At present, nearly 300 million people in China do not have access to safe drinking water.

Bangladesh has some of the most polluted groundwater in the world. In this case, the major contaminant is arsenic, which occurs naturally in soil sediments. Around 85% of the total area of the country has contaminated groundwater. This problem significantly affects in India and other Asian countries. As a matter of fact, this problem is encountered worldwide (see Chapter 2).

Developed Countries

The developed countries have their own set of issues with water quality. For example, the quality of water in Europe's rivers and lakes that are used for swimming and water sports worsened between 2004 and 2005, with 10% of sites not meeting standards. Thirty percent of Ireland's rivers are polluted with sewage or fertilizer. The Sarno, on the continent, is the most polluted river in all of Europe, featuring a nasty mix of sewage, untreated agricultural waste, industrial waste, and chemicals. The Rhine is regarded as being Europe's dirtiest river. Almost one-fifth of all the chemical production in the world takes place along its banks.

The King River is Australia's most polluted river, suffering from a severe acidic condition related to mining operations. Canadian rivers are also polluted. Forty percent of U.S. rivers are too polluted for fishing, swimming, or aquatic life. Even worse are the lakes—46% of them are too polluted for fishing, swimming, or aquatic life. Two-thirds of U.S. estuaries and bays are either moderately or severely degraded from eutrophication (nitrogen and phosphorus pollution). The Mississippi River—which drains nearly 40% of the continental United States, including its central farmlands—carries an estimated 1.5 million metric tons of nitrogen pollution into the Gulf of Mexico each year. The resulting hypoxic coastal dead zone in the Gulf each summer is about the size of Massachusetts. Nearly 1.2 trillion gallons of untreated sewage, storm water, and industrial waste are discharged into U.S. waters annually. The EPA has warned that sewage levels in rivers could be back to the super-polluted levels of the 1970s by the year 2016. In any given year, about 25% of beaches in the United States are under advisories or are closed at least once each summer because of water pollution.

Monitoring Contaminants

Despite the best attempts to purify river water for drinking, it should be recognized that even with well-thought-out purification and reprocessing systems, trace or ultratrace amounts of about every substance present in untreated water are likely to be found in drinking water. To monitor contaminants in water, it is necessary to perform analyses at ultratrace levels (at or below ppb level). In the 1978 Metrochem meeting, I presented a paper, "In Search of Femtogram." For your reference, a femtogram is 10^{-15} g, or 1 part per quadrillion—a phantom quantity at that time. I felt it was essential to analyze very low quantities of various contaminants to fully understand the impact of various chemicals on our body. I also noted that dioxin (2,3,7,8-tetrachloro-dibenzodioxin) can cause abortion in monkeys at the 200 ppt (parts per trillion) level, and polychlorinated biphenyls (PCBs) at 0.43 ppb level can weaken the backbones of trout. It has been known for some time now that water that we call potable may contain many trace and ultratrace contaminants, as exemplified by analysis of Ottawa drinking water as given in Table 2 (Ahuja, 2005).

TABLE 2 GC/MS Analysis of Ottawa Tap Water

Compound	Concentration Detected in Water (ppt)
α-BHC	17
Lindane	1.3
Aldrin	0.70
Chlordane	0.0053
Dibutyl phthalate	29
Di(2-ethylhexyl) phthalate	78

GC/MS: Gas chromatography/mass spectrometry

In 1980, I was invited by CHEMTECH to write an article on ultratrace analyses, in which I explained what the term meant and why it is necessary to perform such analyses (Ahuja, 2006). In 1986, Wiley asked me to write *Ultratrace Analysis of Pharmaceuticals and Other Compounds of Interest*, a book in which I briefly described methods for testing a large variety of compounds, including arsenic, at trace and ultratrace levels (Ahuja, 2008). In 1992, I wrote yet another book, *Trace and Ultratrace Analysis by HPLC*, where I emphasized the need for such analyses in pharmaceuticals, food, cosmetics, and the environment.

What Is Potable Water?

A simple definition of potable water would be any water that is suitable for human consumption. *National Primary Drinking Water Regulations* control the water quality in the United States. However, these regulations vary in the various parts of the world. Table 3 lists what one municipality in the United States does to monitor potable water quality.

This table shows that some other contaminants of concern that are not monitored regularly include:

- MTBE (methyl tertiary butyl ether)
- Herbicides
- Pesticides
- Fertilizers
- Pharmaceuticals
- Perchlorate
- Mercury
- Arsenic

These contaminants and many others are discussed at length in this book. Among pharmaceutical contaminants, the problem of endocrine disruptors is gaining greater importance (see Chapter 14). Recently, it has been reported that liquid formula is the biggest culprit in exposing infants to bisphenol A, a potential hormone-disrupting chemical (*Chemical and Engineering News*, 2008).

DELINEATION OF A MAJOR PROBLEM OF ARSENIC-CONTAMINATED GROUNDWATER

Arsenic contamination of groundwater serves as an excellent example of how water quality and purity problems occur if adequate attention is not paid to detailed testing for all potential contaminants. Millions of wells were installed in Bangladesh in the 1970s to solve the problem of microbial contamination of drinking water. Unfortunately, the well water was not tested for natural contamination of arsenic from the ground. The arsenic problem has now been reported in more than 20 countries, including Argentina, Bangladesh, Canada, Chile,

TABLE 3 Water Quality Results of Potable Water[*]

Substances	EPA's MCL	Amount Detected	Source of Contaminant
TOC		2.6 to 5.2 ppm	Naturally present in the environment
INORGANIC CHEMICALS			
Chlorite	1.0 ppm	0.0–0.94 ppm	By-product of disinfection
Chlorine dioxide	0.8 ppm	0.0–0.23 ppm	Water additive for microbial control
Fluoride	4 ppm	0.0–1.70 ppm	Water additive to promote strong teeth
Nitrate	10 ppm	<1.00 ppm	By-product of disinfection
Sulfate	250 ppm	14.0 ppm	Part of treatment process, erosion of natural deposits
Copper 6/18/04		0.249 ppm	Corrosion of household plumbing
Lead 6/18/04		0.003 ppm	Corrosion of household plumbing
ORGANIC CHEMICALS			
Chloramines	4 ppm	1.5–3.00 ppm	Water additive for microbial control
Trihalomethanes	80 ppb	8.0–90.0 ppb	By-product of disinfection
Haloacetic acids	60 ppb	3.5–19.7 ppb	By-product of disinfection
RADIONUCLIDES			
Beta 1/07/00	10 pCi/l	2.48 pCi/l	Erosion of natural deposits
UNREGULATED CONTAMINANTS			
Sodium	Nonregulated	24.8 ppm	Part of treatment process, erosion of natural deposits
Manganese	Nonregulated	0.051 ppm	Erosion of natural deposits
Iron	Nonregulated	0.10 ppm	Erosion of natural deposits
Bromoform	Nonregulated	0.8 ppb	THM component

(Continued)

TABLE 3 (Continued)

Substances	EPA's MCL	Amount Detected	Source of Contaminant
Chlormethane	Nonregulated	14.0 ppb	THM component
Bromomethane	Nonregulated	8.7 ppb	THM component
Chloroform	Nonregulated	34.0 ppb	THM component

THM—trihalomethane; MCL—maximum contamination level; ppm—parts per million; ppb—parts per billion
**Brunswick County, NC 2004.*

China, France, Ghana, Hungary, India, Mexico, Nepal, Thailand, Taiwan, the United Kingdom, the United States, and Vietnam (see Chapter 2). This problem is most pronounced in Bangladesh, India, and several other countries in South Asia. Arsenicosis results from drinking arsenic-contaminated water. It is seriously affecting the health of over 100 million people in South Asia, where it leads to a slow and painful death for many. Numerous suggestions to rectify this problem were received in response to my worldwide appeal in *Chemical & Engineering News* of June 9, 2003. To fully delineate this problem and seek viable solutions, a workshop in Dhaka in 2005 and several symposia at the Atlanta ACS meeting in 2006 were organized, with the support of American Chemical Society and IUPAC. The lessons learned from South Asia can help solve the problem in other parts of the world. Detailed information is provided in Chapter 2 as to how groundwater is contaminated with arsenic, desirable method(s) for monitoring arsenic contamination at ultratrace levels, and the best options for remediation. Various options presented by inventors vying to win the million-dollar Grainger Prize for remediation of arsenic contamination are evaluated in terms of their suitability for resolution of this problem.

WATER QUALITY IN EASTERN AFRICA

Climate change has added to the urgency of need for water conservation. This is because climate change has altered the temperature and rainfall patterns in Eastern Africa and the remainder of the continent (see Chapter 3). The temperature scenario indicates that global mean surface temperature is projected to increase by between 1.5°C (2.7°F) and 5.8°C (10.8°F) by 2100. Climate change scenarios for Africa reflect future warming across the continent ranging from 0.2°C (0.36°F) per decade (low scenario) to more than 0.5°C (0.9°F) per decade (high scenario). This warming will be greatest over the interior of semiarid margins of the Sahara and central Southern Africa. Many of the changes are now attributed to temperature increases caused by anthropogenic greenhouse gas emissions. Warmer climates increase evaporation of water as well. The quality of blue water in East Africa has experienced drastic changes in the last 40 years, as evidenced by the onset of massive cyanobacteria blooms offshore that took place in Lake Victoria. The poor water quality has led to the collapse of the indigenous fish stock. The collapse has also been correlated to poor agricultural practices that have led to an accelerated rate of deposition and sedimentation of soil rich in nutrients. In addition to activities related to human population and agricultural production is the human abuse of water systems. Today most of the lakes and rivers are choking with wastes that are wantonly dumped into them. One of the major sources of water pollution in East Africa is human waste. It is possible to prevent this by changing the "I don't care attitude" of people by the governments' enacting stringent regulations on water pollution.

EFFECT OF LAND DEVELOPMENT

Conversion of natural landscapes to agriculture or urban areas results in a cascading series of events that degrade the quality of receiving water bodies (see Chapter 4). Removal of the forest cover reduces evapotranspiration of rainfall and increases surface storm-water runoff, causing erosion and suspended sediment pollution of receiving waters. If the land is converted into agricultural use, a number of pollutants may enter the surface and/or groundwater over time, including suspended sediments, nutrients (nitrogen and phosphorus), pathogenic fecal microbes, and pesticides and herbicides. On the other hand, if the cleared land is urbanized, much land will be covered by impervious surfaces that will cause further increases in runoff and the loss of groundwater recharge, erosion of streambeds, and loss of aquatic animals' habitat. When the area is served by septic systems, nutrient and fecal microbial pollution to nearby wells and waterways may result if the soils are porous and there is a high water table; surface runoff may result if soils are too impervious for proper percolation. Urban storm-water runoff will lead to pollution of receiving surface waters by nutrients, fecal microbes, metals, and toxic chemicals including PCBs and PAHs (polycyclic aromatic hydrocarbons). The amount of impervious surface coverage in a watershed is strongly related to degraded fish and invertebrate communities, increases in nutrient and chemical pollution, and increases in fecal microbial pollution. Agricultural and urban runoff can be treated by constructed wetlands, streamside vegetated buffer zones, and the use of proper irrigation techniques. Urban runoff may also be treated by properly designed wet retention ponds and by the use of sand filters and rain gardens. Furthermore, urban runoff may be minimized by reducing runoff at the source through increasing green space and minimizing impervious surface coverage.

SAMPLING AND ANALYSIS OF ARSENIC IN GROUNDWATER IN BANGLADESH AND INDIA

Elevated levels of geogenic arsenic have been detected in the groundwater of Bangladesh and in West Bengal state, in India (see Chapter 5). In this chapter, the findings on arsenic contamination of groundwater are reported, based on the analysis of more than 190,000 water samples from hand-pumped tube wells covering the entire areas of these two regions. The study also presents an overview on the sampling strategy, preservation methods, and analytical techniques used for the determination of arsenic (total and inorganic arsenic species) in water samples. In Bangladesh, arsenic concentrations above 50 µg/L (the national standard level of arsenic in drinking water) have been reported from 50 out of a total of 64 districts, based on the analysis of water samples from 50,515 hand-pumped tube wells. The overall water analyses showed 43% and 27.5% of the samples had arsenic above 10 and 50 µg/L, respectively. The authors analyzed 44,696 hand-pumped tube well water samples from 50 arsenic-affected districts and found that 31% of the samples had arsenic above 50 µg/L and 48.5% above 10 µg/L. The districts situated in the floodplain and the deltaic regions are the most contaminated areas of Bangladesh. Based on the overall water analysis from West Bengal, 48.1% had arsenic above 10 µg/L, 23.9% above 50 µg/L, and 3.4% had levels above 300 µg/L, the concentration that may have arsenical manifestation. Analysis of deep tube wells (>100 m) from Bangladesh and West Bengal showed that arsenic above 50 µg/L is usually not present in depths beyond 350 m.

FORENSIC WATER QUALITY INVESTIGATIONS

As mentioned earlier, the aim of water quality analysis is to determine the presence of pollutants of concern and to estimate their concentrations within acceptable levels of precision. Water pollution can be defined as concentrations of harmful materials or their indicators at or above certain levels that have been established by epidemiological or other methods, or as set by regulation. Remediation or mitigation of water pollution requires that sources be identified and quantified. In the case of pollution caused by human actions, source identification also entails determination of responsibility, which may engender civil actions or even criminal charges that would deter polluters by imposing penalties and/or remediation costs (see Chapter 6). Consequently, polluters may challenge the methods, results, and interpretations of water quality investigations, as well as the skill and veracity of investigators. The term "forensic" is used to describe situations such as trials or administrative hearings, in which adversarial argumentation is used to establish facts, eliminate incorrect observations and interpretations, and test propositions. Clearly, skillful analytical work is a requisite for effective environmental forensic investigation, but a larger set of skills and methods must be employed to yield satisfactory outcomes from good field and laboratory work in a forensic context.

REGULATORY CONSIDERATIONS

Federal drinking water regulations are based on risk assessment of human health effects and research conducted on source water, treatment technologies, residuals, and distribution systems. Chapter 7 focuses on the role that EPA research plays in ensuring pure drinking water in the United States and throughout the world. The first part of this chapter explains the EPA's strategic goals for drinking water, the rulemaking process, and applicable drinking water regulations. The second part of this chapter highlights the EPA's human health and drinking water research. The EPA's strategic goals for clean and safe water have evolved from focusing on contaminants in water to protecting source water and water infrastructure. Objectives include protecting human health, protecting water quality, and enhancing science and research. The EPA's first objective is to protect human health by reducing exposure to contaminants in drinking water (including protecting source waters), in fish and shellfish, and in recreational areas. The EPA's second objective is to protect the quality of rivers, lakes, and streams on a watershed basis and to protect coastal and ocean waters. The EPA's third objective is the enhancement of science and research by conducting sound, leading-edge scientific studies to support the protection of human health through the reduction of human exposure to contaminants in drinking water, fish and shellfish, and recreational waters and to support the protection of aquatic ecosystems, specifically the quality of rivers, lakes, and streams, and coastal and ocean waters. The strategic plan targets the improvement of drinking water quality in community water systems serving 6% of the U.S. population that do not meet all applicable health-based drinking water standards. The EPA plans to accomplish these goals through effective treatment and source water protection and improvements in regulatory monitoring and reporting.

MICROBIAL ANALYSIS

As mentioned earlier, a large number of aquatic microorganisms can infect or parasitize humans, and these pathogens and parasites are responsible for considerable morbidity and mortality worldwide. Most such organisms are problematic when human or animal wastes contaminate surface water supplies used for drinking or body contact, but some occur naturally or can infect humans by other routes of transmission. The magnitude of the public health threat posed by these organisms requires a comprehensive effort to identify, quantify, and remediate these problems. A subset of sentinel or indicator organisms have been identified as representative of the widest and most serious public health threats in water. The strategies and methods for studying these organisms are discussed in Chapter 8, including molecular techniques and microbial source tracking approaches. In addition, the risks posed by microbial biofilms and sediment pathogen reservoirs are discussed as emerging problems.

MONITORING INORGANIC COMPOUNDS

The focus of Chapter 9 is inorganic substances in surface water that must be monitored to assure that it is suitable for drinking. In the United States, potable water must be cleaner than the maximum contaminant levels (MCL) mandated by local, state, and federal guidelines to protect human health. The EPA not only enforces the guidelines but also is required to help communities establish wastewater treatment facilities to ensure compliance. These regulations specify the allowable concentration of microorganisms, disinfectants, and disinfection by-products (DBPs; see Chapters 8 and 12), inorganic chemicals, and organic chemicals discussed throughout the book, and radionuclides (see Chapter 10). Secondary contaminants such as iron and sulfur affect the smell, taste, or color of water but are not known to cause illness.

RADIONUCLIDES IN SURFACE- AND GROUNDWATER

Unique among all the contaminants that adversely affect surface and water quality, radioactive compounds pose a double threat from both toxicity and damaging radiation. The extreme energy potential of many of these materials makes them both useful and toxic. The unique properties of radioactive materials make them invaluable for medical, weapons, and energy applications. However, mining, production, use, and disposal of these compounds provide potential pathways for their release into the environment, posing a risk to both humans and wildlife. Chapter 10 discusses the sources, uses, and regulation of radioactive compounds in the United States, biogeochemical processes that control mobility in the environment, examples of radionuclide contamination, and current work related to contaminated site remediation.

VOLATILE AND SEMIVOLATILE CONTAMINANTS

As mentioned in various chapters, a large number of contaminants can find their way into the water sources we use for drinking water. Of these contaminants, volatile and semivolatile contaminants can enter directly from various spills, by improper disposal, or from the atmosphere in the form of rain, hail, and snow (Chapter 11). Rain is nature's way of providing fresh water; however, now it is generally contaminated with various pollutants that we release into the atmosphere, most of which are volatile contaminants. According to a USGS report, volatile organic compounds (VOCs) are a group of organic compounds with inherent physical and chemical properties that allow these compounds to move between water and air. In general, VOCs have high vapor pressure, low-to-medium water solubilities, and low molecular weights. By contrast, semivolatile organic compounds (SOCs) have a higher boiling point than VOCs; however, they can be volatilized under various environmental conditions and may pollute water.

MONITORING DISINFECTANTS

Disinfection is a process that deliberately reduces the number of pathogenic microorganisms in water to protect public health (Chapter 12). Chemical disinfection has been an integral part of drinking water treatment processes in the United States since the introduction of chlorine as a disinfectant in the early 1900s. Chlorine, along with some other disinfectants such as ozone and chlorine dioxide, was found to provide additional benefits including color, taste, and odor reduction. Therefore, chemical disinfectants were used in as large quantities as required to achieve desired water quality. Although chlorine was known to react with organic material in water, it was only in the early 1970s that scientists were able to identify the formation of chloroform ($CHCl_3$) and other volatile halogen-substituted organics in drinking water. These compounds were related to chlorine and were termed "by-products" of chlorination. These findings led to a large number of studies to learn about the formation of these by-products and their effects. As more became known about the potential by-products, it was also found that alternative chemical disinfectants (such as ozone and chlorine dioxide) form by-products of their own. EPA regulates DBPs and also the amount of disinfectants that can be used in drinking water. The balance between the risk of microbial contamination and DBP formation is still a challenge. Currently, there are several options for the disinfection of drinking water, and each has its own merits however, the by-products must be minimized.

HERBICIDES AND THEIR DEGRADATION PRODUCTS

History shows that it has always been important that water and food be linked to population and quality (Chapter 13). Innovative changes in farming equipment, laboratory instrumentation, and new ideas from around the world have been helpful in meeting the global demands of water and food. Pesticides are not new, but usage has changed over time. The introduction of herbicides is important to agriculture production. At the same time, the need for continued investigation and development is necessary for the proper management of weeds connected with food production. Water samples from around the world have been collected in various studies to help determine what changes have been made with herbicide use and what the results of these herbicide concentrations are. Studies have produced the necessary information to help revolutionize government regulations for better quality water, as well as increased yields of grain; and changes in production practices have been made as a result.

PHARMACEUTICALS IN SEWAGE EFFLUENTS

The growing use of pharmaceuticals worldwide has become a new environmental problem that has awakened great concern among scientists in the last few years (see Chapter 14). Although the first reports on pharmaceuticals in wastewater effluents and in surface waters were published in the United States in the 1970s, pharmaceuticals as environmental contaminants did not receive a great deal of attention until the link was established between a synthetic birth-control pharmaceutical (ethynylestradiol) and impacts on fish. Over 3,000 chemical substances are used in human and veterinary medicines. These pharmaceuticals include antiphlogistics/anti-inflammatory drugs, contraceptives, β-blockers, lipid regulators, tranquilizers, antiepileptics, and antibiotics. Although their toxicity to aquatic and terrestrial organisms is relatively unknown, a number of reported investigations have shown that pharmaceutical compounds pose a real threat to the environment. For example, diclofenac, which is frequently detected in aquatic matrices, has been found to have adverse effects on both rainbow trout and vultures. Diclofenac accumulates, with a concentration factor of up to 2,732, in the liver of rainbow trout and causes histopathological alterations in both the kidneys and gills. In vulture populations this drug has been shown to cause renal failure. This highlights the potential danger to both terrestrial and aquatic life. Moreover, it underlines the latent risk to humans. Wastewater treatment plants (WWTPs) are major contributors of pharmaceuticals in the environment. Because of their high consumption, pharmaceuticals along with their metabolites are continuously introduced to sewage waters, mainly through excreta, disposal of unused or expired drugs, or directly from pharmaceutical discharges. Recently, research has shown that the elimination of some pharmaceutical compounds during wastewater treatment processes is rather low. The compounds that are not removed are released to receiving water bodies from WWTP effluent streams, and as a result, pharmaceuticals may be found in surface, groundwater, and drinking waters.

MONITORING TERRORIST-RELATED CONTAMINATION

The direction of the last 100 years of analytical science, as it pertains to drinking water, took a dramatic shift after terrorist attacks on U.S. soil on September 11, 2001, followed by anthrax attacks. Prior to that date, the analytical emphasis was on the detection and removal of naturally occurring or accidental contaminants that found their way into drinking water supplies. After the terrorist attacks, a new fear dawned in the water supply industry. What if someone were to intentionally introduce a contaminant into the drinking water? The vast array of potential contaminants that could be used by a terrorist, the innumerable sites at which an attack could occur, and the potential consequences of not rapidly detecting such an event demanded a sea change from the old monitoring paradigm of collecting occasional grab samples and monitoring for a small suite of potential contaminants. The challenges entailed in this endeavor and some of the technologies that are becoming available to protect our water supplies from deliberate attack are discussed in Chapter 15.

GROUNDWATER ARSENIC-REMOVAL TECHNOLOGIES BASED ON SORBENTS

Groundwater, a primary source of drinking water in many parts of the world containing toxic level of arsenic and other species, is harming the health of millions of people. Chapter 16 examines small-scale household water filtration systems based on solid sorbents to obtain potable water. Special emphasis is placed on iron-based filters because they appear to be chemically most suitable for arsenic removal, easy to develop, and environmentally benign. Arsenic-removal mechanisms based on surface complexation reactions, sorption dynamics, and kinetics are discussed as they have been reported in the literature. Several promising filters are described from the standpoint of their applicability and sustainability in field use. Finally, an evaluation of technology verification protocols is critically examined.

REFERENCES

Ahuja, S., 1980. Ultratrace analyses. CHEMTECH 11, 702.

Ahuja, S., 1986. Ultratrace Analysis of Pharmaceutical and Other Compounds of Interest. Wiley, NY.

Ahuja, S., 2005. Assuring water purity for human consumption. MARM Meeting, New Brunswick, NJ.

Ahuja S., 2006. Assuring water purity by monitoring water contaminants from arsenic to zinc. American Chemical Society Meeting, Atlanta, March 26–30.

Ahuja, S., 2008. Arsenic Contamination of Groundwater: Mechanism, Analysis, and Remediation. Wiley, NY.

Chemical and Engineering News, 2008. November 17, p. 42.

Chapter 2

Delineation of a Major Worldwide Problem of Arsenic-Contaminated Groundwater

Satinder Ahuja

Ahuja Academy of Water Quality at UNCW, Calabash, NC, USA

INTRODUCTION

The worst case of groundwater contamination was discovered in Bangladesh in the 1980s (Ahuja and Malin, 2004), where a large number of shallow tube wells (10–40 m) installed in the 1970s, with the help of UNICEF, were found contaminated with arsenic. Arsenic contamination has also been found in regional water supplies of many other developing and developed countries in Asia, Africa, Europe, North America, and South America (Table 1). Groundwater contamination of arsenic (As) can occur from various anthropogenic sources such as pesticides, wood preservatives, glass manufacture, and other miscellaneous

Handbook of Water Purity and Quality
Copyright © 2009 by Elsevier Inc. All rights of reproduction in any form reserved.

TABLE 1 Publications from Arsenic-Affected Countries Around the World

Country	Source of Contamination		Publication Date[*]
	Natural	Anthropogenic	
Argentina	X		1938
Germany		X	1940
New Zealand	X		1961
Taiwan, China	X		1968
Sri Lanka	X		1972
Chile	X	X	1974
Czech Republic		X	1977
United Kingdom		X	1978
Sweden		X	1979
Mexico	X		1983
Hungary	X		1989
Japan	X	X	1989
Ghana	X		1992
Bulgaria		X	1993
China (PR)	X	X	1996
Bangladesh	X		1997
Finland	X		1998
Canada	X	X	1999
United States	X	X	1999
Brazil		X	2000
Egypt	X		2001
Romania	X		2001
Thailand		X	2001
Vietnam	X		2001
India	X	X	2002
Switzerland			2002
Myanmar	X		2002
Australia		X	2003
Iran	X		2003
Nepal	X		2003
Afghanistan	X		2004
Greece	X		2004
Cambodia	X		2005
Pakistan	X		2005
Spain	X		2006

[*]*Personal communication from A. Hussam (2008).*

arsenic uses. These sources can be monitored and controlled. However, this is not so easy with naturally occurring arsenic. The natural content of arsenic in soil is mostly in a range below 10 mg/kg; however, it can cause major havoc when it gets into groundwater (Ahuja, 2008a).

Investigations into the Problem

The arsenic contamination problem in Bangladesh was not discovered at the outset. All the attention was focused on providing water free of microbial contamination, a problem that was commonly encountered in surface water. Unfortunately, potential contamination from naturally occurring arsenic was not realized, and the project did not include adequate testing to detect the arsenic contamination. This unfortunate calamity could have been avoided, as analytical methods that can test for arsenic down to the parts per billion (ppb) levels have been available for many years (Ahuja, 1986).

The chronology of publications on arsenic contamination from various countries that are affected can be seen in Table 1. This table also shows whether the source of arsenic contamination is natural or anthropogenic. In most of the countries listed, the source of contamination is natural. Arsenic contamination was reported as early as 1938; however, skin lesions and cancers attributable to arsenic were rare and ignored until new evidence emerged from Taiwan in 1977. The serious health effects of arsenic exposure that include lung, liver, and bladder cancers were confirmed shortly thereafter by studies of exposed populations in Argentina, Chile, and China. In 1984, Dr. K.C. Saha and colleagues at the School of Tropical Medicine in Kolkata, India, attributed lesions observed on the skin of villagers in the state of West Bengal to the elevated arsenic content of groundwater drawn from shallow tube wells. Of the various countries affected by this contamination, Bangladesh and India are experiencing the most serious groundwater arsenic problem, and the situation in Bangladesh has been described as "the worst mass poisoning in human history."

The magnitude of the problem in India has been investigated for the last 18 years by Chakraborti and others (for more details see Chapter 5) who have analyzed 225,000 tube-well water samples from the Ganga–Meghna–Brahamaputra plain, covering an area of 569,749 km^2 and a population of more than 500 million. The investigators found that Bangladesh and a number of states in India (Uttar Pradesh, Bihar, West Bengal, Jharkhand, and Assam) are affected by a concentration of arsenic >50 µg/L. On average, about 50% of the water samples contained arsenic above 10 µg/L and 30% were above 50 µg/L.

It should be noted that groundwater contamination is found even in advanced countries such as Australia, United Kingdom, and the United States. Figure 1 shows groundwater contamination in the United States; over 31,000 samples analyzed over almost a 30-year period reveal that a large number of states are affected by this contamination. In the United States, nearly 10% of groundwater resources exceed the maximum contamination level (MCL) of 10 µg/L.

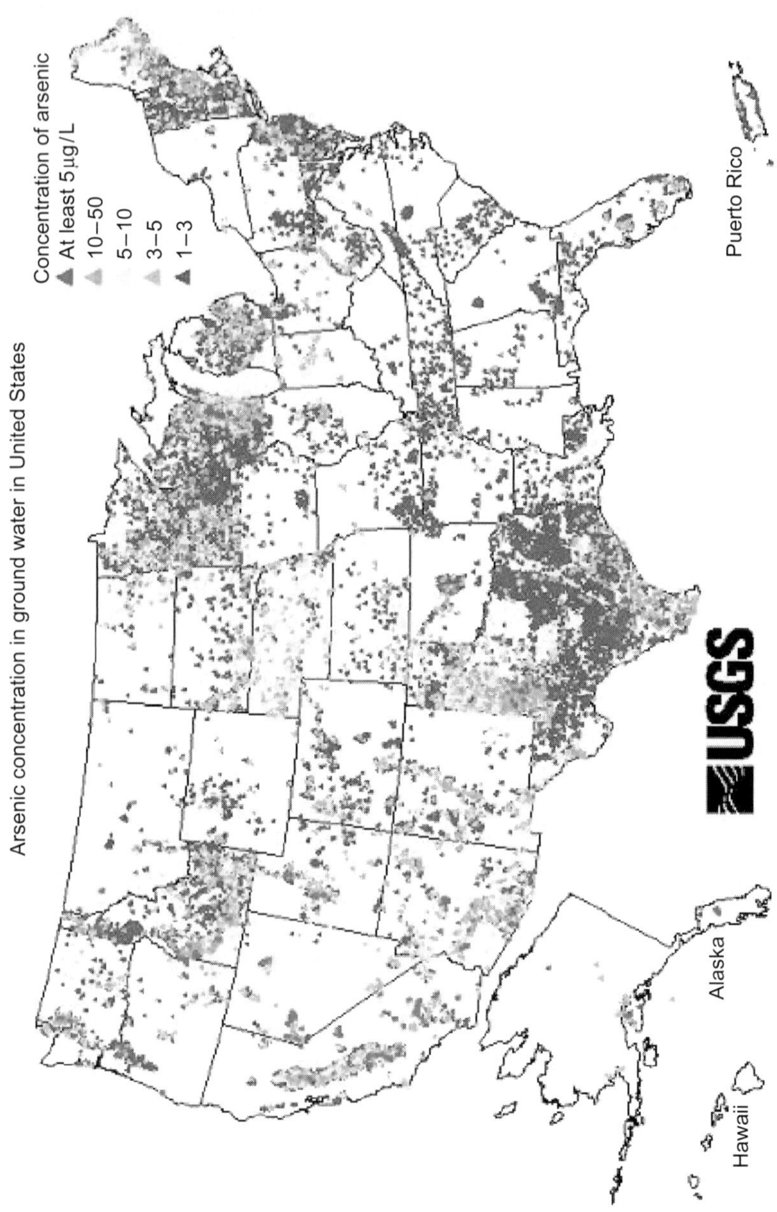

FIGURE 1   Arsenic concentration in groundwater in United States (see Plate 1 of Color Plate secion).

It has been estimated that the population at risk approaches 100 million in Bangladesh at the MCL of 10 ppb set by World Health Organization (WHO), suggesting that as many as 200 million people could be affected by this problem worldwide.

Recognizing the fact that inorganic arsenic is a documented human carcinogen, in 1993 the WHO set a standard at no more than $10\,\mu g/L$ (or 10 ppb) of arsenic in drinking water as the MCL. This standard was finally adopted by the United States in 2006. However, the MCL remains at $50\,\mu g/L$ (or 50 ppb) in Bangladesh and other less developed countries. Furthermore, it should be mentioned that the guidelines do not consider different arsenic species, even though it is well established that the toxicity of arsenic may vary enormously with its speciation, as discussed in the following section.

Toxicity of Various Arsenic Species

Arsenic is a well-known poison, with a lethal dose in humans at about 125 mg. Most of the ingested arsenic is excreted from the body through urine, stools, skin, hair, nails, and breath. In cases of excessive intake, some arsenic is deposited in tissues, causing the inhibition of cellular enzyme activities. In addition to consumption through drinking water, arsenic may also be taken up via the food chain (see section Impact of Arsenic-Laced Irrigation Water on the Food Chain). Direct consumption of rice irrigated with arsenic-rich waters is a significant source of arsenic exposure in areas such as Bangladesh and other countries where rice is the staple food and provides the main caloric intake.

Arsenic is a semimetal or metalloid that is stable in several oxidation states ($-$III, 0, $+$III, $+$V). It is a natural constituent of Earth's crust and ranks 20th in abundance in relation to the other elements. Table 2 shows arsenic concentrations in various environmental media. It should be noted that the $+$III and $+$V states are most common in natural systems. Arsine($-$III), a compound with extremely high toxicity, can be formed under high reducing conditions, but its occurrence in gases emanating from anaerobic environments in nature is relatively rare.

The relative toxicity of arsenic depends mainly on its chemical form and is dictated in part by the valence state. Trivalent arsenic has a high affinity for thiol groups, as it readily forms kinetically stable bonds to sulfur. Hence, reaction with As(III) induces enzyme inactivation, as thiol groups are important to the functions of many enzymes. Arsenic affects the respiratory system by binding to the vicinal thiols in pyruvate dehydrogenase and 2-oxoglutarate dehydrogenase, and it has also been found to affect the function of glucocorticoid receptors. Pentavalent arsenic has a poor affinity toward thiol groups, resulting in more rapid excretion from the body. However, it is a molecular analog of phosphate and can uncouple mitochondrial oxidative phosphorylation, resulting in failure of the energy metabolism system. The effects of the oxidation state on chronic toxicity are confounded by the redox conversion of As(III) and As(V) within human cells and tissues. Methylated arsenicals such as monomethylarsenic acid (MMAA) and dimethylarsenic acid (DMAA) are less harmful than inorganic arsenic compounds.

Clinical symptoms of arsenicosis may take about 6–24 months or more to appear, depending on the quantity of arsenic ingested and also on the nutritional

TABLE 2 Arsenic Concentrations in Environmental Media (U.S. EPA)

Environmental Media	Range of Arsenic Concentrations
Air (ng/m^3)	1.5–53
Rain from unpolluted ocean air (μg/L) (ppb)	0.019
Rain from terrestrial air (mg/L)	0.46
Rivers (μg/L)	0.20–264
Lakes (μg/L)	0.38–1,000
Groundwater (well) (μg/L)	1.0–1,000
Seawater (μg/L)	0.15–6.0
Soil (mg/kg)	0.1–1,000
Stream/river sediment (mg/kg)	5.0–4,000
Lake sediment (mg/kg)	2.0–300
Igneous rock (mg/kg)	0.3–113
Metamorphic rock (mg/kg)	0.0–143
Sedimentary rock (mg/kg)	0.1–490

status and immunity level of the individual. Untreated arsenic poisoning results in several stages; for example, various effects on the skin with melanosis and keratosis; dark spots on the chest, back, limbs, and gums; enlargement of the liver, kidneys, and spleen. Later on, patients may develop nephropathy, hepatopathy, gangrene, or cancers of the skin, lung, or bladder.

It should be noted that a number of toxicologists consider a 10-ppb level of arsenic to be too high because even at 1 ppb, the risk of getting cancer is one in 3,000 (see Figure 2). The fact remains that prolonged drinking of this contaminated water has caused serious illnesses in the form of hyperkeratosis on the palms and feet, fatigue symptoms of arsenicosis, and cancers of the bladder, skin, and other organs. In the long term, one in every 10 people could die of arsenic poisoning if they continue using water with high concentrations of arsenic.

Arsenic toxicity has no known effective treatment, but drinking of arsenic-free water can help the arsenic-affected people who are at the preliminary stage of their illness alleviate the symptoms of arsenic toxicity. Hence, provision of arsenic-free water is urgently needed for the mitigation of arsenic toxicity and the protection of the health and well-being of people living in the areas of these countries where the arsenic problem is acute.

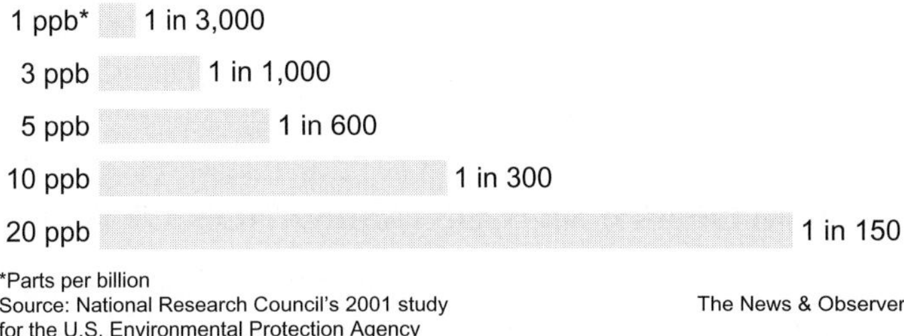

FIGURE 2   Risk of cancer with arsenic contamination of water.

This chapter focuses mainly on groundwater pollution by arsenic from natural sources. The following discussion focuses on how groundwater is contaminated with arsenic, desirable method(s) for monitoring arsenic contamination at ultratrace (ppb or below) levels, and the best options for remediation. In an attempt to improve our understanding of this horrific problem that affects the world, a book has been recently published to improve our understanding of the problem and to offer some meaningful solutions (Ahuja, 2008b).

Impact of Arsenic-Laced Irrigation Water on the Food Chain

The fact that arsenic poisoning in the world's population is not consistent with the level of water intake has raised questions on the possible pathways of arsenic transfer from groundwater to the human system. Even if an arsenic-safe drinking water supply could be ensured, the same groundwater may continue to be used for irrigation purposes, leaving a risk of soil accumulation of this toxic element and eventual exposure to the food chain through plant uptake and animal consumption. Studies on arsenic uptake by crops indicate that there is a great potential for the transfer of groundwater arsenic to crops. The fate of arsenic in irrigation water and its potential impact on the food chain, especially as it occurs in Bangladesh and other similar environments, has been discussed at length (see Chapter 2 in Ahuja, 2008b).

Green leafy vegetables have been found to act as arsenic accumulators, with arum (kochu), gourd leaf, *Amaranthus*, and *Ipomea* (kalmi) at the top of the list. Arum, a green vegetable commonly grown and used in almost every part of Bangladesh, seems to be unique in that the concentration of arsenic can be high in every part of the plant. Arsenic in rice seems to vary widely. Speciation of

Bangladeshi rice shows the presence of As(III), DMAA, and As(V); more than 80% of the recovered arsenic is in the inorganic form. It has been reported that more than 85% of the arsenic in rice is bioavailable, compared to only about 28% of arsenic in leafy vegetables. It is thus pertinent to assess the dietary load of arsenic from various food materials contaminated with arsenic. A person consuming 100 g of arum daily, with an average arsenic content of 2.2 mg/kg, 600 g of rice with an average arsenic content of 0.1 mg/kg, and 3 L of water with an average arsenic content of 0.1 mg/L would ingest 0.56 mg/day, which exceeds the calculated threshold value based on the U.S. Environmental Protection Agency (EPA) model.

MECHANISM OF ARSENIC CONTAMINATION OF WATER

Until recently, it was generally believed that arsenic is released in the soil as a result of weathering of the arsenopyrite or other primary sulfide minerals.

Weathering of Arsenopyrite

Important factors controlling the oxidation–reduction phenomenon of arsenopyrite are listed as follows:

- Moisture (hydrolysis)
- pH
- Temperature
- Solubility
- Redox characteristics of the species
- Reactivity of the species with CO_2/H_2O

It has been reported that weathering of arsenopyrite in the presence of oxygen and water involves oxidation of S to SO_4^{-2} and As(III) to As(V):

$$4FeAsS + 13O_2 + 6H_2O \leftrightarrow 4SO_4^{-2} + 4AsO_4^{-3} + 4Fe^{+3} + 12H^+$$

Although there are both natural and anthropogenic inputs of arsenic to the environment, elevated arsenic concentrations in groundwater are often due to naturally occurring arsenic deposits. While the average abundance of arsenic in Earth's crust is between 2 and 5 mg/kg, enrichment in igneous and sedimentary rocks, such as shale and coal deposits, is not uncommon. Arsenic-containing pyrite (FeS) is probably the most common mineral source of arsenic, although it is often found associated with more weathered phases. Mine tailings can contain substantial amounts of arsenic, and the weathering of these deposits can liberate arsenic into the surface water or groundwater, where numerous chemical and biological transformations can take place. Arsenic may also be directly released into the aquatic environment through geothermal water such as hot springs. Anthropogenic sources of arsenic include pesticide application, coal fly ash, smelting slag, feed additives, semiconductor chips, and arsenic-treated wood, which can cause local water contamination.

In Bangladesh and India (in West Bengal), where the problem has received the most attention, the aquifer sediments are derived from weathered materials from the Himalayas. Arsenic typically occurs at concentrations of 2–100 ppm in these sediments, much of it sorbed onto various mineralogical hosts including hydrated ferric oxides, phyllosilicates, and sulfides. The mechanism of arsenic release from these sediments has been a topic of intense debate, and both microbial and chemical processes have been invoked. The oxidation of arsenic-rich pyrite has been proposed as one possible mechanism. Other studies have suggested that reductive dissolution of arsenic-rich Fe(III) oxyhydroxides deep in the aquifer may lead to the release of arsenic into the groundwater. Additional factors that may further complicate potential arsenic-releasing mechanisms from sediments include the predicted mobilization of sorbed arsenic by phosphate generated from the intensive use of fertilizers, carbonate produced via microbial metabolism, or changes in the sorptive capacity of ferric oxyhydroxides.

Role of Microbes in the Release of Arsenic into Groundwater

Recently, a great deal of support has been found in the role of microbes in the release of arsenic into groundwater. A brief review of high arsenic concentrations in groundwater and also proposed mechanisms for the release of arsenic into groundwater systems, with particular significance to the possible role of metal-reducing bacteria in arsenic mobilization into the shallow aquifers of the Ganges delta, is provided here (for more details, see Chapter 3 in Ahuja, 2008b). The bacterial effects on arsenic behavior in anoxic sediments and the different interactions between minerals, microbes, and arsenic that have a significant impact on arsenic mobilization in groundwater systems are also discussed. Throughout evolution, microorganisms have developed the ability to survive in almost every environmental condition on Earth. Their metabolism depends on the availability of metal ions to catalyze energy-yielding reactions and synthetic reactions and on their ability to protect themselves from toxic amounts of metals by detoxification processes. Furthermore, microorganisms are capable of transforming various elements as a result of (i) assimilatory processes in which an element is taken up into cell biomass and (ii) dissimilatory processes in which transformation results in energy generation or detoxification. Arsenic is called an "essential toxin" because it is required in trace amounts for the growth and metabolism of certain microbes, but is toxic at high concentrations. However, it is now evident that various types of microorganisms gain energy from this toxic element, and, subsequently, these reactions have important environmental implications.

Bacterial reduction of As(V) has been recorded in anoxic sediments, where it proceeds via a dissimilatory process. Dedicated bacteria achieve anaerobic growth by using arsenate as a respiratory electron acceptor for the oxidation of organic substrates, quantitatively forming arsenite as the reduction product.

The reaction is energetically favorable when coupled with the oxidation of organic matter, because arsenate is electrochemically positive; the As(V)/As(III) oxidation–reduction potential is $+135\,mV$. To date, at least 19 species of organisms have been found to respire arsenate anaerobically, and these have been isolated from freshwater sediments, estuaries, hot springs, soda lakes, and gold mines. They are not confined to any particular group of prokaryotes and are distributed throughout the bacterial domain. These microbes are collectively referred to as dissimilatory arsenate-reducing prokaryotes (DARPs), and there are other electron acceptors used by these organisms, which are strain-specific, including elemental sulfur, selenate, nitrate, nitrite, fumarate, Fe(III), dimethyl-sulfoxide, thiosulfate, and trimethylamine oxide. For example, *Sulfurospirillum barnesii* (formerly strain SES-3), a vibrio-shaped Gram-negative bacterium isolated from a selenate-contaminated freshwater marsh in western Nevada, is capable of growing anaerobically, using As(V) as the electron acceptor. It can also support growth from the reduction of various electron acceptors including selenate, Fe(III), nitrate, fumarate, and thiosulfate. The Gram-positive sulfate-reducing bacterium *Desulfotomaculum auripigmentum*, isolated from the surface lake sediments in eastern Massachusetts (United States), has been found to reduce both As(V) and sulfate. DARPs can oxidize various organic and inorganic electron donors including acetate, citrate, lactate, formate, pyruvate, butyrate, fumarate, malate, succinate, glucose, aromatics, hydrogen, and sulfide. Two Gram-positive anaerobic bacteria, *Bacillus arsenicoselenatis* and *B. selenitireducens*, have also been isolated from the anoxic muds of Mono Lake, California (United States). Both grew by dissimilatory reduction of As(V) to As(III), coupled with the oxidation of lactate to acetate plus CO_2.

Arsenic Mobilization and Sequestration

Diagenesis is driven primarily by the mineralization of organic carbon and the subsequent changes in redox potential with depth. As the sediments become more reducing, the redox equilibrium of various chemical species in the sediment shifts (see Chapter 5 in Ahuja, 2008b). Sediment diagenesis involves chemical, physical, and biological processes including deposition, diffusion, reductive dissolution, and secondary mineral precipitation. However, it is important to recognize that the kinetics of these reactions is variable and is sensitive to environmental parameters such as microbial activity. Thus, it is common to observe As(III) and As(V) or Fe(III) and Fe(II) appearing together in sequence or simultaneously under various redox conditions because of kinetic factors. It must be recognized that the interplay of biogeochemical mechanisms makes understanding the processes responsible for arsenic mobilization in the environment inevitably complex.

Microbially mediated reduction of assemblages comprising arsenic sorbed to ferric oxyhydroxides is gaining consensus as the dominant mechanism for the mobilization of arsenic into groundwater. For example, a recent microcosm-based study provided the first direct evidence of the role of indigenous metal-reducing bacteria in the formation of toxic, mobile As(III) in sediments from the Ganges delta (see Chapter 6 in Ahuja, 2008b). This study showed that the addition of acetate to anaerobic sediments, as a proxy for organic matter and a potential electron donor for metal reduction, resulted in the stimulation of microbial reduction of Fe(III), followed by As(V) reduction and the release of As(III). Microbial communities responsible for metal reduction and As(III) mobilization in the stimulated anaerobic sediment were analyzed using molecular polymerase chain reaction (PCR) and cultivation-dependent techniques. Both approaches confirmed an increase in the number of metal-reducing bacteria, principally *Geobacter* species. Further studies have suggested that most *Geobacter* strains in culture do not possess the *arrA* genes required to support the reduction of sorbed As(V) and mobilization of As(III). Interestingly, in the strains lacking the biochemical machinery for As(V) reduction, Fe(II) minerals formed during respiration on Fe(III) have proved to be potent sorbents for arsenic present in microbial cultures, preventing mobilization of arsenic during active iron reduction. However, the genomes of at least two *Geobacter* species (*G. unraniumreducens* and *G. lovleyi*) do contain *arrA* genes, and, interestingly, genes affiliated with the *G. unraniumreducens* and *G. lovleyi arrA* gene sequences have been identified recently in Cambodian sediments stimulated for iron and arsenate reduction by heavy (C-13–labeled) acetate, using a stable isotope-probing technique. Indeed, the type strain of *G. unraniumreducens* has recently been shown to reduce soluble and sorbed As(V), resulting in the mobilization of As(III) in the latter case. A study of North Carolina (United States) wells found that *arrA* genes were closely related to the gene found in *Geobacter uraniumreducens* (see Chapter 4 in Ahuja, 2008b). This suggests that some *Geobacter* species may play a role in arsenate release from sediments. However, other well-known arsenate-reducing bacteria, including *Sulfurospirillum* species, have also been detected in C-13–amended Cambodian sediments and hot spots associated with arsenic release in sediments from West Bengal (India). Although the precise mechanism of arsenic mobilization in Southeast Asian aquifers remains to be identified, the role of As(V)-respiring bacteria in the process is gaining support. Indeed, recent studies with *Shewanella* sp. ANA-3 and sediment collected from the Haiwee Reservoir (Olancha, CA) have suggested that such processes could be widespread, but not necessarily driven by As(V) reduction, following exhaustion of all bioavailable Fe(III).

ANALYTICAL METHODS

It is not very difficult to determine arsenic at $10\,\mu g/L$ or at $10\,ppb$ or at an even lower level in water (Ahuja, 1986; see also Chapter 5). A number of methods can be used for determining arsenic in water at the ppb level.

- Flame atomic absorption spectrometry
- Graphite furnace atomic absorption spectrometry
- Inductively coupled plasma mass spectrometry

- Atomic fluorescence spectrometry
- Neutron activation analysis
- Differential pulse polarography

Very low detection limits for arsenic, down to $0.0006\,\mu g/L$, can be obtained with inductively coupled plasma mass spectrometry (ICP-MS). The speciation of arsenic requires separations based on solvent extraction, chromatography, and selective hydride generation (HG). High-performance liquid chromatography (HPLC) coupled with ICP-MS is currently the best technique available for the determination of inorganic and organic species of arsenic; however, the cost of the instrumentation is prohibitive. For developing countries that confront this problem, the improvement of low-cost, reliable instrumentation (see Chapter 3) and reliable field test kits is very desirable.

Low-Cost Measurement Technologies for Arsenic

Using HG, a method that has been known for many decades, arsenic can be determined by a relatively inexpensive atomic absorption spectrometer or atomic fluorescence spectrometer (AFS) at single digit microgram per liter concentrations (see Chapter 7 in Ahuja, 2008b). Its generation is prone to interference from other matrix components, and as a result, different matrices can present various analytical problems. In this technique, arsenic compounds are converted to volatile derivatives by reaction with a hydride transfer reagent, usually tetrahydroborate III. HG can be quite effective as an interface between HPLC separation and element-specific detection. In fact, it is possible to get the same performance from HG-AFS as from ICP-MS. Therefore, as the former detector represents significant savings in both capital and operation costs compared with the latter, there is considerable interest in this technique in developing countries.

Test Kit Reliability

There are two main approaches currently used for on-site analysis of arsenic (see Chapter 8 in Ahuja, 2008b). (a) The most widely used systems are those based on a colorimetric principle. These systems require few reagents and are easy to use. (b) The electroanalytical approach is based on reduction–oxidation of the arsenic species. Although electroanalysis is more difficult to operate, the detection limits obtained by such devices can be much lower than those obtained by colorimetry. The U.S. EPA supports the environmental technology verification (ETV) program to facilitate the implementation of innovative new technologies for environmental monitoring. The ETV tested several commercially available kits in July 2002 and added four more in August 2003 under field conditions with trained and untrained operators.

Assessments of the "new generation" of on-site testing kits may be found in Table 3. The performance of in-field testing kits for arsenic is overall unsatisfactory,

TABLE 3 ETV Joint Verification (U.S. EPA and Batelle) for Arsenic Field Testing Methods

Products	Accuracy (%)	Precision[1] (%)	Linearity[2]	MDL (μg/L)	Matrix Effects Interferences[3]	Interunit Reproducibility (ub), Operator Bias (ob)	Rate (%) of False Positive (fp) and False Negative (fn) for 10 μg/L	Cost and Time
PeCo (Peters Engineering) with visual testing	For 10 μg/L ±2 to 17 For 23 to 93 μg/L ±1 to 113	0–40 (NTO) 0–26 (TO)	0.977 (NTO)	15–50 (NTO) 20–40 (TO) (given for 25 μg/L)	No significant effects	No performance differences for ob	Fp: 0 (NTO), 3 (TO), Fn: all 0	100 samples for $200
As 75 (Peters Engineering) with electronic testing	For 10 μg/L ±1 to 157 For >10 μg/L ±6 to 310	11–38 (NTO) 12–71 (TO)	0.990	33 (NTO) 28 (TO)	No significant effects	No performance differences for ob	Fp: 2 (NTO), 13 (TO), Fn: all 0	Cost of tester $330, additional 100 cost $60
Quick™ Low-range II color chart	−92 to −8 (TO) −74 to 74 (NTO)	0–55	0.99 (NTO) $R = 0.90$ (TO)	1.2–1.5	No significant effects	Better performance of NTO	Fp: 0, Fn: 62 (TO), 33 (NTO)	15 min analysis, 50 samples for $350
Quick™ Low-range II arsenic scan	−98 to −27 (TO) −76 to 9 (NTO)	0–84	0.96 (NTO) $R = 0.98$ (TO)	0.7–2.1	No significant effects	Better performance of NTO, unit differences not significant	Fp: 0, Fn: 62 (TO), 38 (NTO)	15 min analysis, 50 samples for $350, additional to $1,600
Quick™ Low-range II compuscan	−93 to 104 (TO) −67 to 81 (NTO)	7–91	0.98 (NTO) $R = 0.98$ (TO)	0.5–3.9	No significant effects	Better performance of NTO, unit differences smaller but significant	Fp: 0–3, Fn: 52–67 (TO), 9 (NTO)	15 min analysis, 50 samples for $350, additional to $1,600

(Continued)

TABLE 3 (Continued)

Products	Accuracy (%)	Precision[1] (%)	Linearity[2]	MDL (μg/L)	Matrix Effects Interferences[3]	Interunit Reproducibility (ub), Operator Bias (ob)	Rate (%) of False Positive (fp) and False Negative (fn) for 10 μg/L	Cost and Time
Quick™ II color chart	−61 to 10 (TO) −77 to 96 (NTO)	16–24 (TO) 0–38 (NTO)	$R = 0.98$ (NTO) $R = 0.96$ (TO)	3.6–7	No significant effects	Better performance of NTO	Fp: 0, Fn: 19 (TO), 24 (NTO)	15 min analysis, 50 samples for $220
Quick™ II arsenic scan	−78 to −4 (TO) −85 to −22 (NTO)	11–44 (TO) 13–38 (NTO)	0.93 (NTO) $R =$ 0.92–0.93 (TO)	4.5–6.1	No significant effects	Better performance of NTO, no ub	Fp: 0, Fn: 19–33 (TO), 29 (NTO)	15 min analysis, 50 samples for $220, additional to $1,600
Quick™ II compuscan	−71 to 96 (TO) −82 to 108 (NTO)	10–58 (TO) 16–108 (NTO)	0.91 (NTO) $R = 0.92$ (TO)	3.7–18.2	No significant effects	No significant differences (ob), significant unit biased (ub)	Fp: 3–9 (TO), 0 (TO), Fn: 38–10 (TO), 14 (NTO)	15 min analysis, 50 samples for $220, additional to $1,600
Quick™ Low-range color chart	−38 to 239 (TO) −81 to 579 (NTO)	0–10 (TO) 0–23 (NTO)	0.98 (NTO) $R = 1.00$ (TO)	3.1–6.7	Positive bias when higher levels of sodium chloride, sulfide, and iron	Better performance of NTO	Fp: 3 (TO), Fp: 12.5 (NTO), Fn: 0 (TO), 14 (NTO)	15 min analysis, 50 samples for $180
Quick™ Low-range arsenic scan	−93 to 99 (TO) −86 to 66 (NTO)	5–23 (TO) 0–42 (NTO)	0.966 (NTO) $R = 0.997$ (TO)	4.0–7.2	Positive bias when higher levels of sodium chloride, sulfide, and iron	No significant difference (ob), no ub	Fp: 0–3, Fn: 14–19 (TO), 10 (NTO)	15 min analysis, 50 samples for $220, additional to $1,600

Data from June 2002, August 2003; source: Chapter 8, Ahuja (2008). NTO, non-technical operator; TO, technical operator.
[1]*Relative standard deviation.*
[2]*Linearity = slope X reference value + offset.*
[3]*Tested for high levels of sodium sulfate, iron, or acidity.*

although the new generation of kits has become much more reliable. The report of false-negative and false-positive results of over 30% is not unusual, although the latest figures seem encouraging, and more reliable measurements can be done in the field. However, these studies used a water standard of 50 µg/L as a decisive concentration. If the new WHO guideline of 10 µg/L is adopted as a decision-making criterion, the sensitivity of most arsenic-testing kits based on colorimetric methods will not be sufficient. This is particularly true in the case for kits that are battery powered and also for the electronic systems. Although some reports surprisingly suggest that in a number of cases untrained operators produce more reliable results, the training aspect of the operators should not be underestimated.

To obtain a fast and accurate measurement of arsenic in the field still remains a significant challenge. The quartz crystal microbalance, a device whose interface is more robust than an electrode for stripping voltammetry, holds promise especially as the measurement incorporates an inherent preconcentration step (the accumulation of arsenic at the surface of the oscillating crystal). Voltammetric sensors could be ideally suited for on-site analysis of arsenic. However, the need of the chemical reduction step seems to be the major problem, limiting both potential application in the field and sample throughput. Although promising results have been obtained using voltammetric systems, it is essential to develop methods for determining the arsenic species. The most promising development in direct arsenic speciation is by electrochemical detectors, but they still must be tested in the field.

ARSENIC-FREE WATER SUPPLIES

Two options for a safe water supply are the development of water-supply systems avoiding arsenic-contaminated water sources and the removal of arsenic to acceptable levels (see Chapter 14 in Ahuja, 2008b). Totally arsenic-free water is hard to find in nature; hence, the only viable option for avoiding arsenic is to develop water-supply systems based on sources having very low arsenic content. Rainwater, well-aerated surface water, and groundwater in very shallow wells and in deep aquifers are well-known sources of low-arsenic water. The arsenic content of most surface water sources varies from <1 up to 2 µg/L. Very shallow groundwater replenished by rainwater or surface water and by relatively old, deep aquifers shows arsenic content within acceptable levels.

The technologies for producing drinking water using the sources that are known to have a low arsenic content include the following:

- Treatment of surface water by slow sand filtration, conventional coagulation–sedimentation–filtration, and disinfection is effective.
- Rivers, lakes, and ponds are the main sources of surface water, and the degree of treatment required varies with the level and type of impurities present in water.

- Dug wells/ring wells or very shallow tube wells provide low-arsenic groundwater from very shallow aquifers.
- Deep tube wells (DTW) from deep protected aquifers are a good source of safe drinking water.
- A rainwater harvesting system (RWHS) to collect and store rainwater is another viable system.

Arsenic-removal technologies have improved significantly during the last few years, but reliable, cost-effective, and sustainable treatment technologies have not yet been fully identified.

Remediation of Arsenic-Contaminated Water

A large number of approaches have been investigated for removing arsenic from drinking water. Several useful reviews of the techniques for removing arsenic from water supplies have been published (Bissen and Frimmel, 2003a, b; Ng et al., 2004; Ahuja, 2005, 2006, 2008b; Daus et al., 2005; Ahmed et al., 2006; Ahuja and Malin, 2006). Various existing and emerging arsenic removal technologies are listed here.

- Coagulation with ferric chloride, alum, or natural products
- Sorption on activated alumina
- Sorption on iron oxide–coated sand particles
- Granulated iron oxide particles
- Polymeric ligand exchange
- Nanomagnetite particles
- Sand with zero valent iron
- Hybrid cation-exchange resins
- Hybrid anion-exchange resins
- Polymeric anion exchange
- Reverse osmosis

Reverse osmosis is essentially a nonselective physical process for excluding ions with a semipermeable membrane. The basic chemistry for the rest of the processes includes either one or both of the following interactions (Ahuja and Malin, 2004). As(V) oxyanions are negatively charged in the near-neutral pH range and therefore can undergo coulombic or ion-exchange types of interactions (Ahuja, 2008a). As(V) and As(III) species, being fairly strong ligands or Lewis bases, are capable of donating lone pairs of electrons. They participate in Lewis acid–base interactions and often show high sorption affinity toward the solid surfaces that have Lewis acid properties.

Flocculation of Arsenic

There is a need for a benign and sustainable water purification technology based on natural products because of their inherently renewable character, low cost, and low toxicity. The use of mucilage, derived from the nopal cactus *Opuntia ficus-indica*, can provide reliable methods to treat drinking water supplies that have been contaminated with particulates and toxic metals (see Chapter 9 in Ahuja, 2008b). A study has been conducted to develop an optimized system for rural and underdeveloped communities in Mexico, where drinking water supplies are contaminated with toxic metals and the nopal cactus is readily available and amenable to sustainable agriculture. Comparison with aluminum sulfate (a synthetic flocculant) shows the high efficiency of cactus mucilage to separate particulates and arsenic from drinking water. Further investigations are required to determine the feasibility of implementing this technology for small-scale household units.

Arsenic Removal by Adsorptive Media

Inexpensive, rapid tests are needed to predict the arsenic adsorption capacity of adsorptive media to help communities select the most appropriate technology for meeting compliance with the new arsenic MCL of $10\,\mu g/L$. A study was performed to evaluate alternative methods to predict pilot-scale and full-scale performances from laboratory studies (see Chapter 10 in Ahuja, 2008b). Three innovative adsorptive media that have the potential to reduce the costs of arsenic removal from drinking water were selected. Arsenic-removal performance of these different adsorptive media under constant ambient flow conditions was compared, using a combination of static (batch) and dynamic flow tests. These included batch sorption isotherm and kinetic sorption studies, rapid small-scale column tests (RSSCT), and a pilot test at a domestic water supply well. The media that were studied include a granular ferric oxyhyroxide (E33), a granular titanium oxyhydroxide (Metsorb), and an ion-exchange resin impregnated with iron oxide nanoparticles (ArsenX[np]). They exhibited contrasting physical and chemical properties. The E33 media gave the best performance, based on the volume of water treated until breakthrough at the arsenic MCL (10 ppb) and full capacity at media exhaustion.

Iron Oxide–Coated Coal Ash

A simple technique was developed for removing arsenic from water using fine particles of coal bottom ash that are coated with iron oxide (see Chapter 11 in Ahuja, 2008b). The bottom ash is the ash left at the bottom of a coal-fired boiler after the combustible matter in coal has been burned off. Reduction of the arsenic concentrations to less than the Bangladesh standard of 50 ppb in six of the eight samples of Bangladesh groundwater has been demonstrated. It is believed that a larger dose of coal bottom ash coated with iron oxide would have certainly lowered the concentration to below 50 ppb in those failed samples, because the study also demonstrated the feasibility in some samples of reducing arsenic concentrations in the water to below 10 ppb. Prior to further use of this system, it is necessary to investigate whether any potential contaminants from bottom ash would be released into drinking water.

Composite Iron Matrix Filter

The development and deployment of a water filter based on a specially made composite iron matrix (SONO filter) for the purification of groundwater to safe potable water has been described at length (see Chapter 12 in Ahuja, 2008b). The manufacturer claims that filtered water meets WHO and Bangladesh standards, has no breakthrough, works without any chemical treatment (pre or post), without regeneration, and without producing toxic waste based on EPA guidelines. It costs about \$40, lasts for 5 years, and produces 20–30 L/h for daily drinking and cooking needs of one to two families. Approved by the Bangladesh government, about 35,000 SONO filters are deployed all over Bangladesh and continue to provide more than a billion liters of safe drinking water. This innovation was recognized by the National Academy of Engineering Grainger Challenge Prize for Sustainability with the highest award for its affordability, reliability, ease of maintenance, social acceptability, and environmental friendliness (see Chapter 15). The filter requires the replacement of the upper sand layers when the apparent flow rate decreases. Experiments show that the flow rate may decrease 20–30% per year if the groundwater has high iron levels ($>5\,mg/L$) because of the formation and deposition of natural hydrous ferric oxide (HFO) in sand layers. The sand layers (about an inch thick) can be removed, washed, and reused, or replaced with new sand.

Pathogenic bacteria can still be found in drinking water because of unhygienic handling practices and in many shallow tube wells, possibly because they are located near unsanitary latrines and ponds. A protocol for their elimination must be used once in a week in areas where coliform counts are high. It should be noted that, as with all commercial filters, the consumer needs to be alert to manufacturing defects, quality of water related to natural disasters such as flooding, and mechanical damage because of mishandling and transportation.

Wellhead Arsenic-Removal Units

In many remote villages in West Bengal, India, arsenic-contaminated groundwater remains the only viable source of drinking water. Cost-effective arsenic-removal technology is thus a bare necessity to provide safe drinking water, the groundwater is free of other contaminants and is considered safe for drinking. Over 150 wellhead arsenic-removal units, containing activated alumina as the adsorbent, are currently being operated by local villagers in this Indian state that borders Bangladesh (see Chapter 13 in Ahuja, 2008b). The units are maintained and run by the beneficiaries and do not require any chemical addition, pH adjustment, or electricity for their regular operation. Each of the units serves approximately 250–350 families living within a short distance of the unit, and the flow rate is modest at approximately 10 L/min. Arsenite as well as arsenate from groundwater are effectively removed to render the water safe for drinking and cooking. Regenerateness and durability of the adsorbent allows for a low cost, sustainable solution for the widespread arsenic poisoning in this area. After regeneration, the spent regenerants containing a high concentration of arsenic are converted to a small-volume sludge that is stored under oxidizing conditions to prevent future arsenic leaching. It has been claimed that this process offers superior economic advantages in regard to treatment and management of dangerous treatment residuals, compared with conventional adsorbent-based processes where regeneration and reuse are not practiced. With conventional processes where the adsorbents are treated as garbage, huge amounts of media in landfills leach out dangerous concentrations of arsenic. A global scheme for the overall process of arsenic removal, including the management of treatment residues, has been provided. Input to the process is groundwater contaminated with arsenic and caustic soda and acid for regeneration, whereas the output is treated drinking water and neutralized brine solution. Thus, the technology, besides being appropriate for the rural settings of the affected area in terms of ease of use and economics, also offers considerable ecological sustainability.

It has been estimated from the data of 150 running units that the total volume of water treated by a unit in 1 year, on average, was about 8,000 bed volumes, i.e., 800,000 L. The calculated cost of the water/1000 L is 85¢ U.S. The estimated amount of arsenic-safe water used for a family of six for drinking and cooking purposes in a month at the rate of 5 L *per capita* per day is 900 L. The water tariff for a family of six for 1 month is around 75¢ U.S. or 30 Indian rupees. While regeneration helps reduce the volume of the sludge by about 150 times, reusability of the adsorbent media significantly helps decrease the cost of the treated water.

Reasons for Slow Progress

Since the recognition of the arsenic-contaminated water problem several decades ago, many efforts have been made to solve it. However, the advancement to date has been poor.

The progress in arsenic mitigation has been very slow, as is indicated by the fact that only about 4 million people in Bangladesh have been provided with arsenic-safe water during the last 5 years. A study on the progress of arsenic-mitigation options, the trend in the installation of different arsenic-mitigation technologies and operational monitoring and the evaluation of performance of technologies revealed the following constraints in the progress in arsenic-safe water supplies:

- The problem of the selection of appropriate technology for arsenic mitigation in different areas still remains a major hindrance.
- The trial of prioritized options in the implementation plan before the installation of an appropriate technology in an area is an impractical, time- and resource-consuming approach.

- The overwhelming demand of deep tube wells from communities and local arsenic committees restricts the installation of other technologies prioritized in the implementation plan.
- There are abundant arsenic-removal technologies; however, poor water quality obtained from some of the arsenic-mitigation technologies has deterred the implementing agencies from the deployment of these technologies.
- The implementation of a national policy has received poor support from donor agencies.

While many technologies have been developed to treat arsenic-contaminated water, on scales ranging from individual family filters that sell for approximately $40 to very expensive industry-sized plants, none has yet emerged as optimal for the conditions encountered. In most cases, the materials used are not fully characterized, and the systems sold commercially have not been fully validated. However, while it is relatively easy to remove arsenic by adsorption on supported iron oxides, small point-of-use filters may become clogged after an indeterminate period of time. It should be noted that no provision has been made to assure that systems are working at the time of initial usage or that they remain functional when they have been in use for a period of time. Finally, technologies must be developed to safely dispose of the waste or to recycle the active materials.

Even in advanced countries such as the United States, arsenic-removal technologies are scarce; the few that are available are generally very expensive. They are needed in communities where well water is used for drinking and cooking. It is anticipated that the family- or community-level arsenic-removal technologies that are being developed for Bangladesh, which are also economically and environmentally sustainable, can be replicated or further improved for use in developing and developed countries where arsenic poisoning is a menace.

VIABLE SOLUTIONS

After thorough consideration of the National Policy for Arsenic Mitigation of Bangladesh issued in 2004 and inputs from various participants in a CHEMRAWN conference (Ahuja and Malin, 2004, 2006), Dhaka workshop (Ahuja, 2005), and Atlanta symposia (Ahuja, 2006), and a trip to West Bengal, India, in 2007, the following recommendations appear to be logical for Bangladesh and other countries in South Asia that are most severely affected by this problem:

1. Piped surface water should be the intermediate to long-term goal and should be given the desired priority. This will require total commitment from local governments and the funding agencies that deem this a desirable option. Along these lines, other surface water options such as rainwater harvesting, sand filters, dug wells, etc., should be tapped as much as is reasonably possible.
2. The next best option is safe tube wells. More than likely they would be deep tube wells. It is important to assure that they are located properly and do not

contain other contaminants that can add to the arsenic problem. Furthermore, they should be installed properly such that they are not prone to surface contaminants.

3. Arsenic-removal filtration systems can work on a small scale; however, their reliability initially or over a period of time remains an issue. Other contaminants in water including microbial contamination can affect their performance. Low-priced reliable test kits are needed that can address this issue. There is a need to identify dependable filters that can be scaled up for larger communities. In this manner, both maintenance and reliability issues can be addressed.

4. The education and training of local scientists and technicians need to be encouraged so that local people can address these problems themselves. There is a need for more analytical scientists, low-priced instrumentation, and testing laboratories. The consumers of contaminated water need to be better educated so that they do not continue drinking contaminated water because of their reluctance to either switch wells or take other steps to purify water.

CONCLUSIONS

Arsenic contamination of groundwater can seriously affect as many as 200 million people worldwide. The problem occurred in Bangladesh because of inadequate testing of the wells. Nearly three decades later, the problem still festers; it demands an expeditious solution. A number of viable solutions are offered here. The discussion of the advantages of safe water supply options for Bangladesh, including pond sand filters, river sand filters, rainwater harvesting, dug wells; sharing safe shallow tube wells and deep tube wells; and arsenic-removal technologies must integrate water hygiene and sanitation programs. The application of some of these options depends on local conditions. It is important to remember that local scientists and other well-meaning people are the final arbiters as to what is best for their area.

REFERENCES

Ahmed, M.F.S., Ahuja, S., Alauddin, M., Huq, S.J., Lloyd, J.R., Pfaff, A., 2006. Ensuring Safe Drinking Water in Bangladesh. Science 314, 1687–1688.

Ahuja, S., 1986. Ultratrace Analysis of Pharmaceuticals and Other Compounds of Interest. Wiley, New York.

Ahuja, S. 2005. Origins and Remediation of Groundwater Contamination by Arsenic: Objectives and Recommendations. In: International Workshop on Arsenic Contamination and Safe Water, 11–13 December, Dhaka.

Ahuja, S. 2006. Assuring water purity by monitoring water contaminants from As to Zn. American Chemical Society Meeting, 26–30 March 2006, Atlanta, GA.

Ahuja, S., 2008a. Arsenic Contamination of Groundwater: A Worldwide Problem. UNESCO Conference on Water Scarcity, Global Changes, and Groundwater Management Responses, 1–5 December, 2008, Irvine, CA.

Ahuja, S., 2008b. Arsenic Contamination of Groundwater: Mechanism, Analysis and Remediation. Wiley, New York.

Ahuja, S., Malin, J., 2004. A search for solution to the problem of arsenic contamination of water in Bangladesh. International Conference on Chemistry for Water, 21–23 June 2004 Paris.

Ahuja, S., Malin, J., 2006. Chem. Int. 28 (3), 14–17.

Bissen, M., Frimmel, F., 2003a. Arsenic—A review. Part I: occurrence, toxicity, speciation, mobility. Acta Hydrochim. Hydrobiol. 31, 9–18.

Bissen, M., Frimmel, F., 2003b. Arsenic—A review. Part II: oxidation of arsenic and its removal in water treatment. Acta Hydrochim. Hydrobiol. 31, 97–107.

Daus, B., Weiss, H., Wennrich, R., 2005. Sorption materials for arsenic removal from water: a comparative study. Water Res. 38, 2948–2954.

Ng, K.-S., Le-Clech, P., Ujang, Z., 2004. Arsenic removal technologies for drinking water treatment. Rev. Environ. Sci. Biotechnol. 3, 43–53.

Chapter 3

Water Quality Issues in Eastern Africa

Shem O. Wandiga and Vincent O. Madadi

Department of Chemistry, School of Physical Sciences, College of Biological and Physical Sciences, University of Nairobi, Nairobi, Kenya

INTRODUCTION

At no time has the world faced such acute shortage of potable water than today. This is despite the fact that 75% of the world's surface is covered with water. Unfortunately 97% of this water is saline and therefore undrinkable. Of the available freshwater, 2% is locked up in ice caps and glaciers. Hence there is only about one percent of water available to meet human needs. Further analysis of the available water shows that the blue water locked up in lakes and reservoirs and the green water (rainfall water) have become increasingly polluted with human, industrial and agricultural wastes, and cosmetic chemicals. Scientific challenges exist in determining the quantities of such wastes and their breakdown products, effects on life and environment, and how best to control their distribution in the environment. This chapter describes some of the challenges existing in East Africa.

CLIMATE CHANGE IN EASTERN AFRICA AFFECTS WATER AVAILABILITY

Climate change has added to the urgency of need for water conservation. This is because the climate change has altered the temperature and rainfall patterns in

Eastern Africa and the remainder of the continent. The temperature scenario for Africa described by Hulme et al. (2001), shows that global mean surface temperature is projected to increase between 1.5°C (2.7°F) and 5.8°C (10.8°F) by 2100; climate change scenarios for Africa indicate future warming across the continent ranging from 0.2°C (0.36°F) per decade (low scenario) to more than 0.5°C (0.9°F) per decade (high scenario), and this warming will be greatest over the interior of semiarid margins of the Sahara and central southern Africa. Figure 1 gives mean surface air temperature anomalies for the African continent,

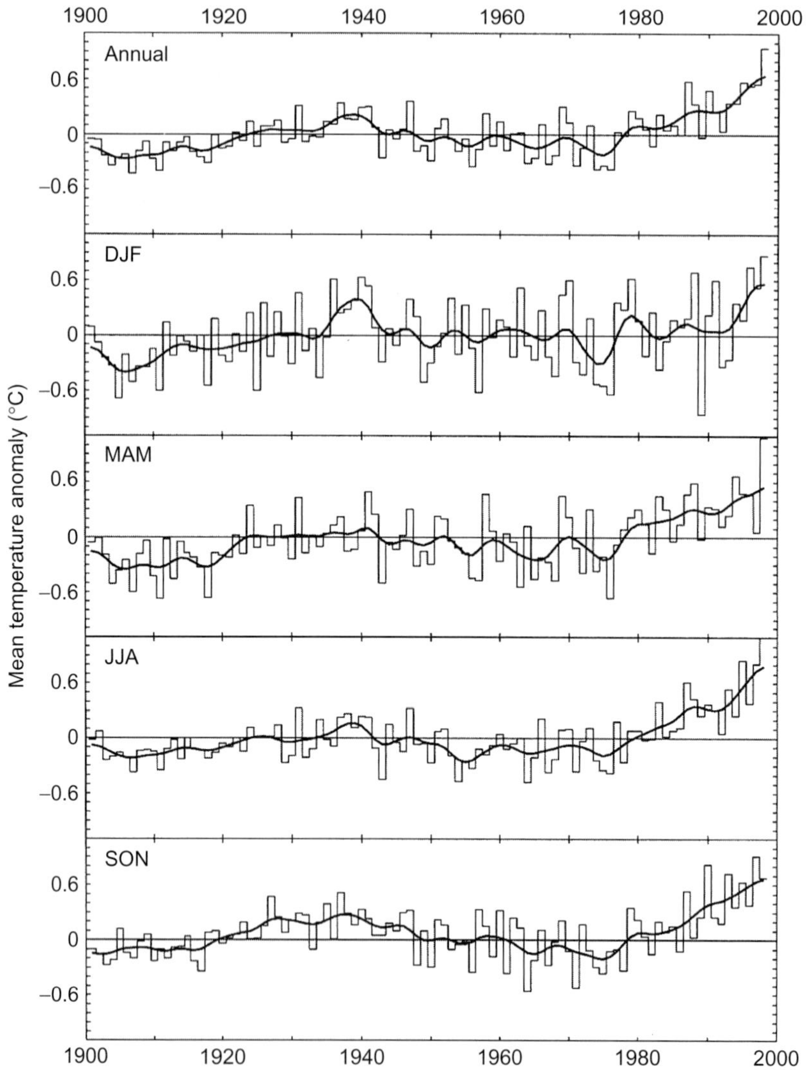

FIGURE 1   Mean surface air temperature anomalies for the African continent, 1901–1998, expressed with respect to the 1961–1990 average; annual and four seasons—DJF, MAM, JJA, SON.

1901–1998, expressed with respect to the 1961–1990 average; annual and four seasons—December, January, February (DJF); March, April, May (MAM); June, July, August (JJA); September, October, November (SON). The smooth curves result from applying a 10-year Gaussian filter (Hulme et al., 2001). The figure shows warming trend in all seasons. The fourth Intergovernmental Panel on Climate Change (IPCC [AR4], 2007) concludes that "changes in climate are now affecting physical and biological systems on every continent. Effects on human systems, although more difficult to discern because of adaptation and non-climatic drivers, are emerging. Over 90% of observed changes in systems and sectors are consistent with regional temperature trends. Many of the changes are now attributed to temperature increase caused by anthropogenic greenhouse gas emissions." Warmer climate increases evaporation of water as well.

Rainfall is similarly affected under climate change scenarios. Figure 2 gives the projected rainfall pattern for various African regions (Hulme et al., 2001).

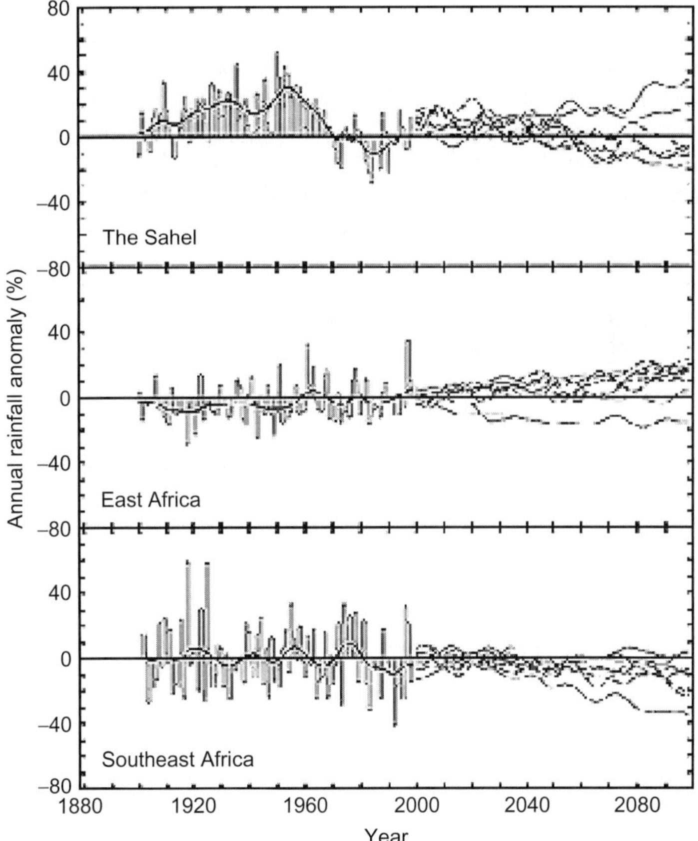

FIGURE 2   The projected trend in rainfall anomalies (in percent) in the Sahel, East, and southern Africa, 1880–1988 actual average annual rainfall, 1999–2090 projected average annual rainfall (Hulme et al., 2001).

The rainfall trend shows decline of rainfall in southern Africa but some probable increase in the Sahel and East Africa. However, the quantity of rainfall will vary even within a country.

Sub-Saharan Africa is likely to experience large impacts from climate change, because of high levels of intrinsic vulnerability stemming from heavy dependence on rain-fed agriculture, propensity to drought, and relative low levels of adaptive capacity. The proportion of the African population at risk of water scarcity could increase from 47% in 2000 to 65% in 2025, when about 370 million African people may experience increased water stress based on a range of Special Report on Emission Scenarios (SRES) of IPCC and other scenarios as observed by IPCC (AR4) (2007). Nonclimatic changes such as water policy and management practice may add significant effects. Changes in primary production of large lakes will have important impacts on local food supplies. For example, Lake Tanganyika currently provides 25–40% of animal protein intake for the population of the surrounding countries and, on the basis of observed and paleoclimate records, it is expected that climate change will reduce catches by 30% (IPCC [AR4], 2007). East African countries already have signs of water scarcity because of various reasons.

Some of the basic problems with water as a resource in Africa include the very high potential evaporation, which occurs throughout the year and is in excess of 2,000 mm p.a. over large tracts; very high aridity indices; a generally low ration of conversion of rainfall to runoff; an often very concentrated seasonality of rainfall, and hence runoff; a strong response to the El Niño Southern Oscillation (ENSO) signal and thus generally high interannual coefficient of variability of rainfall; an amplification of the interannual coefficient of variability of rainfall by the hydrological cycle (Schulze, 2000). Other problems that affect the quantity, quality, and availability of freshwater include land use leading to enhanced erosion/siltation and possible ecological consequences of land-use change on the hydrological cycle.

DRASTIC WATER QUALITY CHANGES IN EAST AFRICA

The quality of blue water in East Africa has experienced drastic changes in the last 40 years as evidenced by the onset of massive cyanobacteria blooms offshore that took place in Lake Victoria. The poor water quality has led to the collapse of the indigenous fish stock. The collapse has also been correlated to the poor agricultural practices that have led to an accelerated rate of deposition and sedimentation of soil rich nutrients (Verschuren et al., 2002) (Figure 3). On top of activities related to human population and agricultural production is added the human abuse of water systems. Today most of our lakes and rivers are choking with wastes that are wantonly dumped into them. It is possible to prevent this "I don't care attitude" of our people by enacting stringent regulations on water pollution by governments. Investment in water cleaning technologies will greatly assist in water purification.

An example of waste pollution of water systems is given in Figure 4, which presents a case of polluted River Ngong in Nairobi. It is evident from the figure

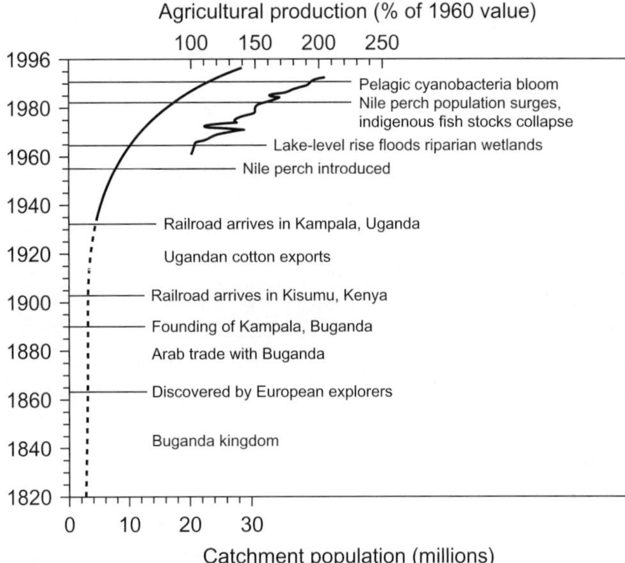

FIGURE 3 Principal events in the recent environmental history of Lake Victoria basin in relation to human population growth and agricultural production (Verschuren et al., 2002).

FIGURE 4 Ngong River pollution, Nairobi, Kenya (Wandiga et al., 2006).

FIGURE 5   Soil erosion, sedimentation, and nutrient loading in Lake Victoria (Photo: ICRAF).

that citizens living along this river use it as their septic tank. Huge investments are required to restore and rehabilitate the river to its original conditions.

It is incomprehensible what dwarfs the human mind to a level of filth glorification. The Ngong River joins the Nairobi River and Athi River, which are sources of water for the communities living downstream. Consideration of the welfare of other water users would dictate better treatment of the water system than is presently the case.

CHALLENGES FACING LAKE VICTORIA

Several challenges face water systems such as Lake Victoria. Some of the recent challenges are human activities contributing to chemical pollution, such as washing clothes with phosphate detergents, car wash, and agricultural sediments.

The International Centre for Research in Agroforestry (ICRAF) has captured an aerial view of nutrients and sediments floating in Lake Victoria (Figure 5). The photo shows a rich variation of different particles of nutrients that have accelerated the eutrophication of the lake. Eutrophication rapidly changes the surface of the water system, and in a few months one finds thick growth of various macrophytes.

The sediments loading on the Lake Victoria and other lakes have been estimated by Verschuren et al. (2002) and are given in Table 1.

The sediment load affects the lakes differently depending on its depth. Lakes Malawi and Tanganyika are deep lakes, and the effect of siltation is not currently observed. However, for a shallow lake such as Lake Victoria, siltation has reduced the depth from about 80 m in the deepest area to an average of 40 m today. The lake is receding and this may be attributed to some extent to siltation and climate change. In addition to the lake's depth, bottom oxygen content has declined and transparency (the Secchi index) decreased from 5 m in 1930 to less than 1 m in

TABLE 1 Sediment Impact on the Lakes as a Result of Human Activities Such as Deforestation and Agriculture

Lake Tanganyika	1,500 mm/1000 year S. Basin 500 mm/1000 year C. Basin 4,700 mm/100 year N. Basin
Lake Turkana	Oromo River Impact (1,600 t/km^2/a)
Lake Victoria	0.032–0.001 g/cm^2/a 2.3 mm/year
Lake Malawi	1000 mm/year

the 1990s, whereas sediment and water phosphorus and nitrogen concentrations have also increased (Hecky, 1993). Poor agricultural practices—deforestation, desegregation of farms, overstocking of livestock, lack of antierosion techniques, degradation of floodplain wetland buffers—have all resulted in massive soil erosions and nutrient loads into the lake that have altered its ecology.

The consequences of these changes have resulted in:

- increased soil erosion and vulnerability to floods and droughts
- loss of soil fertility, declining crop yields
- food insecurity
- loss of dry season gardening, livestock husbandry capability and groundwater recharge, and discharge due to degradation of floodplain wetlands

HUMAN WASTE DISPOSAL: AN IGNORED FACTOR

One of the scientific challenges facing policy makers in equatorial Africa is to ascertain the source of phosphorus that causes eutrophication in the equatorial lakes. The Lake Victoria Environmental Management Project (LVEMP) (2001, 2002) estimated the nutrients input into the lake with atmospheric deposition contributing 102,000 t/year of total N and 24,000 t/year of total P. These values indicated that the atmospheric deposition is by far the most significant contributor to the overall nutrient budget of the lake. Several authors have given various estimates as shown in Table 2.

As Table 2 indicates the atmospheric phosphorus deposition debate has been with us for the last 10 years. Given the fact that there are no phosphorus mines within the Lake Victoria catchments, points to the possibility of a large source of atmospheric P come from human activities such as biomass burning. Evidence of large-scale biomass burning is not there. Further given the fact that the wind regime around the lake is influenced by NW and SW monsoons, it is doubtful

TABLE 2 Various Estimates of Amounts of Atmospheric Deposition of Phosphorus (in %)

Auothor(s)	Agriculture	Urban	Atmosphere
Bullocks et al. (1995)	50	30	20
Scheren (1995)	25	na	75
Lindenschmidst et al. (1998)	57	6	37
LVEMP (2002)	2	18	80

that if these estimates give a clear picture of the source of the P atmospheric load. A new project is being launched under the Global Environment Facility (GEF), United Nations Environment Programme (UNEP)/World Bank execution to explore the origin of phosphorus in equatorial lakes. In the mean time, our studies have started to shed light on the sources of P in Lake Victoria waters and sediments.

One of the major sources of water pollution in East Africa is human waste. The effluents from untreated municipal sewers pose great danger to Lake Victoria sustainable ecological conservation. Municipal sewers contain both feces and urine that are the sources of phosphorus; therefore, let us make two assumptions that will enable us calculate the P contributions from these sources. Let us first assume that each person produces 25–50 kg of feces per year, which contains 0.18 kg P; second, assume that each adult produces about 400 L of urine per year, depending on liquid consumption, and it contains 0.40 kg P. This is because municipal and industrial wastewater treatment plants (WWTPs) are known to be the major source points of phosphorus in urban areas (Smith et al., 1999). Waste disposal sites, construction sites, fertilizers, and farmyards also make substantial contributions to the total phosphorus load (Hooda et al., 2000; Morgan et al., 2000; Sharpley et al., 2000; Tunney ct al., 2000). However, all these have not been adequately evaluated.

Given the number of sewered and unsewered municipalities and their populations in Table 3, one is able to calculate the amount of phosphorus produced.

Using these figures it is therefore possible to calculate the phosphorus load coming from this source as given in Table 4.

If we assume that the sewage–treatment plants are efficiently working, an assumption that we know is not true, then the phosphorus (P) contribution from unsewered municipalities is about 396,865 t/year—a figure which is much larger than that reported in the LVEMP (2001, 2002) reports. If we add the P contribution from sewered municipalities with inefficient operations, then we have even a larger figure of about 500,423 t/year. However, if we assume that only 20% of P from this source reaches the lake, then we calculate a figure of 79,373–100,085 t/year. However, it is not possible that only 20% of P nutrient

TABLE 3 The Population of Sewered and Unsewered Municipalities in the Lake Victoria Catchments

Country	Total Population (1,000 people)	Sewered Urban Population (×1,000)	Unsewered Urban Population (×1,000)	Number of Towns
Kenya	10,200	390	630	18
Uganda	5,600	210	870	9
Tanzania	5,200	27	340	4
Rwanda	5,900	–	400	5
Burundi	2,800	–	140	4
Total	29,700	627	2,380	40

TABLE 4 Calculated Phosphorus Discharge into the Lake from Sewered and Unsewered Municipalities

	Feces				Urine
Country	Sewered Pop*	Phosphorus (t/year)	Unsewered Pop*	Phosphorus (t/year)	Phosphorus (Sewered–Unsewered) (t/year)
Kenya	390	1,755–3,510	630	2,835–5,670	61,400–100,800
Uganda	210	945–1,890	870	3,915–7,830	33,600–139,200
Tanzania	27	122–243	340	1,530–3,060	4,320–54,400
Rwanda	–	–	400	1,800–3,600	–64,000
Burundi	–	–	140	630–1,260	–22,400
Total	627	2,822–5,643	2,380	10,710–21,420	99,320–380,800

Pop = population multiplied by 1,000.

from this source will reach the lake, but even this assumption gives a figure that is still larger than that reported in the LVEMP reports.

We have followed this theoretical calculation of P content with measurements of the element in water, soil, and sediments. Water, soil, and sediments were sampled for 2 years covering four different seasons per year (wet, two dry,

TABLE 5 Pearson Correlation Analysis for Phosphates in Water

		Wet Season	Dry Season 1	Short Rain	Dry Season 2
Wet season	Pearson correlation	1.000	0.938*	0.966*	0.966*
	Sig. (two-tailed)	–	0.000	0.000	0.000
	N	32	32	32	32
Dry season 1	Pearson correlation	–938	1.000	0.926*	0.873*
	Sig. (two-tailed)	0.000	–	0.000	0.000
	N	32	32	32	32
Short rain	Pearson correlation	0.966*	0.926*	1.000	0.969*
	Sig. (two-tailed)	0.000	0.000	–	0.000
	N	32	32	32	32
Dry season 2	Pearson correlation	0.966	0.873	0.969*	1.000
	Sig. (two-tailed)	0.000	0.000	0.000	–
	N	32	32	32	32

Correlation is significant at the 0.01 level (two-tailed).

and short rain seasons) experienced in the Kenyan Lake Victoria catchments. Sampling points covered lakeshores, river mouths, and effluent discharge points; and the parameters analyzed included total reactive phosphates, total hydrolyzable phosphates, total phosphate, sediment exchangeable phosphates, sediment bioavailable phosphate, and soil available phosphates. Soils from the catchments were found to contain 10–100 times higher concentration compared to sediments and water samples. Water from both the rivers and the lake were found to contain phosphate levels much higher than the recommended guidelines for aquatic life, indicating the influence of anthropogenic sources. The seasonal average of total phosphate in the water was 4.61, 3.43, 2.45, and 2.30 mg/L for wet, short rain, and dry seasons 1 and 2, respectively, whereas the total reactive phosphates had means of 2.22, 2.08, 1.12, and 1.61 mg/L in the same seasons. Sediment bioavailable phosphates were higher than exchangeable phosphates, with the highest mean concentrations of 24.45 and 8.22 mg/kg occurring during the dry season, whereas average of soil available phosphorus ranged between 639 and 1,076 mg/kg. Pearson correlation analysis of the data indicated a strong positive correlation between the levels detected in the water for all the seasons, implying increasing accumulation of the nutrients in the drainage basin (Table 5).

Figures 6–8 give the water, sediment, and soil bioavailability phosphates. The highest concentrations were observed at municipalities that have no sewer treatment. Secondly, phosphate levels in the Lake Victoria drainage system vary with seasonal changes. The levels in soils along the shore are higher than at the

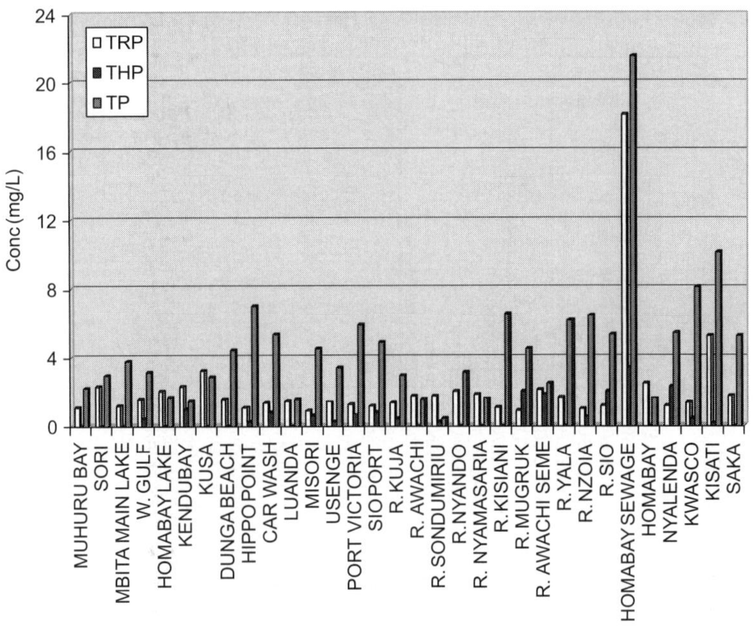

FIGURE 6   Phosphates in water during wet season.

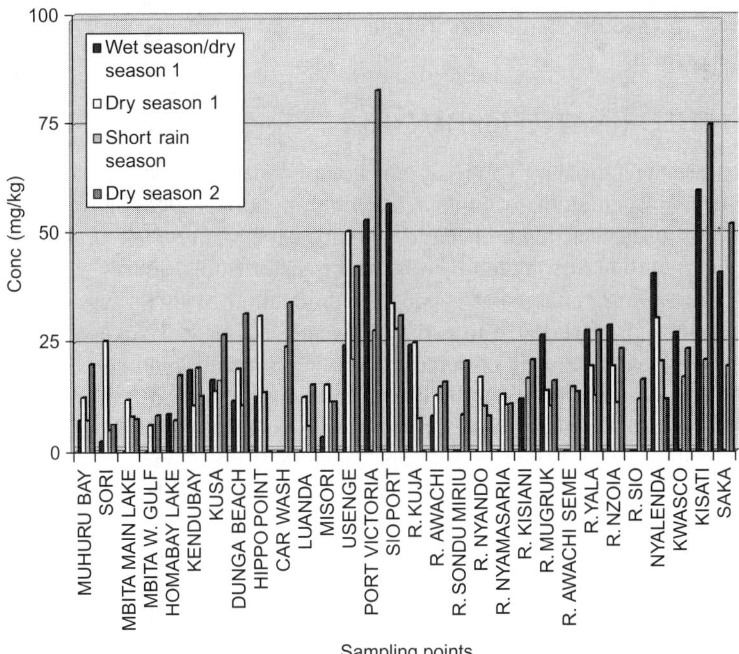

FIGURE 7   Sediment bioavailable phosphates.

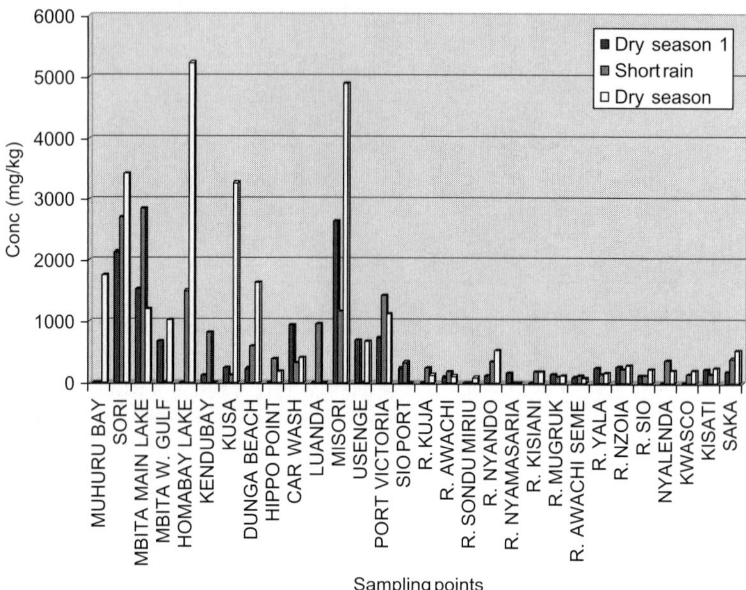

FIGURE 8   Soil available phosphates.

river mouth, and the sites where there is direct sewage discharge are leading in the concentration of phosphates. Furthermore, the high nutrient load at points of direct sewage discharge also stimulates the growth of macrophytes such as water hyacinth.

POLYCHLORINATED BIPHENYLS

Polychlorinated biphenyls (PCBs) are compounds derived from substitution of 1–10 hydrogen atoms of biphenyl by chlorine atoms. PCBs have been produced and used worldwide in large quantities for many years as transformer oils, metal-cutting oils, hydraulic oils, heat transfer fluids, additives in plastics, dyes, and carbonless copying paper. The production of PCBs was terminated worldwide in the late 1970s to early 1980s because of their adverse effects on the environment as a result of persistency, bioaccumulation properties, and toxicity. The amount of PCBs in the global environment has been estimated to be about 3.7×10^8 kg, and further 7.8×10^8 kg was estimated to be still available for utilization or deposited in different ways. PCB poisoning can lead to liver damage, respiratory disorders, various neurological symptoms, and reproduction disorders. Hence, it is of great importance to assess the source of PCBs that may be detrimental to the lake ecosystem and ultimately to man.

We have undertaken analysis of PCBs along the shores of Lake Victoria. Sampling stations were chosen to capture upstream activities, runoffs and sediments brought along the rivers as the source of PCBs. Three composite sites

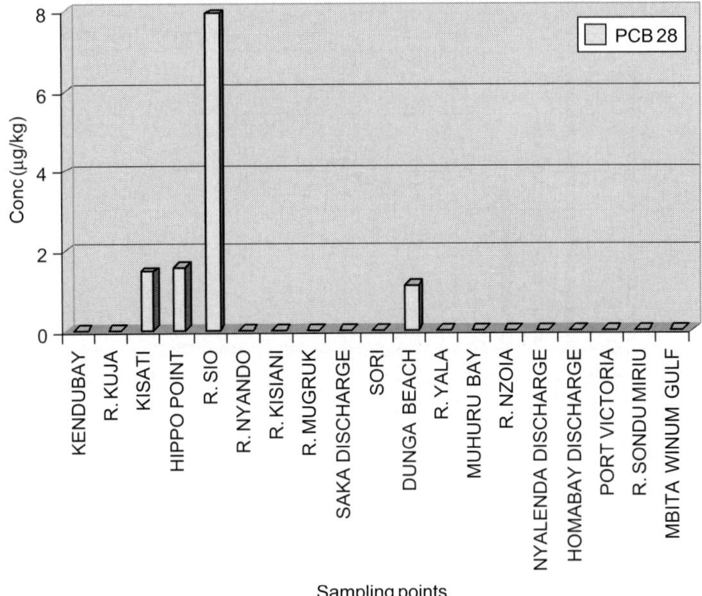

FIGURE 9   Concentrations of PCB 28 in sediments at different sampling stations in Kavirondo Gulf of Lake Victoria catchments.

surface sediments were sampled with a stainless steel shovel and then combined to one pole sediment sample because PCBs in sediments can show a patchy distribution. Sediment samples were wrapped with aluminum foil, labeled, and stored at −25°C. Figure 9 shows the results of PCB 28 analysis along the Kavirondo Gulf (Kenyan side) of Lake Victoria.

High concentration of PCB 28 was found at River Sio, Hippo point, Kisati, and Dunga Beach. However, it is curious to note the presence of this compound at these sites and the environment in general. Similarly, Figure 10 shows the concentrations of PCB 52; the highest concentrations were found at the mouth of River Kuja and River Kisiat. River Kuja drains the Kisii highlands where mainly high agricultural activities take place, whereas River Kisiat drains the industrial area of Kisumu Town. The third highest concentration was found at Hippo point, which is close to the sewage discharge point of Kisumu Town. Low concentrations were observed at other sampling points.

We also carried out analysis for PCB 105 at the shore sampling sites of the Kavirondo Gulf of Lake Victoria. Figure 11 gives the analysis results. High concentration of PCB 105 was found in the sediments of River Sio mouth, followed by Muhuru Bay and Port Victoria. Generally, high concentration of PCB 105 is observed at the river mouths and the bays.

Further analysis of representative congeners of PCB included that of PCB 118. High concentration of PCB 118 was found in the sediments of Muhuru Bay,

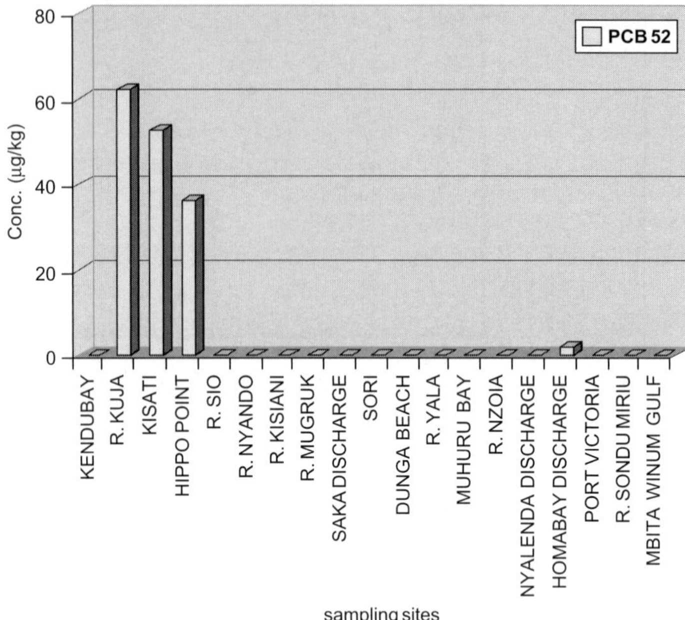

FIGURE 10 Concentrations of PCB 52 in sediments at different sampling stations of Kavirondo Gulf of Lake Victoria catchments.

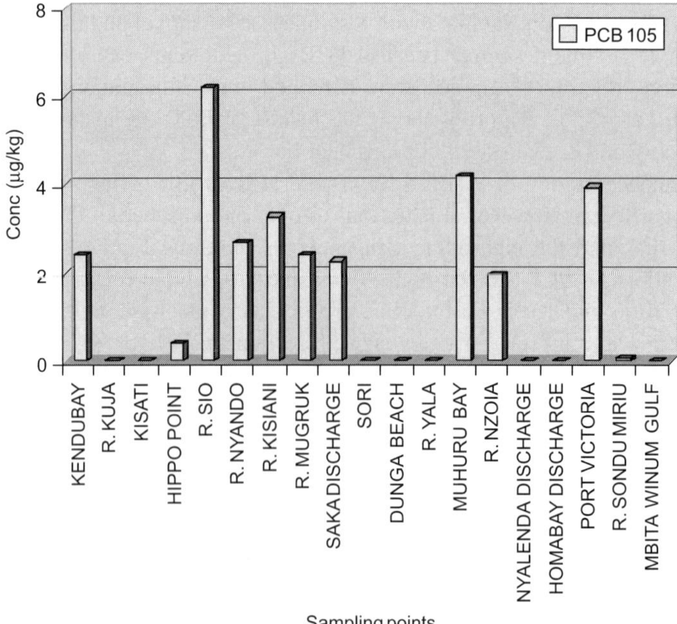

FIGURE 11 Concentrations of PCB 105 in sediments at different sampling stations of Kavirondo Gulf of Lake Victoria catchments.

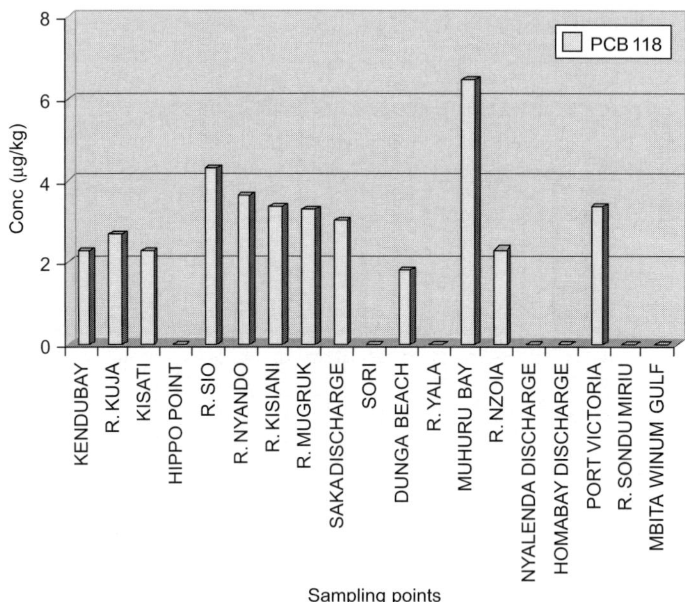

FIGURE 12 Concentrations of PCB 118 in sediments at different sampling stations of Kavirondo Gulf of Lake Victoria catchments.

followed by River Sio. Generally, high concentration of PCB 118 is observed at the river mouths and the Bays. It is of interest to note that no PCB 118 was detected in sediments from Nyalenda in Kisumu Town (this is the open concentration sewer pond for Kisumu Town). Otherwise this compound was detected in most river samples sampling sites (Figure 12).

Analysis for PCB 153 is given in Figure 13. Again it was found that the highest concentration of this compound was observed at River Sio mouth, followed by Muhuru Bay and Sori Towns. Again the sewage discharge points of Homa Bay and Nyalenda concentration pond had low concentration, indicating that the pollution by PCBs does not arise from sewage treatment points.

The last two PCBs analyzed were PCB 156 and PCB 180 (Figures 14 and 15). The concentration of PCB 156 was highest in the sediments of River Sio mouth, followed by River Kisian and Port Victoria, whereas the concentration of PCB 180 was highest in the sediments of Winam Gulf, Hippo point, and Rivers Sio and Kisian.

The observation of PCBs at almost all sampling points poses a challenge as to the source of the compounds. These are rural areas with very little electrification points. At this time it is speculative to attribute the source to any one activity except to speculate that there may be unauthorized dumping of PCBs in the catchments. Further studies are needed to pinpoint the point source of PCBs in the Lake Victoria waters and sediments. However, at present, data indicate that river mouths

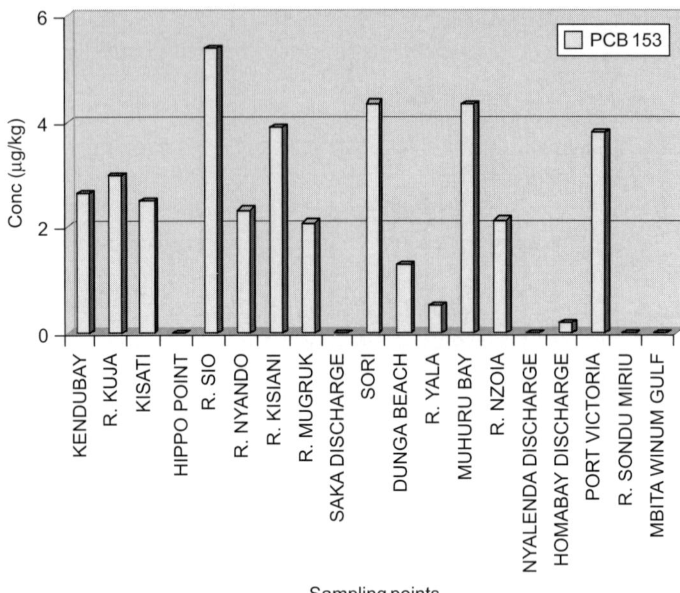

FIGURE 13 Concentrations of PCB 153 in sediments at different sampling stations of Kavirondo Gulf of Lake Victoria catchments.

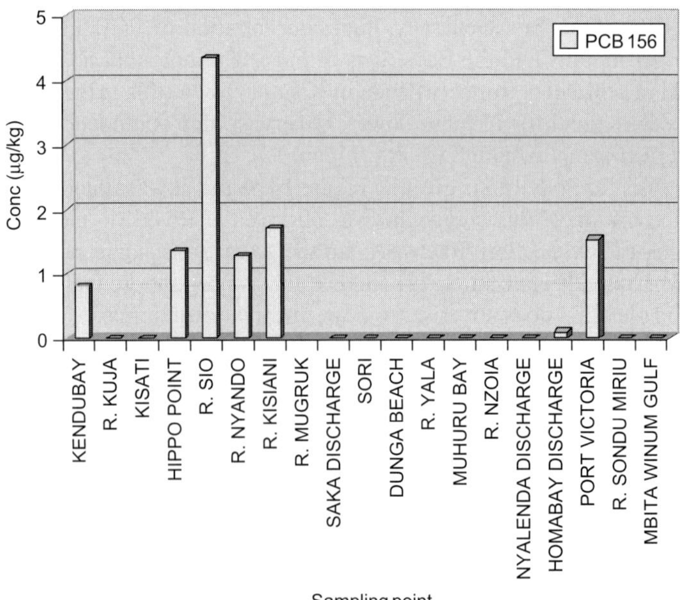

FIGURE 14 Concentrations of PCB 156 in sediments at different sampling stations of Kavirondo Gulf of Lake Victoria catchments.

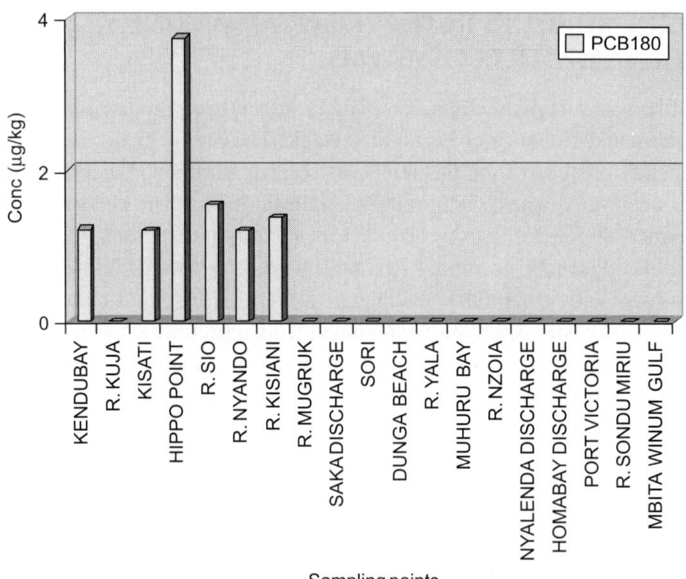

FIGURE 15 Concentrations of PCB 180 in sediments at different sampling stations of Kavirondo Gulf of Lake Victoria catchments.

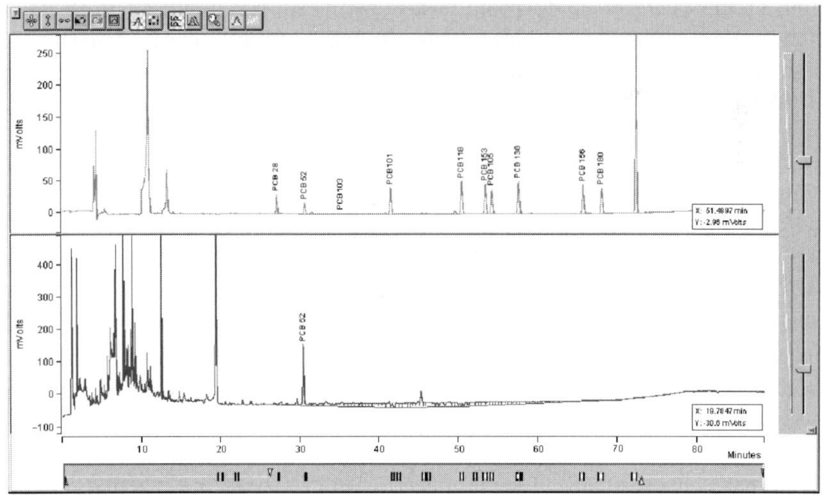

FIGURE 16 Sample chromatogram of PCB standard and the sample.

have higher concentrations of PCBs except for two bays that are unique, such as Hippo point and Muhuru Bay, which is known for active commercial activities.

Figure 16 shows a sample of PCB standard chromatogram and a chromatogram of sediment extract of one of the samples.

PESTICIDE RESIDUES IN THE TROPICAL MARINE AND FRESHWATER ECOSYSTEMS

Kenya, like other tropical countries, heavily depends on the use of pesticides for economic management of crops and livestock. The public health sector in Kenya also depends on the use of pesticides to control vector-borne diseases such as malaria, sleeping sickness, bilharziasis, and fascioliasis. The pesticide spray programs aimed at controlling disease vectors such as mosquitoes, tsetse flies, and water snails succeeded to render Mwea Tabere settlement scheme, Kano Plain, and Lambwe Valley habitable using p,p'-DDT, dieldrin, and endosulfan. Most organochlorine pesticides have been banned from agricultural use because of their high levels of persistence and toxicity to nontarget organisms and replaced by organophosphates, carbamates, and pyrethroids. However, their residues are still in the environment.

Various studies in Kenya have reported organochlorine pesticide residues in a number of river and lake ecosystems at varying concentrations and frequencies, indicating the buildup of pesticide residues in the environment (Wandiga et al., 2002a; Getenga et al., 2004). Earlier studies by Everaarts et al. (1997) also reported the presence of PCBs and cyclic pesticides in benthic organisms from the Kenyan coast and the mouths of Sabaki and Tana Rivers. The bivalve molluscs from the mouth of Sabaki River and Kiwaya Bay had the highest levels of PCBs. Residues of p,p'-DDE were detected in all the samples at levels ranging from 15 to 48 ng/g of lipid in both bivalve and gastropod molluscs. In Wandiga et al. (2002a), sediment samples reported from the Indian Ocean coast contained high presence of lindane, aldrin, p,p'-DDT, and p,p'-DDE.

The presence of pesticides at the top of the food chain has been reported in some studies and may be of critical concern. Pesticide residues have been detected in samples including cows' and human milk, and birds. Kituyi et al. (1997) found contamination of the cows' milk by chlorfenvinphos residues in levels ranging between 0.52 and 3.90 mg/kg in dry season and from 1.58 to 10.69 mg/kg during wet season. The same study showed that milk collected from plunge dipped cows had higher concentrations of the pesticide residues than milk obtained from hand sprayed cows. To understand the behavior of pesticides applied in the tropical environment, a number of simulation studies as well as field experiments carried out on different ecosystems in Kenya have been reviewed (Wandiga et al., 2002b).

The presence of a number of pesticide residues in the aquifer systems and in the food chain is causing great concern. This is due to the fact that most of the synthetic pesticides have high ability to bioconcentrate and bioaccumulate in the food chain and thus cause long-term impact to the sustainability of the ecosystem. As a result, there is a need to monitor the levels and extents of contamination of the environment by these compounds.

We describe here the results of our recent study of pesticide residues in water and sediments from rivers draining into Lake Victoria. The results are presented based on short rain, dry, and wet seasons experienced in the Lake Victoria catchment because the use of pesticides in the region depends on seasons.

The field samples were collected from 12 points located on River Nzoia, River Sio, and Lake Victoria. Field sampling was done during the short rain (October–December, 2002), dry, (January–March, 2003), and wet (April–June, 2003) seasons.

Water was sampled by grab method into 2.5–L amber bottles and preserved with 1 g mercuric chloride to stop microbial degradation of pesticides. In the laboratory, samples were stored in a refrigerator at 4°C prior to extraction. Extraction was done using 2 L of water for all the samples. The water was neutralized to pH 7 using a phosphate buffer, whereas 100 g sodium chloride was added to salt out the pesticides from the aqueous phase. In all cases, extraction was achieved by extracting the water thrice using 60–L portions of triple distilled dichloromethane. The extracts were combined and cleaned by eluting through 20–g florisil column with 200 mL 6%, then 15% and 50% diethyl ether:hexane mixture.

Sediments were sampled using a precleaned stainless steel shovel and mixed on clean aluminum foil. Approximately 500 g each of three representative samples were wrapped in aluminum foil, packed in black polythene bags, and placed in an icebox containing wet ice. In the laboratory, sediment samples were stored in a deep freezer at −20°C prior to extraction. Extraction was done in triplicates using 25 g sample sizes following AOAC method 970.52 for multiresidue analysis. Sample extracts were cleaned by eluting through the florisil column as described for water samples above.

All cleaned water and sediment samples were reduced to 2 mL using LABCON rotary evaporator and analyzed by Varian CP 3800 gas chromatograph equipped with electron capture detector and CP-SIL 8CB capillary column with dimensions 15 m $\times$ 0.25 mm $\times$ 0.25 μm film thickness using a temperature program: 150°C with zero hold time, then increased to 240°C at 4°C/min with a hold time of 4.5 min. Confirmation of pesticide residues was done using second column DB 1701 capillary column with dimensions 15 m $\times$ 0.53 mm $\times$ 0.5 μm film thicknesses, using a temperature program: 150°C with zero hold time, then increased to 200°C at 4°C/min with a hold time of 4.5 min. Identification and quantification of residues was done using high-purity pesticide standards obtained from Dr. Ehrenstorfer GmbH (Ausburg, Germany). The methods used had acceptable recovery rates ranging between 70% and 130% established by spiking blank sediment and distilled water samples and following the same extraction, cleanup, and analysis procedures as for the real samples.

Short Rain Season

Water Samples

Figure 17 shows pesticide residues detected in water samples collected during the short rain season. *p,p'*-DDT, hexachlorocyclohexane (HCH), dieldrin, and heptachlor were the highest detected pesticides.

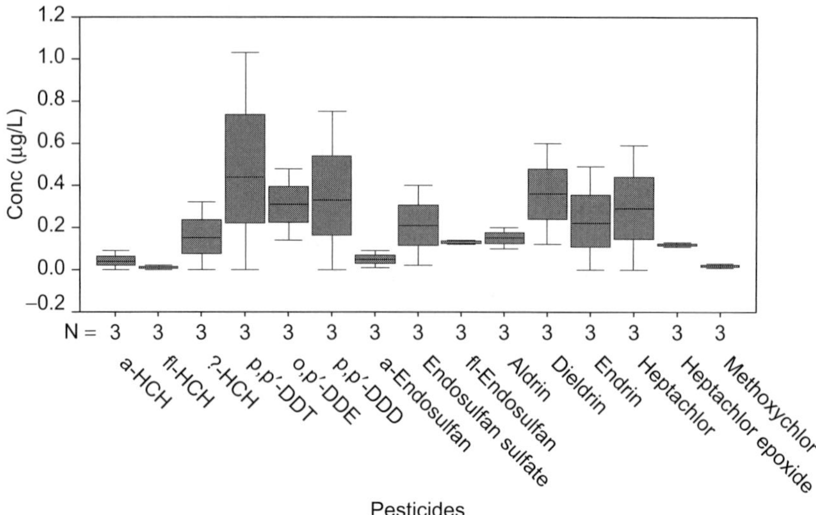

FIGURE 17   Organochlorine pesticide residues in water during short rain season.

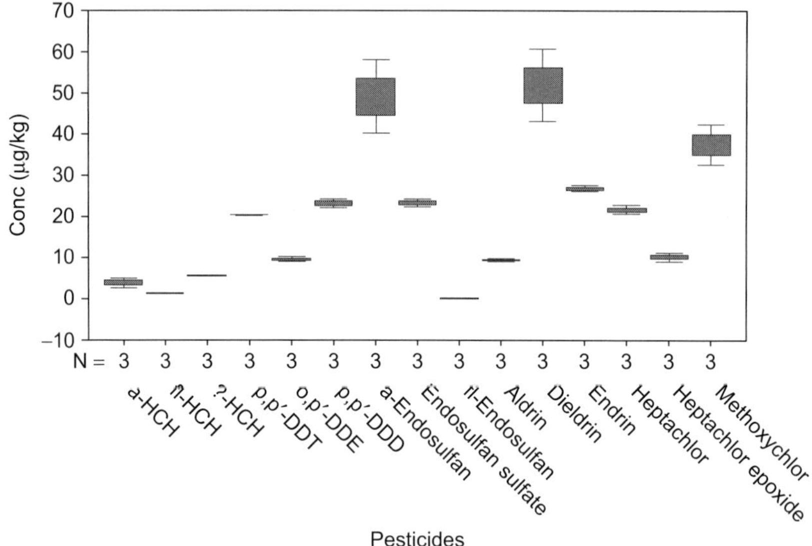

FIGURE 18   Pesticide residues in sediments during short rain season.

The actual values of *p,p'*-DDT ranged from below detection limit (BDL) to 1.13 µg/L, whereas γ-HCH was between BDL and 0.32 µg/L. The levels of endosulfan sulfate were between 0.02 and 0.4 µg/L, compared to alpha endosulfan that was between 0.01 and 0.09 µg/L, indicating a potential degradation process of

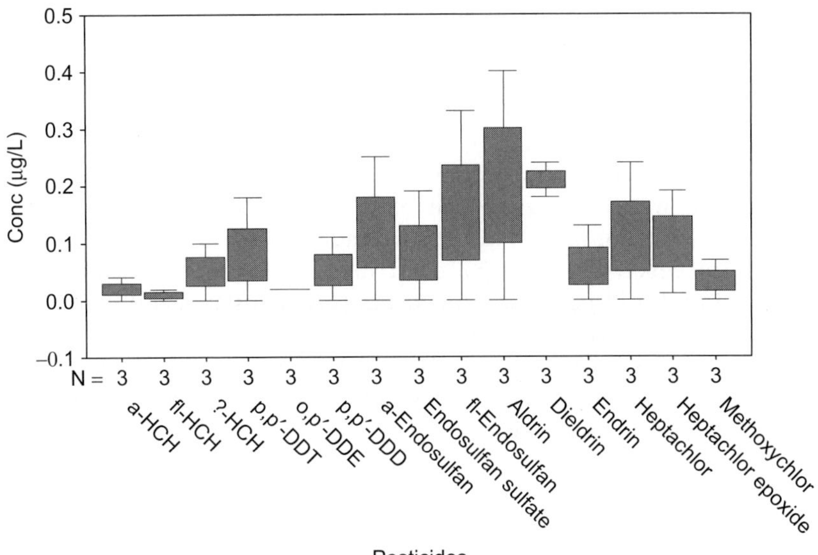

FIGURE 19   Pesticide residues in water during dry season.

the latter. In the same samples, residues of heptachlor ranged between BDL and 0.59 µg/L compared with heptachlor epoxide, whose concentration was between 0.11 and 0.13 µg/L.

Sediment Samples

Figure 18 shows pesticide residues detected in sediments in the same season. It can be noticed that pesticide residues in sediments were 10 times higher than the levels detected in water samples.

In general, α-endosulfan and dieldrin were the highest residues detected in the sediment samples. The levels of *p,p'*-DDD and dieldrin were higher than those of *p,p'*-DDT and aldrin, indicating no recent use of these pesticides in the region. The levels of γ-HCH detected in the sediments were higher than those of α-HCH, and this indicated that HCH might be still in use by some farmers. However, the use of HCH in agriculture was banned in Kenya, and thus, there is need to monitor the sources of these residues in the environment.

Dry Season

Water Samples

Figure 19 shows the range of pesticide residues detected in the water samples collected during the dry season was lower than that detected during the short rain season. This indicated that most of the organochlorine residues are swept from the agricultural land, where they were previously applied, by the storm water into the drainage systems as observed in the short rain season. Dieldrin, o,p'-DDE, and heptachlor epoxide were detected in all the water samples collected during the dry season.

The presence of dieldrin in all the samples was attributed to the breakdown of aldrin that was previously applied in the region. This was confirmed by the dieldrin–aldrin ratio, which was 1.05, indicating no recent use of aldrin in the region. The concentrations of p,p'-DDT ranged between BDL and 0.18 μg/L compared to that of p,p'-DDD, which was between BDL and 0.11 μg/L. The comparison of levels of HCH isomers indicated that γ-HCH (lindane) was leading with concentration between BDL and 0.1 μg/L. Usually, γ-HCH breaks down to α-HCH, therefore the high level of lindane implies that it is still in use by some farmers in the region.

Sediment Samples

The range of pesticide residues detected in the dry season is shown in Figure 20. The results show that the levels were slightly lower than those detected in the same matrix during the short rain season. This could be attributed to the degradation and the transport processes within the stream.

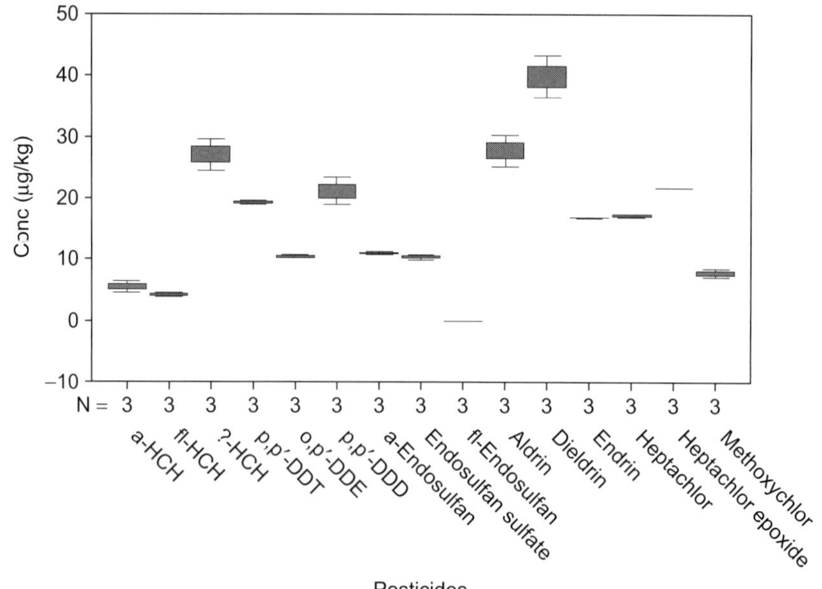

FIGURE 20   Pesticide residues in sediments during dry season.

Dieldrin was the highest of all the pesticides analyzed, followed by aldrin and gamma HCH. The metabolites *p,p'*-DDD, dieldrin, and heptachlor epoxide residues were higher than the parent compounds *p,p'*-DDT, aldrin, and heptachlor. This implied that there was no recent application of DDT, aldrin, and heptachlor in the area. However, the residues of γ-HCH and α-endosulfan were higher than those of α-HCH and endosulfan sulfate, respectively, indicating recent application of these compounds in the environment. The residue levels detected in sediments were in comparable range with those reported by Wandiga et al. (2002a) in sediment samples from the coastal part of Kenya.

Heavy Rain Season

Water Samples

Figure 21 shows residues detected in water samples collected during heavy rain season. The levels of pesticides in the water during the heavy rain season were the lowest of all the three seasons. This was attributed to the large volume of water associated with the heavy rain season, which could have increased the dilution factor of the pesticides and thus lowering the concentrations in the samples.

The levels of DDT, HCH, and methoxychlor were below the WHO limit for all the samples analyzed. However, the levels of dieldrin, aldrin, α-endosulfan,

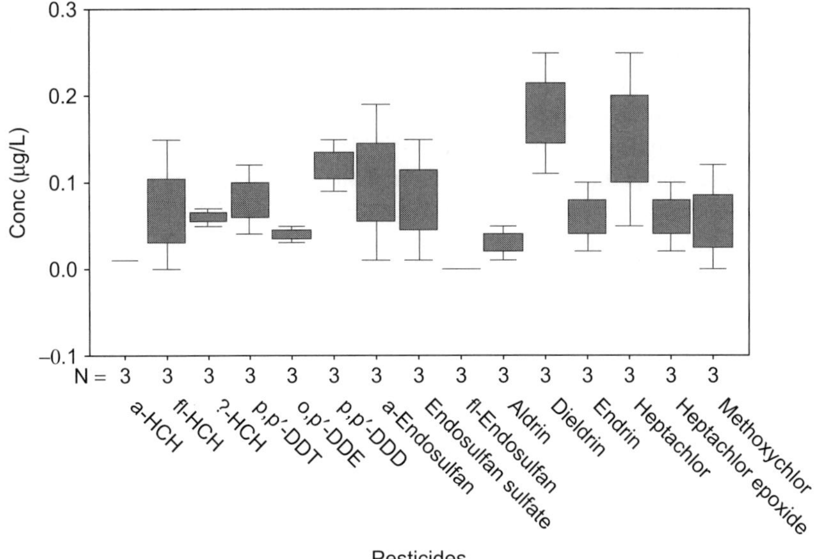

FIGURE 21 Pesticide residues in water during heavy rain season.

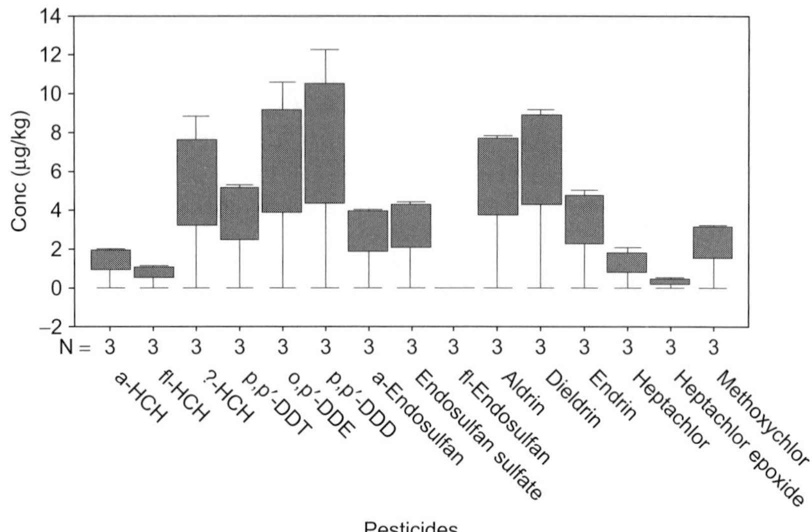

FIGURE 22   Pesticide residues in sediments during heavy rain season.

and heptachlor were still above the WHO limit in some of the samples. The concentration of *p,p′*-DDT detected in samples collected during the heavy rain season ranged between 0.04 and 0.12 µg/L. The level of *p,p′*-DDD was higher, with concentration ranging between 0.09 and 0.15 µg/L. The observed trend was attributed to the degradation of *p,p′*-DDT to *p,p′*-DDD. Generally DDT breaks down in the environment to DDE and DDD. Dieldrin concentration ranged between 0.11 and 0.25 µg/L, whereas aldrin was between 0.01 and 0.19 µg/L. Similar comparison was extended to α-endosulfan and endosulfan sulfate. The analysis of the metabolites showed that the levels of *p,p′*-DDD and dieldrin were higher than those of *p,p′*-DDT and aldrin, indicating no recent use of these compounds. However, the residues of γ-HCH and α-endosulfan were still greater than those of α-HCH and endosulfan sulfate, indicating that those pesticides might be still in use in the region despite the fact that they have been banned from agricultural use.

Sediment Samples

Figure 22 shows the levels of pesticide residues in sediment samples collected during the heavy rain season. *p,p′*-DDD was the highest of all the organochlorines analyzed in the sediments during this season. This could be attributed to the previous use of DDT in the region. Similarly, the levels of dieldrin and endosulfan sulfate were detected in the sediments compared to aldrin and α-endosulfan, indicating the previous use of these compounds.

CONCLUSIONS

Climate change and unsustainable human activities in the Eastern Africa region are contributing to the drastic changes observed in the water quantity and quality of most of the lakes and rivers in the region. Unfortunately, these issues have not been adequately addressed, yet they are already affecting both the aquatic life and the quality of human health.

Our study established high levels of phosphorus in the soils around Lake Victoria especially at the beaches, indicating that the human settlement and related anthropogenic activities strongly contribute to the phosphorus load into the lake waters. It is therefore likely that issues addressing sustainable management of the lake water and the catchment in general should also consider the human settlement matters along the lakeshore to give substantial results.

The presence of PCBs in the Lake Victoria sediment samples presents interesting results that need to be picked by monitoring programs. PCBs are among the least studied chemicals in the region because of the complexity associated with their large number of congeners that demand long and complex procedures in their extraction, cleanup, and analysis. However, their presence in the environmental samples poses issues of great concern because of the biological effects associated with these compounds to both human beings and wildlife.

The use of synthetic pesticides in Kenya is now over eight decades old. The major areas of applications include agricultural pests control and public health control of mosquitoes, snails, and tsetse flies. However, with the rapid development of resistance of pests to the chemicals coupled with environmental persistence, most of the organochlorines have been banned for agricultural use and restricted in use for public health vector control. Nevertheless, there are detected residue levels of these compounds in the amounts of concern for food chain magnification. These levels arise from either previous application or unscrupulous use through illegal means. As a consequence, our research group has been involved in the assessment of the residue levels in both marine and freshwater ecosystems, as well as determination of persistence, toxic effects to fish, and transport of these compounds in different soils in the country. Some comparison of the pesticide residues detected in water, sediments, and biota from marine and freshwater ecosystems and their effect on fish species has been summarized. The summary gives comparative levels of number pesticides studied that include aldrin, dieldrin, α-endosulfan, endrin, DDT, DDE, DDD, and lindane for the period between 1998 and 2004. The detected residue levels in marine samples ranged from 0.503 to 9.025 µg/L in seawater, from 0.584 to 59.0 µg/kg in sediments, and from BDL to 1,011 µg/kg in biota. The residue levels in freshwater ecosystems ranged between BDL and 0.44 µg/L in water, BDL and 65.48 µg/kg in sediments, BDL and 10.07 µg/kg in weeds, and BDL and 481.18 µg/kg in fish samples.

REFERENCES

Bullock, A., Keya, S.O., Muthuri, F.M., Baily-Watts, Williams, R., Waughray, D., 1995. Lake Victoria Environmental Management Programme Task Force 2. Final report by regional consultants on Tasks 11, 16 and 17 (Water quality, land use and wetlands) Centre for Ecology and Hydrology, Wallingford, UK, and FAO, Rome, Italy.

Everaarts, J.M., vanWeerlee, E.M., Fischer, C.V., Hillebrand, M.T.J., 1997. Polychlorinated biphenyls and cyclic pesticides in sediments and micro invertebrates from the coastal regions of different climatological zones. In: Environmental Behaviour of Crop Protection Chemicals. Proceedings of an International Symposium on the Use of Nuclear and Related Techniques for Studying Environmental Behaviour of Crop Protection Chemicals, July 1–July 5, 1996. IAEA-SM-343/45, Vienna: IAEA/FAO, 407–431.

Getenga, Z.M., Kengara, F.O., Wandiga, S.O., 2004. Determination of organochlorine pesticides residues in soil and water from River Nyando drainage system within Lake Victoria basin, Kenya. Bull. Environ. Contam. Toxicol. 72 (2), 335–342.

Hecky, R.E., 1993. The eutrophication of Lake Victoria. Proc. Int. Assoc. Theor. Appl. Limnol. 25, 39–48.

Hooda, P.S., Edwards, A.C., Anderson, H.A., Miller, A., 2000. A review of water quality concerns in livestock farming areas. Sci. Total Environ. 250, 143–167.

Hulme, M., Doherty, R.M., Ngara, T., New, M.G., Lister, D., 2001. African climate change: 1900–2100. Clim. Res. 17, 145–168.

Intergovernmental Panel on Climate Change (IPCC). Fourth assessment report. Climate change 2007: synthesis report. Summary for policymakers. http://www.ipcc.ch/pdf/assessment-report/ar4/syr/ar4_syr_spm.pdf

Kituyi, E., Jumba, I.O., Wandiga, S.O., 1997. Occurrence of chlorfenvinphos residues in cow milk in Kenya. Bull. Environ. Contam. Toxicol. 58 (6), 969–975.

Lindenschmidt, K.-E., Suhr, M., Magumba, M.K., Hecky, R.E., Bugenyi, F.W.B., 1998. Loading of solute and suspended solids from rural catchment areas flowing into Lake Victoria in Uganda. Wat. Res. 32, 2776–2786.

LVEMP, 2001. Study on toxic chemicals/oil products spill contingency plan for Lake Victoria RFP#LVEMP/RCON/003 Volume V of the Toxic Chemical/ Oil Products Spill Contingency Plan Waste Water Report (2001).

LVEMP, 2002. Lake Victoria Environmental Management Project Phase, Water Quality and Ecosystem Management Component, Preliminary Findings of Studies Conducted on Lake Victoria

Morgan, G., Xie, Q., Devins, M., 2000. Small catchments—NMP Dripsey—water quality aspects. In: Tunney, H. (Ed.), Quantification of Phosphorus Loss from Soil to Water. EPA, Wexford, pp. 24–37.

Schulze, R.E., 2000. Modelling hydrological responses to land use and climate change: a southern African perspective. Ambio 29 (1), 12–22.

Scheren, P.A.G.M., 1995. A systematic approach to lake water pollution assessment: water pollution in Lake Victoria, East Africa. Eindhoven Univ. Tech. pp. 71.

Sharpley, A.N., Foy, B., Withers, P., 2000. Practical and innovative measures for the control of agricultural phosphorus losses to water; an overview. J. Environ. Qual. 29, 1–9.

Smith, V.H., Tilman, G.D., Nekola, J.C., 1999. Eutrophication: impacts of excess nutrient inputs on freshwater, marine, and terrestrial ecosystems. Environ. Poll. 100, 179–196.

Tunney, H., Copulter, B., Daly, K., Kurz, I., Coxon, C., Jeffrey, D., et al., 2000. Quantification of Phosphorus Loss from Soil to Water. EPA, Waxford.

Verschuren, D.T.J., Johnson, Hedy J., King, D.N., Edington, P.K., Leavitt, E.T., Brown, M.R., et al., 2002. History and timing of human impact of Lake Victoria, East Africa. Proc. R. Soc. Lond., B, Biol. Sci. 269, 289–294.

Wandiga, S.O., Yugi, P.O., Barasa, M.W., Jumba, I.O., Lalah, J.O., 2002a. The distribution of organochlorine pesticides in marine samples along the Indian Ocean coast of Kenya. Environ. Toxicol. 23, 1235–1246.

Wandiga, S.O., Ongeri, D.M.K., Mbuvi, L., Lalah, J.O., Jumba, I.O., 2002b. Accumulation, distribution and metabolism of ^{14}C-1,1,1-trichloro-2,2-,bis(p-Chlorophenyl)ethane(p,p-DDT) residues in a model tropical marine ecosystem. Environ. Toxicol. 23, 1285–1292.

Wandiga, S.O., Kariuki, D.K., Oduor, F.D.O., Madadi, V.O., 2006. Nairobi river basin programme III: desk study of the Nairobi river basin published literature. http://www.unep.org/roa/Nairobi_River_Basin/Publications/?case=WQ

Chapter 4

Effect of Human Land Development on Water Quality

Michael A. Mallin

Center for Marine Science, University of North Carolina Wilmington, Wilmington, NC 28409, USA

INTRODUCTION

When humans move into a natural area, major changes are initiated to the landscape and associated ecosystems that usually result in various negative environmental impacts. The native wildlife undergo species changes, reductions in abundance and/or diversity, or virtual elimination, whereas the native flora is either reduced in abundance, eliminated, or replaced by introduced food crops or ornamental species. The natural land morphology may be altered to fit its new

Handbook of Water Purity and Quality
Copyright © 2009 by Elsevier Inc. All rights of reproduction in any form reserved.

uses and the native soils may be amended by fertilizers, replaced by other soils, or covered by roads and structures. All of these processes will change the physical, chemical, and ecological structure of the aquatic systems the land drains into. These changes will likely lead to a number of negative impacts including increased flooding and erosion, pollution by sediments, nutrients and chemical contaminants, algal blooms, and loss of native aquatic species. Pollution by pathogenic microbes and toxic compounds will also be a direct threat to human health, particularly if it impacts drinking well water, recreational bathing beaches, and harvestable shellfish beds. The objectives of this chapter are to describe the various hydrological, physical, chemical, biological, and ecological impacts that occur in receiving water systems following removal of the native vegetation and subsequent conversion of the land to agriculture or urbanized areas.

IMPACTS OF LAND CLEARING ON RECEIVING WATERS

When rainfall comes to earth, it has one of three fates. One, it can enter the ground through percolation or infiltration, and from there enters the water table, or upper aquifer. It most easily does this in porous soils such as sandy soils or limestone and dolomite "karst" soils. Soils of smaller size and smaller interstitial spaces percolate less freely, and at the far end of the spectrum some clays provide little or no percolation. Beneath the surface, water table water can move laterally in a downslope manner until it encounters a stream bed, or, if the underlying occluding layer has some porosity or channels, it may migrate deeper into deep groundwater. The second fate rainwater can have is to return to the air as vapor, either through evaporation or by plant uptake and subsequent transpiration. Together, these processes are generally known as evapotranspiration. Third, the water that is not infiltrated or evapotranspired becomes surface runoff, also called stormwater runoff. Stormwater runoff flows downhill until a water body is encountered, which it becomes part of. As it moves downhill, it carries with it all manner of physical, chemical, and biological pollutants. It is stormwater runoff that leads to much of the pollution of surface waters, including streams, rivers, lakes, reservoirs, estuaries, and the coastal ocean (National Oceanic and Atmospheric Administration, 1998; Schueler and Holland, 2000; Dorfman and Stoner, 2007).

A wooded area has little runoff associated with rain events. Most water is retained on-site and is infiltrated or evapotranspired. When such a site is to be developed for housing, commerce, silviculture, or agriculture it is often clear-cut to remove the trees. This removes much of the evapotranspiration potential of the site, and causes a large increase in surface runoff. This runoff increase causes erosion of the nearby and downslope land, with steeper slopes more susceptible to erosion than gentler slopes. The runoff picks up and transports dirt that become suspended sediments upon reaching a receiving water body. Suspended sediments are measured by gravimetry as milligrams per liter, and are usually reported as total suspended solids (TSS). In the water the TSS load increases the cloudiness of the water, which is termed turbidity. Turbidity is commonly quantified by instruments that project a beam of light into a small volume of water, with the amount measured that is reflected at a 90° angle. This process is called nephelometry, and the units are called nephelometric turbidity units (NTUs). States and provinces often have pollution standards for turbidity; for example, in the State of North Carolina, there is a marine and brackish water standard of 25 NTUs, a general freshwater standard of 50 NTUs, and a trout water standard of 10 NTUs. Significant increases in stream-suspended solids and/or turbidity often occur following clear-cuts (Waters, 1995; Ensign and Mallin, 2001).

Besides degrading the clarity of the water, stormwater runoff that enters a stream greatly increases the stream volume and discharge. This causes more erosion, primarily within the stream itself along the banks and at the bottom. This results in the widening, deepening, and effective straightening of a stream (Paul and Meyer, 2001). This scouring of the channel adversely impacts the stream fauna by destroying habitat for fish and benthic organisms (including shellfish), and by covering the bottom with upland-derived sediments, further altering the habitat (Waters, 1995). Besides the suspended sediments, other pollutants that are carried into the stream in runoff include organic and inorganic nitrogen and phosphorus, animal manure, other large or small organic material [biochemical oxygen demand (BOD)] and anything that has been deposited on the landscape from the air including metals and toxic chemical compounds. Streams passing through clear-cuts are also subject to increased sunlight from the newly opened canopy, and combined with nutrient inputs from increased runoff, the increase in solar irradiance may stimulate nuisance algal blooms (Ensign and Mallin, 2001). Clear-cut upland areas may be consequently converted to agriculture or urban areas.

AGRICULTURE AND WATER POLLUTION

Historically agriculture has been a major use of clear-cut landscapes. Agriculture is a broad category that includes crop agriculture, pastureland, and more recently, concentrated animal feeding operations (CAFOs). Forest clearing for agricultural usage is rare presently in Europe and North America, but is a critical issue in South America and Southeast Asia especially in rain forests. Crop agriculture and pasturing have been with us for millennia; however, CAFOs have become a major means of animal production only in recent decades (Mallin, 2000; Burkholder et al., 2007) and a brief introduction is required here. CAFOs are systems wherein cattle, swine, or poultry are closely confined in buildings where they are fed, grown, and defecate with little or no contact with the outdoors. Cattle CAFOs contain dozens to hundreds of animals, swine CAFOs hundreds to many thousands of animals, and poultry CAFOs thousands to millions of birds. CAFOs create vast amounts of feces and urine as waste (Mallin and Cahoon, 2003). Waste from cattle CAFOs is often spread on fields and disked into the soil; waste from swine CAFOs is pumped into large outdoor ponds called lagoons, from where it is periodically sprayed out on nearby fields; and waste from poultry CAFOs is sometimes pumped into lagoons but more often spread as dry litter on fields. Waste lagoons in particular have been subject to major accidents resulting in catastrophic pollution to streams, rivers, and estuaries (Mallin, 2000; Burkholder et al., 2007). Poultry waste stored uncovered outdoors is subject to rainfall-induced runoff, but when it is stored under sheds this practice of course removes the runoff threat until it is field applied. In colder climates, swine and cattle manure may be stored indoors in enclosed facilities which present less pollution risk than open lagoons. However, leakage or seepage from such facilities can endanger nearby surface waters as well as ground waters. Each of the agriculture types (crop, pastureland, CAFO) thus present pollution problems to downslope waters and groundwaters, including pollution by suspended sediments, nutrients, fecal microbes, pesticides, and herbicides.

Human Health Pollutants from Agriculture

Animal manure is commonly used as a fertilizer for various crops. As mentioned, in the case of CAFOs, it is also sprayed, spread onto, or disked into rural land as an inexpensive "treatment" process. However, raw manure contains many microbial pathogens that can infect humans if it enters nearby surface water bodies or well water (Table 1). For example, runoff from cattle feedlots caused hundreds of illnesses and several deaths to residents of Washington County, New York (in 1999) and Walkertown, Ontario (in 2000) by infecting drinking well water with the pathogenic bacteria *Escherichia coli* 0157:H7 and *Campylobacter jejuni*.

Another human health issue that can result from agriculture is methemoglobinemia, commonly known as "blue-baby syndrome", a potentially fatal condition (mainly to infants) that is caused by ingestion of elevated nitrate concentrations in drinking water or food. The nitrate is reduced to nitrite by gut microflora,

TABLE 1 Human Pathogenic Microbes that are Found in Animal Waste (Hinton and Bale, 1991; Berger and Oshiro, 2002)

Bacteria	Protozoa	Viruses
Aeromonas spp.	*Cryptosporidium parvum*	Reoviruses
Campylobacter jejuni	*Giardia lamblia*	Hepatitis E virus
Clostridium spp.	*Balantidium coli*	
Escherichia coli 0157:H7	*Encephalitozoon intestinalis*	
Nocardia spp.	*Enterocytozoon bieneusi*	
Salmonella spp.		
Yersenia enterocolitica		

which then reacts with hemoglobin (which carries oxygen in the blood) and produces methemoglobin, which cannot transport oxygen and can lead to infant death (Johnson and Kross, 1990). Both the United States and Canada use a drinking water standard of 10 mg nitrate-N/L to protect against methemoglobinemia. In the United States, all documented cases of methemoglobinemia have been from consumption of water with nitrate concentrations in excess of this standard (Fan and Steinberg, 1996).

The fertilization of crops over long periods, as well as spraying of CAFO wastes on fields has resulted in excessive nitrate entering groundwater, in some cases exceeding the Environmental Protection Agency (EPA) standard. In Maryland, elevated nitrate in drinking water wells has been positively correlated with area corn production and with the numbers of chickens (broilers) produced in area CAFOs (Lichtenberg and Shapiro, 1997). Groundwater nitrate has been documented to exceed the EPA drinking water standard downslope of swine waste CAFOs (Dukes and Evans, 2006). In a survey of 1595 North Carolina wells located adjacent to CAFOs (Rudo, 1999), 34.2% were found with nitrate-N concentrations exceeding 2 mg/L, and 10.2% with nitrate exceeding 10 mg/L, which was three times the statewide average for nitrate contamination of well water based on historical surveys.

Ecosystem Pollution from Agriculture

Major ecosystem-impacting pollutants from agriculture include suspended sediments from surface runoff, nutrients (nitrogen and phosphorus) from surface runoff, airborne travel and subsurface (groundwater) movement, and a variety of pesticides and herbicides that enter water bodies via runoff, airborne deposition, and groundwater.

Agriculture, particularly row crop and grazing, is considered to be a very important source of sediment pollution to stream ecosystems (Waters, 1995). Suspended sediments (TSS) directly impact the ecosystem by siltation, which is covering the natural stream sediments with material from fields. This can render the stream bottom unhabitable to its natural fauna (Waters, 1995). Examples include the covering of spawning fish habitat (such as salmon) in freshwater and the covering of shellfish habitat (which need hard substrates to settle on) in estuarine and nearshore marine waters. Another direct impact of TSS loading to water bodies is creating sufficient turbidity that it reduces photosynthesis for submersed aquatic vegetation and benthic microalgae. Finally, numerous substances (including pollutants such as fecal bacteria, metals, ammonium, phosphate, and organic pollutants) become physically or chemically bound to suspended materials (particularly clays) and can be transported far downstream to impact distant estuarine areas, including shellfish beds. Thus, keeping soil particles from running off of agricultural landscapes has numerous benefits.

Agricultural sources of nutrients (nitrogen and phosphorus) include inorganic and organic fertilizers on crop fields, runoff from grazing lands, and

TABLE 2 Nutrient Concentrations (as milligrams per liter) in Drainage Waters from Swine CAFOs [Revised from Mallin (2000) and References Within]

Water body	Nitrate-N	Phosphorus-P	Ammonium-N
Surface runoff	4.6	4.0	
Subsurface	21.0	0.6	
Receiving stream	5.4	1.3	
Receiving stream	7.7	na	
Drainage ditch	2.1	3.1	7.1

manure waste from poultry, swine and cattle CAFOs. Nitrogen and phosphorus can enter streams either via surface runoff (Table 2), groundwater contamination and subsequent lateral movement, or atmospheric deposition (Walker et al., 2000; Dukes and Evans, 2006). Studies have shown that in agricultural landscapes 60–90% of the phosphorus movement toward water bodies moves with eroded soils (Sharpley et al., 1993). A study of 17 watersheds in Chesapeake Bay found that discharge of phosphorus into receiving estuaries was positively correlated with the concentration of suspended sediments entering the estuaries (Jordan et al., 1997). Phosphorus bound to particulates that are deposited in the sediments of water bodies can then be released as bioavailable orthophosphate into the water column through biological and chemical means (Correll, 1998). However, dissolved phosphorus can also move through groundwater to waterways if the soil is already saturated with phosphorus, the soil is sandy, or where there is a high organic content of the soils (Sims et al., 1998).

The study of the 17 Chesapeake Bay watersheds (Jordan et al., 1997) determined that discharge of nitrogen, particularly nitrate, was strongly correlated with the percent of cropland within the watersheds, and discharge of nitrogen from cropland was 6 times that of forested watersheds. Nitrate is mobile in soils and can also be found in high concentrations in ground and surface waters near CAFOs (Table 2). The reduced inorganic form of nitrogen, ammonia, is a major byproduct of animal waste, volatilizes from CAFOs and becomes airborne, settles back to earth usually within 80 km of the source CAFO (Walker et al., 2000) and is a likely contributor to nutrient loading of water bodies downwind.

Impacts of Nutrient Loading on Receiving Waters

What impacts result from agriculture-sourced nutrient loading to streams, lakes, and estuaries? A group of major ecosystem impacts caused by nutrient loading fall under the blanket term eutrophication, a collection of symptoms caused by an overabundance of nutrients entering fresh, estuarine, or marine waters (Burkholder, 2001). The most noticeable is the growth of algal blooms, which are usually caused by phosphorus loading in freshwater and upper estuaries, by nitrogen in lower estuaries and marine systems (Hecky and Kilham, 1988; Howarth, 1988), and by nitrogen in some freshwater (blackwater) systems (Mallin et al., 2004). Algal blooms do occur naturally and in some cases they can be relatively benign. However, they are usually a symptom of nutrient overloading and can cause a variety of deleterious effects on water bodies (Paerl, 1988). In freshwater some of the worst nuisance blooms are of cyanobacteria (or the Cyanophyceae), commonly called blue green algae. These blooms make poor food for grazers such as zooplankton and benthic invertebrates, and make the food web less efficient (Paerl, 1988; Burkholder, 2002). Such blooms also result in taste and odor problems for drinking water supplies. In freshwater and oligohaline waters, cyanobacteria can form blooms that can be toxic to fish and mammals (Paerl, 1988; Burkholder, 2002). In estuarine and nearshore marine waters, nutrient loading can stimulate the growth of toxic dinoflagellates and other harmful phytoplankton (Burkholder, 1998; Rabelais, 2002), one of the best studied being the dinoflagellate *Pfiesteria* spp. that is toxic to fish and causes illness in humans (Burkholder and Glasgow, 1997; Burkholder, 1998). The growth of algal blooms, in general, has been strongly correlated with increases in BOD in a variety of habitats including rivers, lakes, and estuaries (Mallin et al., 2006a); elevated BOD then causes reductions in water column dissolved oxygen. Lack of dissolved oxygen (anoxia) and low dissolved oxygen <2.0 mg/L (hypoxia) can cause fish and invertebrate kills and habitat loss (Diaz and Rosenberg, 1995). Nutrient induced blooms of marine macroalgae also disrupt marine ecosystems (Lapointe, 1997; Burkholder, 2001; Rabelais, 2002). Nutrient loading has led to overgrowths of epiphytic algae that will depress the growth of several species of seagrass (Tomasko and Lapointe, 1991; Rabelais, 2002).

Critical Nutrient Concentrations in Receiving Waters

Total phosphorus (TP) concentrations exceeding 100 μg-P/L (3.2 μM) are sometimes considered problematic in fresh and estuarine receiving waters (Correll, 1998). The U.S. EPA has determined that geological and hydrological characteristics of an area strongly impact a given water body's susceptibility to eutrophication-associated problems. Thus, it presently utilizes suggested critical nutrient criteria for total nitrogen (TN) and TP that are tailored for 14 individual ecoregions within the continental United States (www.epa.gov/waterscience/criteria/nutrient/ecoregions/).

In nearshore marine waters, elevated nutrient loading from terrestrial sources can decimate seagrass beds, important habitats for many species of fish and their prey. Nitrate from agricultural sources (or other anthropogenic sources) has a direct toxic effect on a major species of seagrass, *Zostera marina* at relatively modest water column concentrations of 100–200 μg-N/L or 7.1–14.3 μM

(Burkholder et al., 1992; Touchette and Burkholder, 2000). In marine waters, concentrations of inorganic nitrogen exceeding $14\mu g$-N/L ($1\mu M$) and concentrations of inorganic phosphorus exceeding $3.1\mu g$-P/L ($0.1\mu M$) have been found to lead to problematic nuisance algal overgrowths of coral reefs in a variety of systems (Lapointe, 1997). In the Chesapeake Bay, a statistical analysis of factors influencing seagrass bed health in 101 subestuaries showed that a rapid decline in seagrass coverage occurred where watershed TN loading exceeded $16.7\,kg$ N/m^2/d and TP loading exceeded $1.3\,kg$ P/m^2/d [37]; this same analysis showed that seagrass beds in subestuaries draining agricultural watersheds had lower density than beds in subestuaries draining forested watersheds.

Best Management Practices to Reduce Agricultural Runoff

Reducing nutrient loading to water bodies is a major factor in reducing eutrophication and associated aquatic problems. Nutrient losses from agricultural areas are especially likely in areas where soil concentrations are already high, where soils are porous and have low sorption capacities such as sandy or organic soils, or where artificial drainage systems such as ditches or tile drains move runoff quickly off the site (Sims et al., 1998). Soil analysis of the fields where manure is deposited or sprayed can demonstrate when the soil is saturated by N or P and further amendments will lead to excessive runoff (Daniel et al., 1998). Proper management of irrigation water can reduce the amount of runoff from the fields (Sharpley et al., 1993; Gilliam et al., 1997). Vegetated buffer zones protect streams from runoff from crop fields and CAFO spray fields by retaining suspended sediments and their associated pollutants (Gilliam et al., 1997; Han et al., 2000), allowing uptake of nutrients from the runoff by the resident vegetation or binding them to soils, and in the case of nitrate by helping to induce denitrification (Young and Briggs, 2007). For crop fields, techniques useful for reducing off-site runoff of nutrients include conservation tillage, contour plowing, terracing, and runoff collection in on-site ponds or small reservoirs (Sharpley et al., 1993; Daniel et al., 1998).

Agriculture and Pesticides

Agricultural fields, underlying groundwater, and adjoining streams are subject to loading of numerous pesticides and herbicides, some of which are harmful to biota. The U.S. EPA has compiled comprehensive information on effects of these substances (US EPA, 2000a, b). Besides impacting fish and invertebrates, pesticides entering water bodies can have deleterious effects on the base of the aquatic food web, including phytoplankton, periphyton, and zooplankton (DeLorenzo et al., 2001). In addition to currently used compounds, a number of pesticides and herbicides that have been banned for many years (including DDT, dieldrin, and others) can be found at harmful levels in animal tissues or the environment (US EPA, 2000a, b).

URBANIZATION AND HYDROLOGICAL IMPACTS

As we saw at the beginning of this chapter, land clearing removes the native vegetative cover, reduces transpiration of water, exposes the soil to erosion, and increases surface runoff. As urbanization occurs, soils are covered by increasing quantities of impervious surfaces such as parking lots, roads, sidewalks, and rooftops. This greatly reduces the ability of the earth to infiltrate rainwater, turning much of it into stormwater runoff. Streams located downslope from urbanizing areas are forced to accept ever increasing inputs of surface runoff, causing more streambank erosion as well as deepening of the stream. Rain events then cause scouring of the fish and invertebrate habitats, leading to reductions in species richness and diversity (Paul and Meyer, 2001). The widening of the stream also leads to increased water temperature through loss of streamside shade, with potential species displacement. The widening and deepening causes a loss of sinuosity to the stream channel, which leads to loss of habitat and also reduces nutrient processing. With a wider and deeper channel, pollutants that are washed into the stream are rapidly carried downstream.

Urbanization can also have major impacts on ground water hydrology. The covering of the natural ground with impervious surfaces greatly reduces infiltration of rainwater, reducing groundwater recharge and base flow of streams (Klein, 1979). With a storm drain system in place, water that would normally be infiltrated in place is conveyed elsewhere to surface water bodies, like wet detention ponds or into streams or lakes or estuaries. This can reduce the base flow of natural stream systems. On Long Island researchers found that the base flow of streams in an urbanized, sewered area was reduced 80% from normal, whereas base flow of streams in a control undeveloped area was not reduced (Simmons and Reynolds, 1982). Septic systems return water to the ground on-site (although not necessarily clean water—see Sections Septic Systems and Fecal Microbial Pollution and Septic Systems and Nutrient Pollution), so recharge does occur in such areas. However, in the aforementioned Long Island study, a nearby urbanized area serviced by septic systems had a 16% reduction in stream base flow, likely through impervious cover infiltration blockage, stormwater removal, and perhaps some effect from the neighboring sewered region (Simmons and Reynolds, 1982).

Urban Pollution—On-Site Wastewater Treatment Issues

During the urbanization of a formerly natural area, a major pollution issue arises in the form of human waste disposal. Especially since the early 1970s centralized sewage collection and wastewater treatment plants have vastly improved the quality of surface waters in terms of both human health risks and ecological soundness. Problems still occur with incomplete treatment, outdated delivery systems, mismanagement, and accidents (Mallin et al., 2007). The types of environmental problems associated with centralized sewage treatment are generally traceable and point source in nature, thus they will not be detailed in this chapter. However, much human sewage in developed nations is treated by on-site waste-water treatment systems (OWTS), commonly called septic systems. They are used to treat human sewage from individual homes, multifamily structures, businesses, and even hotels in both urbanized and rural areas. In the United States, approximately 23% of homes utilize septic systems, they are particularly abundant in North and South Carolina, Georgia, Alabama, Kentucky, West Virginia, Maine, Vermont, and New Hampshire (US EPA, 2002). Wastewater entering septic systems contains elevated concentrations of fecal bacteria and other microbes (Table 3), nutrients, BOD, and potentially toxic chemicals and metals, depending on the source of the wastewater (US EPA, 2002). In its basic form, the septic system consists of a septic tank, a drainfield (also known as a soil absorption field or subsurface wastewater infiltration system) and the underlying soil (Cogger, 1988; US EPA, 2002). Waste enters the septic tank (closed at the bottom) where solids are settled out and some anaerobic digestion occurs. The supernatant liquid is then piped into infiltration trenches (the drainfield) through perforated pipes, where it percolates through the soil to receive pollutant treatment. The infiltration trenches themselves are underlain by gravel or other porous media before being covered up by the native soil. A biological mat forms along the bottom of the trenches where active treatment occurs including filtration, microstraining, aerobic decomposition, and protozoan predation of fecal bacteria. Further treatment occurs as the wastewater then percolates downward through several feet of non-saturated (i.e., aerated) soil, called the vadose zone, before encountering the saturated zone, or water table. The greater the contact between wastewater and the soil particles in this aerated zone the greater is the

TABLE 3 Pathogenic Microbes Found in Human Sewage Effluents (West, 1991; Smith and Perdek, 2004)

Bacteria	Viruses	Protozoa
Campylobacter jejuni	Adenoviruses	*Balanidium coli*
Escherichia coli 0157:H7	Coxsackie virus	*Cryptosporidium parvum*
Salmonella spp.	Echovirus	*Entamoeba histolytica*
Shigella spp.	Hepatitis A virus	*Giardia lamblia*
Yersenia enterolitica	Human caliciviruses	*Toxoplasma gongii*
Vibrio cholera	Noroviruses	
	Reovirus	
	Rotovirus	

degree of treatment (Cogger, 1988). Under ideal circumstances, septic systems can achieve near complete removal of fecal bacteria and BOD (US EPA, 2002).

Septic Systems and Fecal Microbial Pollution

Excessive septic system density in a given area, especially in areas of poor soils can lead to the microbial contamination of surface and groundwater, and nutrient loading to surface waters that contributes toward eutrophication (Duda and Cromartie, 1982; Yates, 1985; Cahoon et al., 2006). When one considers that average water use in the United States ranges from 40 to 70 gallons per day (150–265 L per day), in areas of high septic system density this can account for considerable pollutant loading to ground and surface waters if it is not effectively treated. A number of documented disease outbreaks have been traced to drinking well contamination by fecal bacteria or viruses from septic system drainfields in the United States (Yates, 1985). Besides the direct contamination of drinking waters, microbial pollution from septic system drainfields has been implicated as contributing to 32% of the shellfish bed closures in a 1995 survey of U.S. state shellfish managers (National Oceanic and Atmospheric Administration, 1998). Regionally in U.S. waters, this type of shellfish bed pollution was noted as particularly problematic in coastal areas of the Gulf of Mexico and the West Coast (National Oceanic and Atmospheric Administration, 1998).

Siting of septic systems in improper soils is a major pollution issue. A widespread type of improper septic system siting occurs where drainfields are situated on porous soils (such as sand or karst) with a high water table. This becomes a problem when the seasonal water table is less than 2 ft (0.6 m) vertically below the infiltration trenches. This is an insufficient depth of the vadose zone to achieve necessary treatment; research has demonstrated that at least 2 ft of aerated soil is needed for proper treatment of fecal microbes (Cogger, 1988; Cogger et al., 1988; Bicki and Brown, 1990), and the U. S. EPA (US EPA, 2002) states that between 2 and 5 ft (0.6–1.5 m) of aerated soil are needed to achieve near complete treatment of the wastewater before it enters the groundwater table. Where soils are sandy, porous, and waterlogged, microbial pollutants such as fecal bacteria and viruses can flow through the soils laterally via the surficial groundwater to enter surface waters. An example of this occurs in the sandy coastal soils in south Brunswick County, North Carolina where there are excessive densities of septic systems (8 per acre or 20 per hectare) that cause bacterial and nutrient pollution in stormwater outfalls and marine waters adjacent to the shorelines (Cahoon et al., 2006). In the northern Cape Hatteras area on North Carolina's Outer Banks, pollutants (including nitrate and fecal bacteria) from septic systems in areas of the Village of Nags Head enter the groundwater and move laterally through sandy porous soils into surface waters within Cape Hatteras National Seashore (Mallin et al., 2006b).

In west Florida, (Charlotte Harbor and Sarasota Bay) estuarine canals and bays receive fecal microbial pollution from an abundance of septic systems sited in porous soils with high water tables (Lipp et al., 2001a, b). In such coastal communities, tides can influence groundwater table height and the outgoing tide actually draws polluted groundwater and associated fecal microbes into the estuarine waters (Lipp et al., 1999). The Florida Keys contain tens of thousands of septic systems and injection wells into which raw sewage is disposed. However, the soils are karst (limestone) and very porous. Experiments have demonstrated that fecal viruses injected into the wells flow out through the porous soils into coastal waters, sometimes within hours of being injected (Paul et al., 1997). These Florida Keys septic systems also serve as conduits to deliver elevated concentrations of nutrients into coastal waters where they can impact sensitive seagrass beds and coral reefs (Lapointe et al., 1990). In Florida, it has been estimated that 74% of the soils have severe limitations to conventional septic system usage (US EPA, 2002).

A second major siting problem occurs when septic systems are placed in soils that are too impermeable to permit proper percolation. The polluted liquid from the drainfield will seep to the surface, called ponding. With the high fecal microbial concentrations there, it is an immediate health hazard to humans or animals that contact it. While ponded on the soil surface, it is subject to rainfall and thus can lead to stormwater runoff containing very high fecal bacteria concentrations (Reneau et al., 1975).

Septic Systems and Nutrient Pollution

Nutrient pollution can be a human health and/or ecological problem where septic systems are placed near drinking water wells or nutrient-sensitive surface waters. In the aerated soil beneath the drainfield, nitrogen in the wastewater is nitrified to nitrate, which moves readily through soils. High concentrations of nitrate build up in groundwater that can exceed the U.S. EPA's and Canadian human health "blue-baby" syndrome standard of 10 mg-N/L. Studies have shown nitrate concentrations well in excess of this standard in groundwater plumes draining septic system drainfields (Cogger, 1988; Cogger et al., 1988; Postma et al., 1992; Robertson et al., 1998). Drinking wells have been contaminated by nitrate from septic systems (Johnson and Kross, 1990) and in Maryland elevated nitrate concentrations in drinking well waters have been positively correlated with the number of septic systems in the area (Lichtenberg and Shapiro, 1997). Under reducing conditions where nitrification is suppressed, elevated ammonia concentrations will occur in septic plumes (Robertson et al., 1998). Although not as mobile in the soil as nitrate, under sandy porous soil and waterlogged conditions, groundwater ammonia plumes may also impact nearby surface waters and increase eutrophication. Where plumes containing high nitrate or ammonium concentrations enter nitrogen-sensitive surface waters, such as coastal lagoons or coastal blackwater streams, algal blooms may be stimulated with associated hypoxia issues (Rabelais, 2002; Mallin et al., 2004). In the Chesapeake Bay, a statistical analysis of factors influencing seagrass bed health in 101 subestuaries (Li et al., 2007) showed that a sharp decline in seagrass coverage occurred where watershed septic system density exceeded 39/km^2.

Phosphate in the wastewater plume tends to bind readily to soils and is much less mobile than nitrate (Cogger, 1988). Considerable phosphate sorption occurs in the vadose zone (Robertson et al., 1998). Even so, under sandy soil conditions or conditions where long usage has led to saturation of phosphate sorption capacity in soils, phosphate concentrations exceeding 1.0 mg-P/L in septic system plumes have been documented as far as 70 m from the point of origin (Robertson et al., 1998). For septic systems serving homes or cottages along lake shores, this can be problematic in that it can contribute to algal blooms and the eutrophication of freshwaters. Karst regions and coarse textured soils low in aluminum, calcium, and iron present the biggest risk of phosphate movement and water contamination (US EPA, 2002).

Minimizing Problems from Septic Systems

Under proper circumstances, septic systems serve as efficient and safe means of disposal of human waste. Proper placement of septic systems is the key. Movement of fecal microbes off-site will occur if the soils are too impervious and ponding occurs, or if the soils are too porous (sandy or karst) and the water table is too high for an appropriately aerated vadose zone (at least 1 m of aerated soil is preferred). Also, even under the best circumstances pollution can move off-site if there are excessive densities of septic systems in a given area. Finally, risks to the environment and human health are increased when septic system use is prevalent near drinking water wells, surface water bodies that are nutrient-sensitive, and coastal waters where shellfishing occurs.

STORMWATER RUNOFF

Urban and Suburban Stormwater Runoff

To reiterate, in a naturally vegetated landscape rainfall is largely removed either by percolation through the soil into the groundwater or by absorption and later transpiration back to the atmosphere by trees and other vegetation. Any remaining water from the rain that is not percolated, transpired, or evaporated becomes surface (or stormwater) runoff. Pollutants entrained in runoff water that is flowing over a vegetated landscape are normally filtered by the natural landscape. Nitrogen and phosphorus are taken up by plants, some nitrogen is turned into atmospheric nitrogen by natural microbial activity (denitrification), and phosphate, metals, toxins, and fecal bacteria are adsorbed by the soils as the rainwater percolates downward.

Whether an urbanized or urbanizing area is serviced by a centralized sewer collection and treatment system or by septic systems, covering the natural land cover with impervious surfaces causes a series of cascading impacts to the downslope (and receiving water) environments. First, rainfall can no longer percolate through the soil, forcing the rain to become surface stormwater runoff. This leads to less recharging of the groundwater aquifer, which is a source of well water and irrigation water, and stream base flow (Klein, 1979; Arnold and Gibbons, 1996). The increased surface runoff causes increased flooding in downslope areas and erosion of the landscape, causing pollution of the water by eroded and suspended sediments (Schueler, 1994). These suspended sediments also adsorb many pollutants, including fecal bacteria, and help transport them downstream or downslope. Between rains, the impervious surfaces concentrate many kinds of pollutants on the pavement including nutrients (nitrogen and phosphorus), metals, organic toxicants such as polycyclic aromatic hydrocarbons (PAHs)—common urban toxic compounds, and of course fecal bacteria. When it rains, impervious parking lots, roads, drives, and sidewalks provide a rapid and direct conduit of polluted stormwater runoff into ditches and streams, and in coastal areas into shellfish beds and beach areas (Mallin et al., 2000; Holland et al., 2004). This is non-point source pollution, commonly referred to in urban areas as stormwater runoff.

Impacts of Urban Stormwater Runoff on Aquatic Ecosystems

A good metric for the degree of urbanization in a watershed is the percent coverage by impervious surfaces (Arnold and Gibbons, 1996). Studies on the impacts of impervious surface coverage on stream health was initiated by researchers working in freshwater systems (Klein, 1979; Griffin et al., 1980). Concentrations of a variety of the aforementioned pollutants were found to increase along with increasing impervious surface coverage (Griffin et al., 1980), with negative responses noted in the fish and invertebrate communities (Klein, 1979). Compilations of published and unpublished data indicated that such negative impacts on freshwater stream biota begin at about the 10–15% impervious surface coverage level (Klein, 1979; Schueler, 1994), with sharp increases in the degree of chemical pollution occurring at the 30–50% impervious surface coverage range (Griffin ct al., 1980). In large-scale analyses of tidal creek estuarine systems, once watershed impervious surface coverage exceeded 10% of a variety of responses occurred including altered hydrography and salinity regimes, altered sediment characteristics, and increased chemical contamination (Holland et al., 2004). When impervious coverage exceeded the 20–30% range, changes in the benthic community occurred including reductions in diversity, loss of pollution-sensitive species, and reduced abundance of commercially important species (Lerberg et al., 2000; Holland et al., 2004).

The loading of suspended sediments and consequent increases in water column turbidity in urban streams have similar impacts to those discussed in Section Ecosystem Pollution from Agriculture, with changes in bottom habitat, decreases in photosynthesis of rooted macrophytes and periphyton, interference with finfish and shellfish feeding, and enhanced transport of sediment-associated pollutants (Waters, 1995; Paul and Meyer, 2001). As with streams in agricultural areas (section *Impacts of Nutrient Loading on Receiving Waters*), urban streams, lakes, and estuaries will respond to increased nutrient loading with eutrophication symptoms, including nuisance algal blooms, increases in BOD and subsequent decreases in dissolved oxygen (Mallin et al., 2006a), and enhancement of toxic and potentially toxic algal blooms (Burkholder, 1998; Lewitus et al., 2003). In estuarine areas draining urban landscapes, the water column and sediments demonstrate higher concentrations of pollutants including nitrate, phosphate, various metals, PAHs, and other organic contaminants than estuaries draining forested regions (Vernberg et al., 1992; Comeleo et al., 1996). The benthic community largely has limited mobility in escaping pollutant loadings, thus benthic community indices can provide appropriate response diagnostic tools in measuring impacts of sediment contamination from urban watershed sources. Such indices show significant negative responses to increasingly polluted estuarine habitats (Hyland et al., 2003). In the Chesapeake Bay, the seagrass coverage within subestuaries draining urbanized watersheds was lower than coverage in subestuaries draining forested watersheds, with beds in subestuaries draining agricultural watersheds between those two in terms of seagrass coverage (Li et al., 2007).

An in-depth analysis of the numerous metals and toxic chemical pollutants generated from urbanized areas is beyond the scope of this chapter. Briefly, such polluting metals include, among others, arsenic, cadmium, copper, lead, mercury, selenium, and zinc, whereas toxic compounds include polychlorinated biphenyls (PCBs) and PAHs. The U.S. EPA provides comprehensive information on sources and effects of these metals and toxicants in several publications (US EPA, 2000a, b, 2004). Concentrations of metals and toxicants in estuarine and marine sediments that are likely to cause negative impacts on invertebrates are provided in Long et al. (1995).

Urban Stormwater Runoff and Human Health

In terms of human health, the most important pollutants washed into streams are fecal bacteria, viruses, and protozoans, some of which are pathogenic. Sources of fecal microbes to stormwater runoff include manure deposited on the landscape from domesticated animals such as dogs, cats, and horses; manure from urban wildlife including raccoons, deer, other small mammals, and waterfowl; and sewage from leaking distribution systems, ponded sewage from improperly sited septic systems. Table 1 provides a list of such pathogens that have been isolated from domestic or wild animals; Table 3 lists pathogens commonly found in human sewage.

Humans may become infected by fecal microbes in contaminated water directly by ingesting them through the mouth (swallowing), nose, eyes, or open wounds. This can occur at beaches, rivers, urban lakes, creeks, sounds—anywhere people recreate and contact water physically. Contact situations include swimming, surfing, diving, water skiing, boating, etc. Excessive fecal microbial pollution from stormwater runoff can lead to closure of bathing beaches by regulatory authorities. During 2006 in the United States, there were over 25,640 marine and freshwater beach closing and advisory days, of which approximately 10,600 were attributed to polluted stormwater runoff (Dorfman and Stoner, 2007). Governments use various standards to ensure beach water safety. In the United States, individual states normally set their own standards, but the EPA recommends using fecal enterococcus as a marine beach water standard, with 104/100 ml the instantaneous standard and a geometric mean of 35 CFU/100 mL from a set of five samples within 3 weeks. Some freshwater areas in the United States utilize a human contact standard of 200 CFU/100 mL of fecal coliform bacteria. Other areas in the United States as well as other nations may utilize other indicator organisms such as total coliforms and *Escherichia coli*.

The clearest impacts of non-point source runoff are visible in coastal waters. Coastal waters support the production of shellfish, a major commercial and recreational target for harvest and (avid!) human consumption. Shellfishing beds that are close to shore are the easiest to access, and require minimal capital to harvest. Unfortunately, the beds located closest to shore are also the beds that receive the greatest impact from stormwater runoff from developed coastlines. As shellfish are filter feeders, they concentrate pollutants in their bodies, especially fecal bacteria derived from land sources such as stormwater runoff and septic system leachate. In the United States, shellfish water standards are set by the U.S. Public Health Service because shellfish are integral to interstate commerce (USFDA, 1995). The current standard is 14 CFU/100 mL for fecal coliform bacteria. Thus, a critical challenge is thus to manage coastal development while allowing for the continued and improved propagation of the traditional commercial fishery.

Clearly, coastal urbanization has a major deleterious impact on shellfish bed availability. As an example the State of North Carolina, along the southeast coast of the United States has undergone major population growth in recent decades. Shellfishing historically has comprised an important segment of both commercial and recreational coastal livelihoods, requiring a relatively modest investment in equipment. Fishermen could utilize local shallow waters in small, inexpensive craft to harvest local beds. Hard clams and oysters are currently among North Carolina's highest valued seafood items, with hard clam meat yielding approximately $6.73/lb and oysters approximately $4.45/lb. However, the once viable and vibrant shellfishing industry has suffered along with this population explosion. Over the past 25 years the shellfish catch has decreased dramatically, having a major impact on this historical coastal way of life. In the early1980s, the commercial harvest of clams and oysters in North Carolina yielded over $14 million of revenue (normalized to 2005 dollars); this income dwindled to less than $4.5 million by 2005 (Figure 1).

The author obtained shellfish closure data from the five southernmost North Carolina coastal counties, Carteret, Onslow, Pender, New Hanover, and

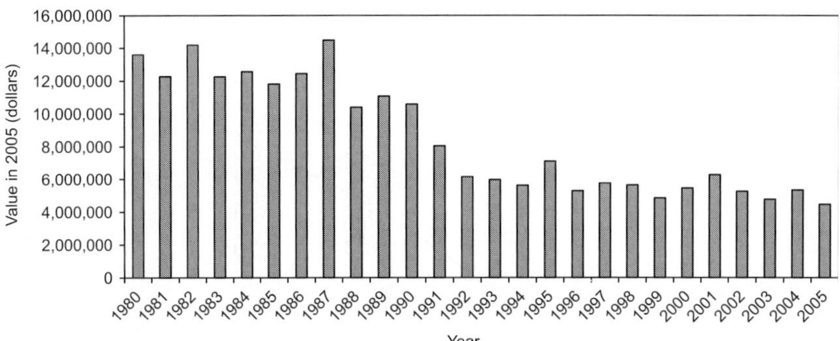

FIGURE 1   Loss of income (in 2005 dollars) for combined clam and oyster harvest from 1980–2005 for coastal North Carolina, USA.

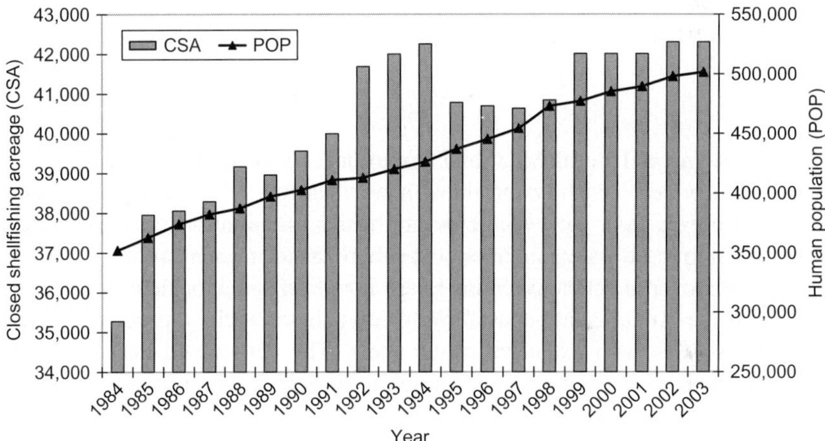

FIGURE 2   Relationship between increases in population and increases in shellfish water closures for five counties in southeastern North Carolina, USA ($r^2 = 0.71$; $p < 0.001$).

Brunswick Counties, and compared those data with human population increases in those same counties. Because, prior to 1980, a number of areas had raw or poorly treated sewage discharges directly entering coastal waters, the analysis was confined to the 20-year period from 1984–2004, when these direct discharges had been halted due to sewage treatment plant improvements (Mallin et al., 2001). Linear regression analyses (Figure 2) showed that the loss in usable shellfishing acreage is directly related to coastal development ($r^2 = 0.71$, $p < 0.001$). Thus, on a broad scale, coastal human population growth appears to be a major impact factor leading to the losses. This leads to the question of what human activities specifically are most responsible for the shellfish bed closures. In more rural areas as well as barrier island areas, the usage of septic systems in unsuitable soils certainly contributes to shellfish bed pollution (Section

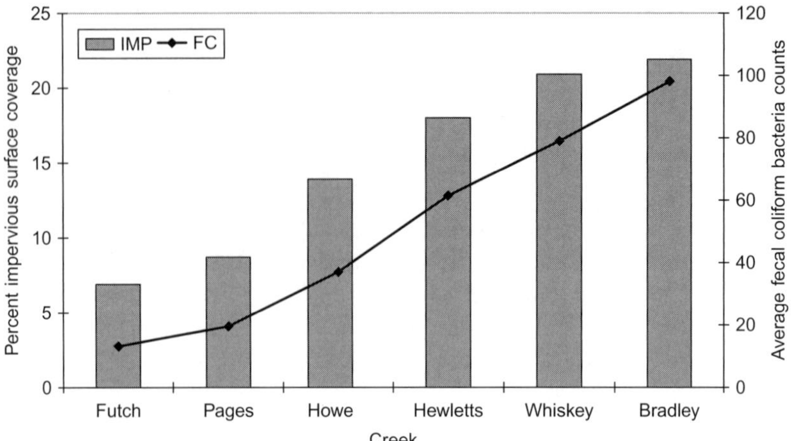

FIGURE 3 Relationship between percent watershed impervious surface coverage and mean fecal coliform bacteria counts for six tidal creeks in New Hanover County, North Carolina, USA ($r = 0.975$; $p = 0.005$).

Septic Systems and Fecal Microbial Pollution). However, large areas of this coast are now serviced by centralized sewer systems that, barring leaks and spills, effectively treat most human waste, which is the most important source of pollution to these waters in urban and suburban stormwater runoff.

New Hanover and Pender counties are host to a series of tidal creeks that drain into the Atlantic Intracoastal Waterway; these creeks are normally rich shellfishing beds with extensive oyster reefs. The author and his laboratory conducted extensive fecal coliform sampling throughout six of these estuarine watersheds and statistically analyzed the results in terms of a series of land use and demographic factors (Mallin et al., 2000, 2001). The results demonstrated that the magnitudes of the fecal bacteria counts in these estuarine creeks were strongly correlated with the total land area draining into the creeks ($r = 0.879$, $p = 0.039$), the human population of the watersheds ($r - 0.922$, $p = 0.026$), the percent of developed land in the watersheds ($r = 0.945$, $p = 0.015$), and especially the percent of impervious surface coverage (roads, roofs, sidewalks, driveways, and parking lots) within the watersheds ($r = 0.975$, $p = 0.005$; Figure 3).

The study also showed that the watersheds with less than 10% impervious surface coverage (Futch and Pages Creeks) still had areas open to shellfishing, whereas those with greater coverage (Bradley, Hewletts, Howe, and Whiskey Creeks) were entirely closed to shellfishing because of high fecal bacterial counts. A later study performed on a set of 22 tidal creeks by a group in Charleston, South Carolina (Holland et al., 2004) showed a similar statistical relationship between impervious surface coverage and fecal bacteria counts, as well as verifying that the 10% impervious coverage percentage is a key level impacting a variety of water-quality parameters. These coastal results of the impact of impervious surfaces on fecal bacterial pollution add a major human health facet to the physical and ecological impacts described by authors addressing the effects of urbanization on freshwater systems (Klein, 1979; Schueler, 1994; Arnold and Gibbons, 1996).

Stormwater Runoff Solutions

Reducing the environmental impacts of stormwater runoff will involve utilizing a variety of simultaneous tactics leading toward the encouragement and establishment of environmentally sound (called "green") coastal development—a concept that has been demonstrated in a growing number of locales. These tactics include reducing stormwater runoff at the source by reducing impervious surface coverage, conservation, and enhancement of natural filtration areas, treatment of stormwater runoff, reducing septic system uses in at-risk coastal areas, public relations efforts, and financial incentives to developers, and public education. The following discusses several tactics helpful in reducing pollution from urban stormwater runoff. Some commonly used websites that also provide much information useful in the reduction of stormwater impacts: the Center for Watershed Protection http://www.cwp.org/, the Water Environment Federation http://www.wef.org/, and the North American Lake Management Society www.nalms.org.

1.   Reduce the amount of stormwater runoff at the source to protect receiving water quality. Less stormwater runoff generated at the source reduces flooding pressure on the receiving stream, reduces the erosion of land and streambed, reduces the total load of pollutants delivered downstream, and leads to lower treatment costs of the runoff. When building a housing development, commercial area, golf course, or even a park, the developer should plan to minimize inputs of runoff to the streams, estuaries, and beaches and minimize physical changes to the natural landscape. This can be accomplished both by conservation practices and technological means.

1a.   Make maximizing greenspace a part of the site planning process. Retention of large trees on a site is very important because trees remove large quantities of water via uptake and transpiration of water vapor to the atmosphere. When landscaping a site, either retain native species or plant mixed native vegetation. Non-native imports usually require more watering and more fertilizer usage; that is a greater cost to the landowner. Trees and green areas are aesthetically pleasing and an advantage to home value.

1b.   Utilize reduced width of roads within new developments, and place sidewalks on only one side of the street. This reduces the amount of impervious surface in the development and the consequent generation of stormwater runoff, and significantly reduces the cost of materials and labor to the developer, which reduces the costs to the homebuyer.

1c.   Use pervious pavement whenever possible in place of traditional pavements. There are several types of pervious pavement available, including porous concrete, porous asphalt, concrete grid pavers, and permeable interlocking concrete pavers, and the soils beneath them need to be able to drain sufficiently for their proper use. Porous concrete is currently about 10–25% more expensive than standard concrete. Porous concrete requires annual vacuuming to maintain its efficiency. On a more basic level, outlying grassed areas associated with parking lots of shopping areas can be used for holiday parking, so the developer can minimize paved areas to amounts that are used on normal shopping days.

1d.   Water conservation and reuse can reduce generation of runoff in both small-and large-scale systems. On the small scale, individual homeowners can install rain barrels so they can utilize roof runoff for irrigation of flowers, shrubs, and vegetable gardens. This reduces watering costs of the homeowner. On a larger scale, the reuse of treated wastewater for golf courses and development of common areas saves treatment costs and recharges the aquifer, while releasing less-treated wastewater into receiving water bodies.

2.   Treating stormwater runoff to remove pollutants can be an effective tactic to reduce pollution of streams, estuaries, beaches. This treatment includes a mix of both passive and engineered means.

2a.   Rain and stormwater runoff is naturally treated for pollutant removal by flowing into and through greenspace. In green areas, pollutants such as suspended sediments settle out and nutrients (nitrogen and phosphorus) are absorbed by plants. Nitrate can be taken up by microbes associated with plant roots and denitrified (turned into atmospheric nitrogen and removed from the system). Soil particles adsorb fecal bacteria, ammonium, orthophosphate, and metals (although some toxic organic contaminants can pass through soils). Trees not only take up vast quantities of water and transpire it away, reducing flooding, but also take up and sequester nutrients. The stormwater runoff is not only cleaned but enters and recharges the groundwater aquifer instead of entering the stream or estuary.

2b.   Runoff also receives treatment when it percolates downward through pervious pavement. This has been found to be effective in removing fecal bacteria, suspended sediments, nitrogen, and phosphorus from stormwater (Pennington et al., 2003). As with greenspace, the water ends up recharging the aquifer rather than polluting the stream or estuary. Pervious pavement can by employed by developers of residential or commercial areas, private and government owners of roadways, sidewalks and parking lots, and individual homeowners.

2c.   Conservation of natural wetlands is critical to protecting coastal waters from pollution. Wetlands absorb runoff and reduce downstream flooding, and the wetland plants and soils absorb or sequester nutrients and other pollutants. A moderate amount of wetlands coverage in watersheds buffers the stream from the input of pollutants such as fecal coliform bacteria and turbidity (Mallin et al., 2001). All wetlands serve as wildlife habitat, and wetlands associated with estuarine areas serve as nursery areas for young finfish and shellfish of numerous species by providing food for them and protection from larger predators. Any individual or private or government entity that owns land containing wetlands needs to practice wetland conservation.

2d.   Wet detention ponds are the most commonly used engineered device for treatment of stormwater runoff. They are seen everywhere, near the parking lots of apartment complexes, shopping centers, municipal facilities, and other complexes, and exist in all manner of shapes and sizes. However, their pollution removal requirements are limited—generally wet detention ponds are required only to remove a proscribed portion (often 85%) of the TSS from the incoming water. There are often no removal standards for the other pollutants. Some wet detention ponds discharge nutrients, algae, and fecal bacteria downstream (Mallin et al., 2002). However, when properly designed and managed, wet detention ponds can achieve removal rates exceeding 80% for TSS, BOD, TN, TP, and selected metals (Livingston, 1995). The efficacy of stormwater ponds can be improved by having a forebay to settle most TSS where the runoff enters (US EPA, 1999). The use of rooted aquatic plants in the pond (native species) greatly improves nutrient removal via plant uptake, and nitrogen removal also occurs from bacterial denitrification. Thus, a shallow shelf around the edge of the pond planted with native rooted aquatic plants (called macrophytes) increases the effectiveness of pond pollutant removal (US EPA, 1999). Plants living and dying within the pond also increases the organic content of the pond sediments, which increases their capacity to absorb metals and nutrients (Schueler and Holland, 2000). The pond should have a deep middle of 6–10 ft (2–3 m) so macrophytes don't fill the pond and restrict flow. Wet detention ponds should be designed with an increased length/width ratio (>2/1), with runoff water (and pollutants) flowing into one end and being discharged at the other end, so short-circuiting is prevented and the entire pond can be used for treatment (US EPA, 1999; Mallin et al., 2002). When ponds have inflow areas located near the outflow areas, this circumvents much of the treatment process (Mallin et al., 2002). Accumulated (polluted) sediments should be removed from the bottom periodically (Livingston, 1995). Wet detention ponds have moderate to high costs associated with their construction and maintenance (Wossink and Hunt, 2003).

2e.   Use constructed wetlands, either alone or in series with wet detention ponds to improve pollution removal of stormwater runoff, although their performance for individual pollutant removal is variable and more research is needed on refining these systems (Livingston, 1995). Constructed wetlands serve many of the functions as natural wetlands (Gilliam et al., 1997), and if placed to take in effluent from a detention pond they can serve as a polishing system to reduce nutrients and other pollutants that exit the wet detention pond (Schueler and Holland, 2000). Constructed wetlands are considerably cheaper to construct and maintain than wet detention ponds (Wossink and Hunt, 2003). Constructed wetlands can be utilized wherever it is geographically and hydrologically feasible by developers of residential or commercial areas or industrial parks, subsequent owners of such entities, and by governments at the local, regional, or national level (whoever needs to treat runoff from impervious areas such as parking lots, extensive roof coverage, roadways, etc.).

2f.   Individual parking lots can control and treat stormwater runoff using engineered filtration systems, which are either sand filters or bioretention areas (also called rain gardens), existing in a wide variety of sizes and designs. These are collection systems with layers of sand or sand with organic matter and vegetation into which runoff is directed, so the runoff can be filtered for fecal bacteria and other pollutants. These systems can be underlain by a drain, so treated runoff then enters the surficial aquifer. These systems are often proprietary and vary considerably in design (this is an active area of stormwater treatment research and development). Sand filters are more heavily engineered and cost more to construct, and can serve urbanized areas up to several acres. Sand filters can also be placed under parking lots. Rain gardens are less costly with less engineering, and utilize plant uptake and microbial activity in the uptake or transformation of pollutants. Rain gardens can be attractively landscaped as well. These systems have a seasonal component, being more effective in summer when biological activity is greater. They can serve only a limited area, such as a parking lot, but are more effective than sand filters for removal of certain nutrients (Hunt, 2003).

2g. Curbside treatment can reduce both runoff entering the storm drains and improve its quality. The standard curb and gutter systems are built to remove road runoff as quickly as possible and send it somewhere else, sometimes into a wet detention pond and sometimes directly into an urban stream, a tidal creek, or urban lake. Instead of directing runoff into gutters and storm drains, road runoff can be directed through gaps in the curb into a median. There, it should flow over a rock riprap (depositing suspended sediments it carried in from the road) then flow sideways through a grassed median before it finally encounters a storm drain set several meters away (as done in a demonstration project by the City of Wilmington, North Carolina, USA). This allows for water percolation through the soil to reduce runoff, settling of suspended sediments and other pollutants in the riprap and grass, and filtration of pollutants through the grass and through the soil. This also allows for some groundwater recharge to occur via percolation of water.

2h. Maintenance of vegetated buffer zones will reduce pollution impacts to streams and drainage ditches (Schueler and Holland, 2000). Although the width of the buffer needed depends on the slope of the land and the soil material, in general, the most effective model is at least 50 ft (15.2 m) buffer. The buffer should be planted with mixed, native vegetation (Han et al., 2000). Within the buffer, overland runoff is slowed down and pollutants such as suspended sediments settle out and nutrients (nitrogen and phosphorus) are absorbed by plants. By using different plant species with mixed root depths, groundwater-borne nitrate can be intercepted before reaching the stream. Nitrate can be taken up by microbes associated with plant roots and denitrified into atmospheric nitrogen (Young and Briggs, 2007). Soil particles adsorb fecal bacteria, ammonium, orthophosphate, and metals, and soil-associated pollutants are controlled more effectively than dissolved pollutants (Han et al., 2000; Mallin et al., 2002). Vegetative buffer zones can be employed by home or business owners, or government entities; that is whoever is the riparian landowner of a given stream.

3a. A non-technical way to protect streams and estuaries from pollution is to obtain conservation easements on shoreline properties. By this approach a government, a private conservancy, or other group can purchase streamside property to keep it from being developed, or buy a long-term lease agreement (a conservation easement) in which the landowner retains the ownership of the land, but his uses are limited by the terms of the agreement. By doing this, landowners can get property tax breaks in many locations.

3b.   Another powerful incentive can be financial. In some cases developers can obtain pollution and stormwater credits for adapting environmentally sound building techniques. For example, developers within the City of Wilmington, North Carolina can reduce their stormwater fees to the city if they use pervious pavement instead of standard pavement. In watersheds considered by the government to be nutrient-sensitive waters, developers may be able to receive pollution credits for using techniques that reduce their runoff load, which over time allows for recouping some of the higher construction costs.

SUMMARY AND CONCLUSIONS

When humans decide to develop land areas that are pristine or near pristine, a cascading series of events occur that impact the quality of nearby (receiving) water bodies. Removal of the forest cover, often by clear-cutting, reduces evapotranspiration of captured rainfall and increases the amount of surface stormwater runoff. This runoff causes erosion of the landscape and carries suspended sediments into receiving waters. If the land is used for agriculture, a number of pollutants may subsequently enter surface and/or groundwaters including suspended sediments, nutrients (nitrogen and phosphorus), pathogenic fecal microbes, and pesticides and herbicides. If the cleared land is converted to urbanized areas, much land will be covered by impervious surfaces which will cause hydrological changes, including further increases in runoff and loss of groundwater recharge, enhanced erosion of streambeds, and loss of aquatic animal habitat. If the area is to be served by septic systems, nutrient and fecal microbial pollution to nearby wells and waterways may result if the soils are porous (sandy or karst) and there is a high water table; surface runoff may result if soils are too impervious for proper percolation. Chances of off-site pollution are increased if septic systems density is excessive for the area. Sewered areas will avoid those pollution problems, but surface stormwater runoff will lead to several problems. These include pollution of receiving surface waters by nutrients, fecal microbes, metals, and toxic chemicals including PCBs and PAHs. The amount of impervious surface coverage in a watershed is strongly related to receiving water impacts, including degraded fish and invertebrate communities, decreases in dissolved oxygen, increases in nutrients and chemical pollution, and increases in fecal microbial pollution.

Suspended sediment loading to water bodies increases turbidity and changes bottom habitat, reduces photosynthesis of aquatic plants, interferes with fish and shellfish feeding, and enhances transport of nutrients and other pollutants. Increased nutrient loading to streams from agriculture, on-site wastewater treatment, or stormwater runoff will cause eutrophication symptoms in receiving streams, lakes, and estuaries. These include stimulation of noxious algal blooms that can lead to low dissolved oxygen and fish kills, toxic algal blooms that can kill fish and sicken humans, overgrowths of nuisance algae on beneficial aquatic plants, and toxicity to seagrass. Excessive nitrate ($>10\,$mg/L) in drinking wells can cause illness to humans. High concentrations of fecal microbial pathogens (bacteria, protozoans, and viruses) can sicken and even kill humans if ingested through water contact or shellfish consumption. Closing of shellfish beds by authorities because of microbial contamination results in considerable economic losses to commercial fishermen as well.

A variety of preventative tactics can be employed to reduce the pollution of surface waters from agricultural and urban runoff. Runoff itself can be treated by the use of constructed wetlands, streamside buffer zones with mixed vegetation, and the use of proper irrigation techniques. In urban situations, runoff can also be treated by properly designed wet detention ponds and the use of sand filters and rain gardens. Urban stormwater runoff can be minimized by reducing runoff at the source through increasing greenspace, minimizing impervious surface coverage, and using pervious pavement. Finally, a variety of non-technical means to protect water quality may be available including the use of conservation easements and financial incentives to landowners.

REFERENCES

Arnold Jr., C.L., Gibbons, C.J., 1996. Impervious surface coverage: the emergence of a key environmental indicator. J. Am. Plann. Assoc. 62, 243–258.

Berger, P.S., Oshiro, R.K., 2002. Source water protection: microbiology of source water. In: Bitton, G. (Ed.), Encyclopedia of Environmental Biology, vol. 5. John Wiley & Sons, New York, pp. 2967–2978.

Bicki, T.J., Brown, R.B., 1990. On-site sewage disposal: the importance of the wet season water table. J. Environ. Health 52, 277–279.

Burkholder, J.M., 1998. Implications of harmful microalgal and heterotrophic dinoflagellates in management of sustainable marine fisheries. Ecol. Appl. 8, S37–S62.

Burkholder, J.M., 2001. Eutrophication and oligotrophication. In: Encyclopedia of Biodiversity, vol. 2. Academic Press, New York, pp. 649–670.

Burkholder, J.M., 2002. Cyanobacteria. In: Bitton, G. (Ed.), Encyclopedia of Environmental Microbiology. Wiley Publishers, New York, pp. 952–982.

Burkholder, J.M., Glasgow, H.B., 1997. *Pfiesteria piscicida* and other *Pfiesteria*-like dinoflagellates: behavior, impacts, and environmental controls. Limnol. Oceanogr. 42, 1052–1075.

Burkholder, J.M., Mason, K.M., Glasgow, H.B., 1992. Water-column nitrate enrichment promotes decline of eelgrass *Zostera marina*: evidence from seasonal mesocosm experiments. Mar. Ecol. Prog. Ser. 81, 163–178.

Burkholder, J.M., Libra, B., Weyer, P., Heathcote, S., Kolpin, D., Thorne, P.S., 2007. Impacts of waste from concentrated animal feeding operations on water quality. Environ. Health Persp. 115, 308–312.

Cahoon, L.B., Hales, J.C., Carey, E.S., Loucaides, S., Rowland, K.R., Nearhoof, J.E., 2006. Shellfish closures in southwest Brunswick County, North Carolina: septic tanks vs. stormwater runoff as fecal coliform sources. J. Coastal Res. 22, 319–327.

Cogger, C.T., 1988. On-site septic systems: the risk of groundwater contamination. J. Environ. Health 51, 12–16.

Cogger, C.G., Hajjar, L.M., Moe, C.L., Sobsey, M.D., 1988. Septic system performance on a coastal barrier island. J. Environ. Qual. 17, 401–408.

Comeleo, R.L., Paul, J.F., August, P.V., Copeland, J., Baker, C., Hale, S.S., et al., 1996. Relationships between watershed stressors and sediment contamination in Chesapeake Bay estuaries. Landsc. Ecol. 11, 307–319.

Correll, D.L., 1998. The role of phosphorus in the eutrophication of receiving waters: a review. J. Environ. Qual. 27, 261–266.

Daniel, T.C., Sharpley, A.N., Lemonyon, J.L., 1998. Agricultural phosphorus and eutrophication: a symposium review. J. Environ. Qual. 27, 251–257.

DeLorenzo, M.E., Scott, G.I., Ross, P.E., 2001. Toxicity of pesticides to aquatic microorganisms: a review. Environ. Toxicol. Chem. 20, 84–98.

Diaz, R.J., Rosenberg, R., 1995. Marine benthic hypoxia: a review of its ecological effects and the behavioural responses of benthic macrofauna. Oceanogr. Mar. Biol. Annu. Rev. 33, 245–303.

Dorfman, M., Stoner, N., 2007. Testing the Waters: A Guide to Water Quality at Vacation Beaches, seventeenth ed. Natural Resources Defense Council. New York. www.nrdc.org

Duda, A.M., Cromartie, K.D., 1982. Coastal pollution from septic tank drainfields. J. Environ. Eng. Div., Proc. Am. Soc. Civil Eng., ASCE 108, 1265–1279.

Dukes, M.D., Evans, R.O., 2006. Impact of agriculture on water quality in the North Carolina coastal plain. J. Irrigat. Drain. Eng. 132, 250–262.

Ensign, S.E., Mallin, M.A., 2001. Stream water quality following timber harvest in a coastal plain swamp forest. Water Res. 35, 3381–3390.

Fan, A.M., Steinberg, V.E., 1996. Health implications of nitrate and nitrite in drinking water: an update on methemoglobinemia occurrence and reproductive and developmental toxicity. Regul. Toxicol. Pharmacol. 23, 35–43.

Gilliam, J.M., Osmond, D.L., Evans, R.O., 1997. Selected agricultural best management practices to control nitrogen in the Neuse River Basin, North Carolina. NC Agric. Res. Serv. Tech. Bull. 311, p. 53.

Griffin Jr., D.M., Grizzard, T.J., Randall, C.W., Helsel, D.R., Hartigan, J.P., 1980. Analysis of non-point pollution export from small catchments. J. Water Pollut. Control Fed. 52, 780–789.

Han, K-H., Isenhart, T.M., Schultz, R.C., Mickelson, S.K., 2000. Multispecies riparian buffers trap sediment and nutrients during rainfall simulations. J. Environ. Qual. 29, 1200–1205.

Hecky, R.E., Kilham, P., 1988. Nutrient limitation of phytoplankton in freshwater and marine environments: a review of recent evidence on the effects of enrichment. Limnol. Oceanogr. 33, 796–822.

Hinton, M., Bale, M.J., 1991. Bacterial pathogens in domesticated animals and their environment. J. Appl. Bacteriol. Symp. Suppl. 70, 81S–90S.

Holland, A.F., Sanger, D.M., Gawle, C.P., Lerberg, S.B., Santiago, M.S., Riekerk, G.H.M., et al., 2004. Linkages between tidal creek ecosystems and the landscape and demographic attributes of their watersheds. J. Exp. Mar. Biol. Ecol. 298, 151–178.

Howarth, R.W., 1988. Nutrient limitation of net primary production in marine ecosystems. Annu. Rev. Ecol. Syst. 19, 89–110.

Hunt, B., 2003. Bioretention use and research in North Carolina and other mid-Atlantic states. NWQEP Notes, Number 109, North Carolina State University Cooperative Extension Service, Campus Box 7637, Raleigh, NC.

Hyland, J.L., Balthis, W.L., Engle, V.D., Long, E.R., Paul, J.F., Summers, J.K., et al., 2003. Incidence of stress in benthic communities along the U.S. Atlantic and Gulf of Mexico coasts within different ranges of sediment contamination from chemical mixtures. Environ. Monit. Assess. 81, 149–161.

Johnson, C.J., Kross, B.C., 1990. Continuing importance of nitrate contamination of groundwater and wells in rural areas. Am. J. Ind. Med. 18, 449–456.

Jordan, T.E., Correll, D.L., Weller, D.E., 1997. Effects of agriculture on discharges of nutrients from coastal plain watersheds of Chesapeake Bay. J. Environ. Qual. 26, 836–848.

Klein, R.D., 1979. Urbanization and stream quality impairment. Water Resour. Bull. 15, 948–963.

Lapointe, B.E., 1997. Nutrient thresholds for bottom-up control of microalgal blooms on coral reefs in Jamaica and southeast Florida. Limnol. Oceanogr. 42, 1119–1131.

Lapointe, B.E., O'Connell, J.D., Garrett, G.S., 1990. Nutrient couplings between on-site sewage disposal systems, groundwaters, and nearshore surface waters of the Florida Keys. Biogeochemistry 10, 289–307.

Lerberg, S.B., Holland, A.F., Sanger, D.M., 2000. Responses of tidal creek macrobenthic communities to the effects of watershed development. Estuaries 23, 838–853.

Lewitus, A.J., Schmidt, L.B., Mason, L.J., Kempton, J.W., Wilde, S.B., Wolny, J.L., et al., 2003. Harmful algal blooms in South Carolina residential and golf course ponds. Popul. Environ. 24, 387–413.

Li, X., Weller, D.E., Gallegos, G.L., Jordan, T.E., Kim, H-C., 2007. Effects of watershed and estuarine characteristics on the abundance of submerged aquatic vegetation in Chesapeake Bay subestuaries. Estuaries and Coasts 30, 840–854.

Lichtenberg, E., Shapiro, L.K., 1997. Agriculture and nitrate concentrations in Maryland community water system wells. J. Environ. Qual. 26, 145–153.

Lipp, E.K., Rose, J.B., Vincent, R., Kurz, R.C., Rodriquez-Palacios, C., 1999. Diel variability of microbial indicators of fecal pollution in a tidally influenced canal: Charlotte Harbor, Florida. Southwest Florida Water Management District, Technical Report.

Lipp, E.K., Farrah, S.A., Rose, J.B., 2001a. Assessment and impact of fecal pollution and human enteric pathogens in a coastal community. Mar. Pollut. Bull. 42, 286–293.

Lipp, E.K., Kurz, R., Vincent, R., Rodriguez-Palacios, C., Farrah, S.K., Rose, J.B., 2001b. The effects of seasonal variability and weather on microbial fecal pollution and enteric pathogens in a subtropical estuary. Estuaries 24, 266–276.

Livingston, E.H., 1995. Lessons learned from a decade of stormwater treatment in Florida. In: Herricks, E.E. (Ed.), Stormwater Runoff and Receiving Systems: Impact, Monitoring and Assessment. Lewis Publishers, Boca Raton, FL, pp. 339–363.

Long, E.R., McDonald, D.D., Smith, S.L., Calder, F.D., 1995. Incidence of adverse biological effects within ranges of chemical concentrations in marine and estuarine sediments. Environ. Manag. 19, 81–97.

Mallin, M.A., 2000. Impacts of industrial-scale swine and poultry production on rivers and estuaries. Am. Sci. 88, 26–37.

Mallin, M.A., Cahoon, L.B., 2003. Industrialized animal production – a major source of nutrient and microbial pollution to aquatic ecosystems. Popul. Environ. 24, 369–385.

Mallin, M.A., Williams, K.E., Esham, E.C., Lowe, R.P., 2000. Effect of human development on bacteriological water quality in coastal watersheds. Ecol. Appl. 10, 1047–1056.

Mallin, M.A., Ensign, S.H., McIver, M.R., Shank, G.C., Fowler, P.K., 2001. Demographic, landscape, and meteorological factors controlling the microbial pollution of coastal waters. Hydrobiologia 460, 185–193.

Mallin, M.A., Ensign, S.H., Wheeler, T.L., Mayes, D.B., 2002. Pollutant removal efficacy of three wet detention ponds. J. Environ. Qual. 31, 654–660.

Mallin, M.A., McIver, M.R., Ensign, S.H., Cahoon, L.B., 2004. Photosynthetic and heterotrophic impacts of nutrient loading to blackwater streams. Ecol. Appl. 14, 823–838.

Mallin, M.A., Johnson, V.L., Ensign, S.H., MacPherson, T.A., 2006a. Factors contributing to hypoxia in rivers, lakes and streams. Limnol. Oceanogr. 51, 690–701.

Mallin, M.A., McIver, M.R., Johnson, V.L., 2006b. Assessment of coastal water resources and watershed conditions at Cape Hatteras National Seashore, North Carolina. Technical Report NPS/NRWRD/NRTR-2006/351, Water Resources Division, National Park Service, Fort Collins, CO.

Mallin, M.A., Cahoon, L.B., Toothman, B.R., Parsons, D.C., McIver, M.R., Ortwine, M.L., et al., 2007. Impacts of a raw sewage spill on water and sediment quality in an urbanized estuary. Mar. Pollut. Bull. 54, 81–88.

National Oceanic and Atmospheric Administration (NOAA), 1998 (on-line). Classified shellfish growing waters, by C.E. Alexander. NOAA's State of the Coast Report. Silver Spring. MD: NOAA. http://state_of_coast.noaa.gov/bulletins/html/sgw_04/sgw.html

Paerl, H.W., 1988. Nuisance phytoplankton blooms in coastal, estuarine and inland waters. Limnol. Oceanogr. 33, 823–847.

Paul, M.J., Meyer, J.L., 2001. Streams in the urban landscape. Ann. Rev. Ecol. Syst. 32, 333–365.

Paul, J.H., Rose, J.B., Jiang, S.C., Zhou, X., Cochran, P., Kellogg, C., et al., 1997. Evidence for groundwater and surface marine water contamination by waste disposal wells in the Florida Keys. Water Res. 31, 1448–1454.

Pennington, S.R., Kaplowitz, M.D., Witter, S.G., 2003. Reexamining best management practices for improving water quality in urban watersheds. J. Am. Water Resour. Assoc. 39, 1027–1041.

Postma, F.B., Gold, A.J., Loomis, G.W., 1992. Nutrient and microbial movement from seasonally-used septic systems. J. Environ. Health 55, 5–10.

Rabelais, N.N., 2002. Nitrogen in aquatic ecosystems. Ambio 31, 102–112.

Reneau Jr., R.B., Elder Jr., J.H., Pettry, D.E., Weston, C.W., 1975. Influence of soils on bacterial contamination of a watershed from septic sources. J. Environ. Qual. 4, 249–252.

Robertson, W.D., Schiff, S.L., Ptacek, C.J., 1998. Review of phosphate mobility and persistence in 10 septic system plumes. Ground Water 36, 1000–1010.

Rudo, K., 1999. Groundwater contamination of private drinking well water by nitrates adjacent to intensive livestock operations (ILOs). In: Aneja, V.P., Murray, G., Southerland, J. (Eds.), Proceedings of Workshop on Atmospheric Nitrogen Compounds II: Emissions, Transport, Transformation, Deposition and Assessment. North Carolina Department of Environment and Natural Resources, Division of Air Quality, Raleigh, NC 27626, pp. 413–418.

Schueler, T.R., 1994. The importance of imperviousness. Watershed Prot. Tech. 1, 100–111.

Schueler, T.R., Holland, H.K., 2000. The Practice of Watershed Protection. Center for Watershed Protection, Ellicott City, MD.

Sharpley, A.N., Daniel, T.C., Edwards, D.R., 1993. Phosphorus movement in the landscape. J. Prod. Agric. 6, 492–500.

Simmons, D.L., Reynolds, R.J., 1982. Effects of urbanization on base flow of selected south-shore streams, Long Island, New York. Water Resour. Bull. 18, 797–806.

Sims, J.T., Simard, R.R., Joern, B.C., 1998. Phosphorus loss in agricultural drainage: historical perspective and current research. J. Environ. Qual. 27, 277–293.

Smith, J.E., Perdek, J.M., 2004. Assessment and management of watershed microbial contaminants. Crit. Rev. Environ. Sci. Technol. 34, 109–139.

Tomasko, D.A., Lapointe, B.E., 1991. Productivity and biomass of *Thalassia testudinum* as related to water column nutrient availability and epiphyte levels: field observations and experimental studies. Mar. Ecol. Prog. Ser. 75, 9–17.

Touchette, B.W., Burkholder, J.M., 2000. Review of nitrogen and phosphorus metabolism in seagrass. J. Exp. Mar. Biol. Ecol. 250, 133–167.

US EPA., 1999. Storm water technology fact sheet: wet detention ponds. EPA 832-F-99-048. United States Environmental Protection Agency, Office of Water, Washington, D.C.

US EPA., 2000a. Guidance for assessing chemical contaminant data for use in fish advisories, vol. 2. Risk assessment and fish consumption limits. EPA-823-B-00-008. United States Environmental Protection Agency, Office of Water, Washington, D.C.

US EPA., 2000b. Guidance for assessing chemical contaminant data for use in fish advisories, vol. 1. Fish sampling and analysis. EPA-823-B-00-007. United States Environmental Protection Agency, Office of Water, Washington, D.C.

US EPA., 2002. Onsite wastewater treatment systems manual. EPA/625/R-00/008. Office of Water, Office of Research and Development, U.S. Environmental Protection Agency, Washington, D.C.

US EPA., 2004. National coastal condition report II. EPA-620/R-03/002. United States Environmental Protection Agency, Office of Research and Development, Office of Water, Washington, D.C.

USFDA., 1995. Sanitation of shellfish growing areas. National shellfish sanitation program manual of operations, Part I. United States Department of Health and Human Services, Food and Drug Administration, Office of Seafood, Washington, DC. 20204.

Vernberg, F.J., Vernberg, W.B., Blood, E., Fortner, A., Fulton, M., McKellar, H., et al., 1992. Impact of urbanization on high-salinity estuaries in the southeastern United States. Neth. J. Sea Res. 30, 239–248.

Walker, J.T., Aneja, V.P., Dickey, D.A., 2000. Atmospheric transport and wet deposition of ammonium in North Carolina. Atmos. Environ. 34, 3407–3418.

Waters, T.F., 1995. Sediment in streams: sources, biological effects and control. American Fisheries Society Monograph 7.

West, P.A., 1991. Human pathogenic viruses and parasites: emerging pathogens in the water cycle. J. Appl. Bacteriol. Symp. Suppl. 70, 107S–114S.

Wossink, A., Hunt, B., 2003. An evaluation of cost and benefits of structural stormwater best management practices in North Carolina. North Carolina Cooperative Extension Service, Raleigh, N.C., US.

Yates, M.V., 1985. Septic tank density and ground-water contamination. Ground Water 23, 586–591.

Young, E.O., Briggs, R.D., 2007. Nitrogen dynamics among cropland and riparian buffers: soil-landscape influences. J. Environ. Qual. 36, 801–814.

Chapter 5

Sampling and Analysis of Arsenic in Groundwater in West Bengal, India, and Bangladesh

Mohammad M. Rahman, Bhaskar Das, and Dipankar Chakraborti*

School of Environmental Studies, Jadavpur University, Kolkata 700 032, West Bengal, India

INTRODUCTION

Before the year 2000, five major incidents of groundwater arsenic contamination have been reported from Asian countries (Mukherjee et al., 2006). These are Bangladesh, West Bengal of India, and three provinces (Taiwan, Inner Mongolia, and Xin-jiang) of China. In the next few years, arsenic contamination in groundwater has emerged from other Asian countries including new sites in China, Lao People's Democratic Republic, Nepal, Cambodia, Myanmar, Afghanistan, DPR Korea, several states (Bihar, Uttar Pradesh [UP], Jharkhand, Assam, and Manipur) of India, Kurdistan province of Iran, Vietnam, and Pakistan (Berg et al., 2001, 2007;Chakraborti et al., 2003, 2008a, 2009; Mosaferi et al., 2003; Ahamed et al., 2006; Mukherjee et al., 2006). Based on the survey conducted in West Bengal

---

*Corresponding author

Handbook of Water Purity and Quality
Copyright © 2009 by Elsevier Inc. All rights of reproduction in any form reserved.

and other states (Jharkhand, Bihar, UP, and Assam) of India and Bangladesh, it was predicted that a significant portion of the Ganga–Meghna–Brahamaputra (GMB) plain in India and Bangladesh comprising an area of $569,749\,km^2$ with a population of over 500 million is potentially at risk from groundwater arsenic contamination (Chakraborti et al., 2004).

The first case of arsenic poisoning of West Bengal was discovered in 1983 when an arsenic patient with skin lesions was identified by Dr. K.C. Saha at the School of Tropical Medicine, Kolkata, on 6th July 1983 (Chakraborti et al., 2002). The first available published article on arsenic contamination in West Bengal reported that 16 patients in three families were affected with arsenical poisoning with the symptoms of hyperpigmentation, hyperkeratosis, edema, and ascites from a village of 24-Parganas district of West Bengal (Garai et al., 1984). The highest arsenic concentration recorded in a tube well was $1250\,\mu g/L$. In 1984, an additional 127 patients of 25 families (total members, 139) were reported to have arsenical skin lesions in five villages covering the districts of 24-Parganas, Nadia, and Bardhaman in West Bengal (Saha, 1984).

When we initiated our survey in arsenic-affected villages of West Bengal in 1988, we knew of only 22 affected villages in 12 blocks (Habra, Barasat, Baruipur, Karimpur, Tehatta, Nabadwip, Chakdaha, Kaliachak, Bhagowangola, Raninagar, Domkal, and Jalangi) of five districts (North 24-Paraganas, South 24-Paraganas, Nadia, Malda, and Murshidabad). In 1995, it was reported that 312 villages from 37 blocks in 6 districts in West Bengal were affected from arsenic groundwater contamination, and from extrapolation of the data, it was predicted that more than 800,000 people were drinking arsenic-contaminated water from these affected districts and about 175,000 people could be suffering from arsenical skin lesions (Chatterjee et al., 1995). However, arsenic contamination in groundwater of West Bengal came into the limelight only after the conference (International Conference on Arsenic, 1995) on arsenic held in Kolkata in February 1995. In the year 2002, 2,700 arsenic-affected villages were further reported from 74 blocks in 9 affected districts in West Bengal after analyzing 105,000 water samples (Chakraborti et al., 2002). From the extrapolation of the water analysis data, it was predicted that about 6 million people were drinking contaminated water with arsenic above $50\,\mu g/L$ and 8 million people with arsenic above $10\,\mu g/L$ (Chakraborti et al., 2002). Based on the analysis of 140,150 hand tube-well water samples, it was recently reported that 48.1% of the samples contained arsenic above $10\,\mu g/L$ and 23.8% above $50\,\mu g/L$ (Chakraborti et al., 2009). Even after working for 20 years in the affected areas of West Bengal, we realized that we are merely seeing the tip of the iceberg of the actual calamity in the nine affected districts of West Bengal.

While working in arsenic-affected Gobindapur village of North 24-Parganas district in West Bengal in 1992, we noticed that in one family, none of the members was showing arsenical skin lesions except a woman who came to West Bengal from Bangladesh (village: Bansdoha, district: Satkhira) after her marriage (Dhar et al., 1997). She informed us that many of her relatives residing in Bangladesh had similar symptoms. She further informed us that she had witnessed similar skin lesions among a few of her neighbors and also in some people living in two neighboring villages (Uttar Sripur and Tona). Besides, during our survey in the arsenic-affected areas of West Bengal close to the border of Bangladesh, we also identified people with arsenical skin lesions who had lived in the district of Nawabganj in Bangladesh but now living in West Bengal. In due course, we began to obtain more information about the arsenic problem in those parts of Bangladesh that border the arsenic-affected areas of West Bengal. Later we analyzed the biological samples including hair, nail, skin scale, and urine of some of these patients who came from Bangladesh to Kolkata for treatment and found that most of the samples to be highly contaminated with arsenic. Realizing the severity of arsenic situation in Bangladesh, we informed World Health Organization (WHO), United Nations International Children Emergency Fund (UNICEF) and Bangladesh government about possible extensive arsenic contamination in Bangladesh (Dhar et al., 1997; Chakraborti et al., 2002, 2004). Although the International Conference on Arsenic (1995) held in Kolkata during February 1995 was attended by the representatives of WHO, UNICEF–Bangladesh, and the government officials of Bangladesh, none of them reported any arsenic groundwater contamination and the resulting suffering of people in Bangladesh (Chakraborti et al., 2002). Immediately after the international conference, medical personnel from Bangladesh's hospitals contacted and informed us that for some years, they had been treating patients with similar skin lesions at the outpatient department of their hospitals. However, the doctors also admitted that at that time they were unaware of the symptoms as that of arsenical skin lesions. After the conference, increasing numbers of people suffering from arsenical skin lesions in Bangladesh started coming to Kolkata for treatment particularly from the districts Faridpur, Narayanganj, Bagerhat, etc. Some of these patients brought water samples from their villages for arsenic analysis, and we determined elevated level of arsenic in most of the samples (Dhar et al., 1997).

During 1996, the Geology Department of Rajshahi University, Bangladesh, sent 600 water samples to our institute (School of Environmental Studies [SOES]) for arsenic analysis from a few districts (Rajshahi, Nawabganj, Kushtia, Jessore, etc.) of Bangladesh that bordered the arsenic-affected districts of West Bengal (Dhar et al., 1997). Many of those samples were found to be arsenic contaminated. WHO–Bangladesh also sent two doctors (Dr. Sheikh Abdul Hadi and Dr. Sheikh Akhter Ahmed) from the National Institute for Preventive and Social Medicine (NIPSOM) to SOES for a 2-week (17–28 June 1996) training to understand the signs and symptoms of arsenicosis. These doctors also brought hand tube-well water and biological samples from the affected districts of Bangladesh, and we found them to be contaminated. After that, SOES and NIPSOM worked jointly for 3 months (August–October 1996) in Bangladesh, covering a few districts. Water samples from 750 tube wells and about 300 each of hair, nail, and some skin scale samples were collected from the affected areas and analyzed for arsenic. The samples tested positive for high arsenic contamination (Dhar et al., 1997).

From December 1996, SOES is jointly working with Dhaka Community Hospital (DCH), Bangladesh, on groundwater arsenic contamination and its health effects in Bangladesh. In 1997, 889 patients were reported with arsenical dermatological features from 45 villages in 18 districts in Bangladesh, and on the basis of the analysis of 3,427 hand tube-wells, it was predicted that 16.7 million people from 18 districts in Bangladesh were drinking arsenic-contaminated water above 50 μg/L (Dhar et al., 1997). Arsenic contamination in groundwater of Bangladesh came into the limelight during international conference on arsenic in Dhaka during 1998 (International Conference on Arsenic, 1998). In 2001, on the basis of the analysis of 34,000 hand tube-wells from Bangladesh, it was reported that in 50 districts, groundwater contained arsenic above 50 μg/L (Rahman et al., 2001). Extrapolation of the water analysis data showed that about 32 million people in Bangladesh were drinking arsenic-contaminated water of above 50 μg/L (Chakraborti et al., 2004). The British Geological Survey (BGS) and Department of Public Health Engineering (DPHE), Bangladesh on the basis of 3,534 hand tube-well water samples throughout Bangladesh excluding Chittagong Hill Tract estimated that 35 million and 57 million people may be exposed to concentrations exceeding 50 and 10 μg/L, respectively (BGS–DPHE, 2001).

COLLECTION, PRESERVATION OF WATER SAMPLES, AND ANALYTICAL METHODS FOR THE DETERMINATION OF TOTAL ARSENIC

To understand the contamination situation in an area, the sampling could be hotspot sampling, blanket sampling, and total screening of samples. Collection and preservation of samples are as important as analysis. Sampling technique is very crucial for the determination of arsenic in water samples. The major concern for sampling and storage are to prevent contamination and minimize the loss of trace amounts of analytes for assessing the total concentration of any element (IARC, 2004). For storage, high-density polyethylene containers are usually preferred than glass containers as the plastic containers are less adsorptive for arsenic (IARC, 2004). For groundwater sampling, tube wells were purged for 5 min prior to collection. Usually the water samples are acidified with strong acid such as concentrated nitric acid or hydrochloric acid to stop precipitation, reduce adsorption of trace metals onto the container walls, and prevent bacterial proliferation (IARC, 2004). Groundwater samples can be kept in a refrigerator or at room temperature and preferably analyzed within a week (Rahman et al., 2002; IARC, 2004). A recent publication investigated the mode of sampling, sample storage, and the time interval study on arsenic concentration in groundwater samples from West Bengal (Roychowdhury, 2008). The study reported that arsenic concentration decreased gradually with time in groundwater samples without treatment. The study also reported higher arsenic loss in the presence of higher concentration of iron for nonacidified samples. Lesser arsenic loss was observed under refrigerated condition (4°C) than at room temperature (Roychowdhury, 2008). The study showed that approximately 91–98% and 96–100% of arsenic was recovered within the first 3 days for acidified samples stored in room temperature and under refrigerated condition, respectively. A significant amount of arsenic was lost with time because of the adsorption on plastic container. The study also reported that a good amount of arsenic was coming out through a large number of small particles, containing mainly colloidal iron hydroxides from newly installed tube wells that increased arsenic level of unfiltered water samples (Roychowdhury, 2008). The study also revealed that if the water samples were not filtered through Millipore filter, then an average 12% higher value of arsenic is expected due to the presence of arsenic-bearing particles. Finally, the study recommended on-site filtration through Millipore filter ($0.45\,\mu m$) followed by acidification and then refrigeration at 4°C (Roychowdhury, 2008). However, to know the arsenic concentration in drinking water to those drinking tube-well water, collection of water samples for arsenic analysis after filtration is not needed as villagers will not drink tube-well water after filtration.

Several analytical methods are currently used for the determination of total arsenic in water samples. The widely used analytical methods for the determination of arsenic in water are colorimetric/spectrophotometric/silver-diethyldithiocarbamate (Ag-DDTC) methods, atomic absorption spectrometry (hydride generation and graphite furnace) methods, and inductively coupled plasma mass spectrometry methods. Flow injection hydride generation atomic absorption spectrometric (FI-HG-AAS) method is the most widely used method to assess total arsenic concentration in water, as this method is characterized by high efficiency, low sample volume and reagent consumption, improved tolerance of interference, and rapid determination (Samanta et al., 1999). The most common analytical methods for total arsenic determination are given in Table 1. Although colorimetric–Ag-DDTC method is widely used for the determination of arsenic, U.S. EPA criticized the method below $50\,\mu g/L$ (U.S. EPA, 1999). Recently, Nickson et al. (2008) reported the current arsenic contamination scenario in groundwater of five states (West Bengal, Bihar, UP, Jharkhand, and Assam) in India. Nickson et al. (2008) used the Ag-DDTC method for the analysis of arsenic in more than 132,000 groundwater samples from eight districts of West Bengal. Field kit methods are also used for the determination of arsenic in drinking water samples in arsenic-impacted regions of Bangladesh, India, and other arsenic-affected Asian countries. Recently, Nickson et al. (2008) employed the field test kit method for the analysis of arsenic in groundwater samples of Bihar, UP, and Jharkhand states of India. National Chemical Laboratories (NCL, Pune, India) kit was used for arsenic testing in Bihar and UP (Nickson et al., 2008). Merck field kit was used for the determination of arsenic in Jharkhand state (Nickson et al., 2008). Several studies have evaluated the effectiveness and reliability of commercially available field kit methods (Rahman et al., 2002; Erickson, 2003; Khandaker, 2004; Cherukurii and Anjaneyulu, 2005; Deshpande and Pande, 2005; van Geen et al., 2005; Steinmaus et al., 2006; Jakariya et al., 2007). A number of field kits are subject to question due to its poor accuracy and uncertain results (Rahman et al., 2002). The field kit methods have several advantages such as no need to transport samples to the laboratory, no need to add preservative, and no storage required, which can reduce the cost of the analysis (IARC, 2004). The main disadvantage

TABLE 1 Commonly Used Analytical Methods for the Determination of Arsenic in Water Samples (IARC, 2004)

Methodology	Detection Limit	Advantages	Disadvantages	References
Colorimetric/spectrophotometric methods	$2\,\mu g/L$ $\sim\!40\,\mu g/L$	Low cost, very simple, uses a simple spectrophotometer		Dhar et al. (2004); Agrawal et al. (1999) Dhar et al. (1997) Pillai et al. (2000) Goessler and Kuehnelt (2002)
Inductively coupled plasma atomic emission spectrometry (ICP-AES)	$\sim\!30\,\mu g/L$			SM 3120 (1999) Goessler and Kuehnelt (2002)
Inductively coupled plasma mass spectrometry (ICP-MS)	$0.1\,\mu g/L$	Analytical method approved by U.S. EPA	Spectral and matrix interferences	Goessler and Kuehnelt (2002)
High resolution-ICP-MS	$0.01\,\mu g/L$	Solves spectral interferences in samples with complex matrices		Gallagher et al. (2001) Karagas et al. (2001, 2002)
Graphite furnace atomic absorption spectrometry (GF-AAS)	$\sim\!0.025\,\mu g/g$	Analytical method approved by U.S. EPA	Preatomization losses, requires the use of matrix modifiers	WHO (2001) SM 3114 (1999)

Hydride generation (FI-HG)-AAS	0.6–6 µg/L	Analytical method approved by U.S. EPA	Chatterjee et al. (1995) Samanta and Chakraborti (1997); Samanta et al. (1999) Shraim et al. (1999, 2000) SM 3114 (1999)
Hydride generation quartz furnace (HG-QT)-AAS	0.003–0.015 µg/L	Inexpensive	IARC (2004)
High performance liquid chromatography (HPLC)-ICP-MS	0.01 µg/L	No need for sample pretreatment	Shibata and Morita (1989)
Polarographic method	~5 µg/L		Whitnack et al. (1972); Whitnack and Brophy (1969)

of the field kit method is that it provides semiquantitative results. Another important limitation of this method is the visual identification of the color in the lower range. Due to the uncertainty of the results of field testing kit, UNICEF–West Bengal discarded the use of field testing kit in West Bengal (Rahman et al., 2002).

In our study, arsenic in hand tube-well water samples was measured by the FI-HG-AAS method. The details of the instrument, flow injection system, and analytical procedure for water were reported earlier (Chatterjee et al., 1995; Samanta et al., 1999). Details of the sample collection procedures have been described in our earlier publications (Chatterjee et al., 1995; Samanta et al., 1999).

QUALITY ASSURANCE AND QUALITY CONTROL PROGRAM

For quality control, interlaboratory tests were performed for arsenic in water samples. Sixteen hand tube-well water samples were collected from arsenic-affected areas of West Bengal and analyzed for arsenic in our laboratory by the FI-HG-AAS method. After analysis, aliquots of the samples were sent to the Intronics Technology Centre (ITC), Dhaka, Bangladesh, and the Central Food Laboratory (CFL), Kolkata, India, for arsenic analysis by FI-HG-AAS. The regression lines and the correlation coefficient values of ITC, CFL, and our laboratory (SOES) are shown in Figure 1. No significant difference was observed from the interlaboratory tests. U.S. EPA standard water samples were also analyzed to check the accuracy of the FI-HG-AAS method (Samanta et al., 1999; Rahman et al., 2002).

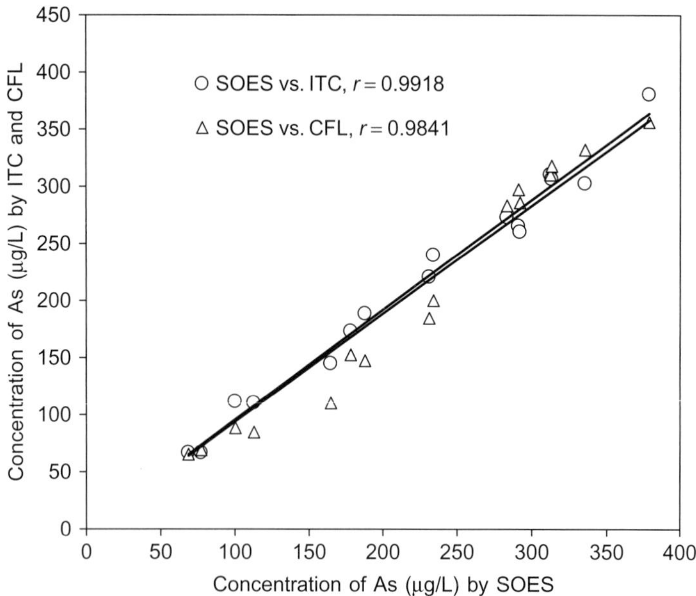

FIGURE 1   Correlation of arsenic analyses (µg/L) by FI-HG-AAS method by three laboratories (SOES, ITC, and CFL).

RESULTS AND DISCUSSION

Arsenic Contamination in Groundwater of West Bengal

We have been conducting analytical, clinical, epidemiological, and hydrogeological studies in arsenic-affected areas of West Bengal since 1988 to determine the magnitude of arsenic contamination and its health effects. Till now, 140,150 hand tube-well water samples were analyzed for arsenic from all 19 districts of West Bengal. Table 2 represents the summary of the arsenic contamination scenario in West Bengal. Figure 2 shows arsenic contamination status in groundwater in all 19 districts of West Bengal. Figure 3 shows the bar diagram for the distribution of water samples according to different arsenic concentration

TABLE 2 Summary of the Arsenic Contamination Scenario in West Bengal

Parameters	West Bengal
Area (km^2)	88,750
Population in million	80.2
Total number of districts (number of district surveyed)	19 (19)
Total number of water samples analyzed	140,150
Percentage of samples having arsenic $>10\,\mu g/L$	48.1
Percentage of samples having arsenic $>50\,\mu g/L$	23.8
Maximum arsenic concentration so far we analyzed ($\mu g/L$)	3,700
Number of severely arsenic-affected districts	9
Number of mildly arsenic-affected districts	5
Number of arsenic-safe districts	5
Total population of highly contaminated nine districts in million	50.4
Total area of highly contaminated nine districts (km^2)	38,861
Total number of blocks/police stations	341
Total number of blocks/police stations surveyed	241
Number of blocks/police stations having arsenic $>50\,\mu g/L$	111
Number of blocks/police stations having arsenic $>10\,\mu g/L$	148
Total number of villages	37,910
Total number of villages surveyed	7,823

(Continued)

TABLE 2 (Continued)

Parameters	West Bengal
Number of villages/paras having arsenic above 50 μg/L	3,417
People drinking arsenic-contaminated water >10 μg/L (in million)	9.5
People drinking arsenic-contaminated water >50 μg/L (in million)	4.6
Population potentially at risk from arsenic contamination > 10 μg/L (in million)	26
Number of districts surveyed for arsenic patients (preliminary survey)	9
Number of districts where arsenic patients found	7
Villages surveyed for arsenic patients	602
Number of villages where we have identified people with arsenical skin lesions (preliminary suvey)	488
People screened for arsenic patients including children (preliminary survey)	96,000
Number of adults screened for arsenic patients	82,000
Number of registered patients with clinical manifestations	9,356 (9.7%)
Number of children screened for arsenic patients	14,000
Number of children showing arsenical manifestations	778 (5.6%)
Total hair, nail, and urine analyzed (20% samples from arsenical skin lesions people)	39,624
Average arsenic concentration above normal level of arsenic in hair, nail, and urine	94%

ranges from all 19 districts of West Bengal. The water analysis results of West Bengal showed that 48.1% had arsenic above 10 μg/L and 23.8% above 50 μg/L (Chakraborti et al., 2009). Surprisingly, 3.3% samples contained arsenic above 300 μg/L. From our field experience in arsenic-affected areas of Bangladesh, West Bengal, and other affected states and finding around 16,000 arsenicosis patients, it was observed that ingestion of 300 μg/L of arsenic and above in drinking water for couple of years may produce arsenical skin manifestations (Chakraborti et al., 2002, 2004). Arsenic concentration above 1000 μg/L was detected in 187 (0.13%) hand tube-well water samples. The highest arsenic concentration was observed as 3,700 μg/L from Ramnagar village of South-24 Parganas district of West Bengal.

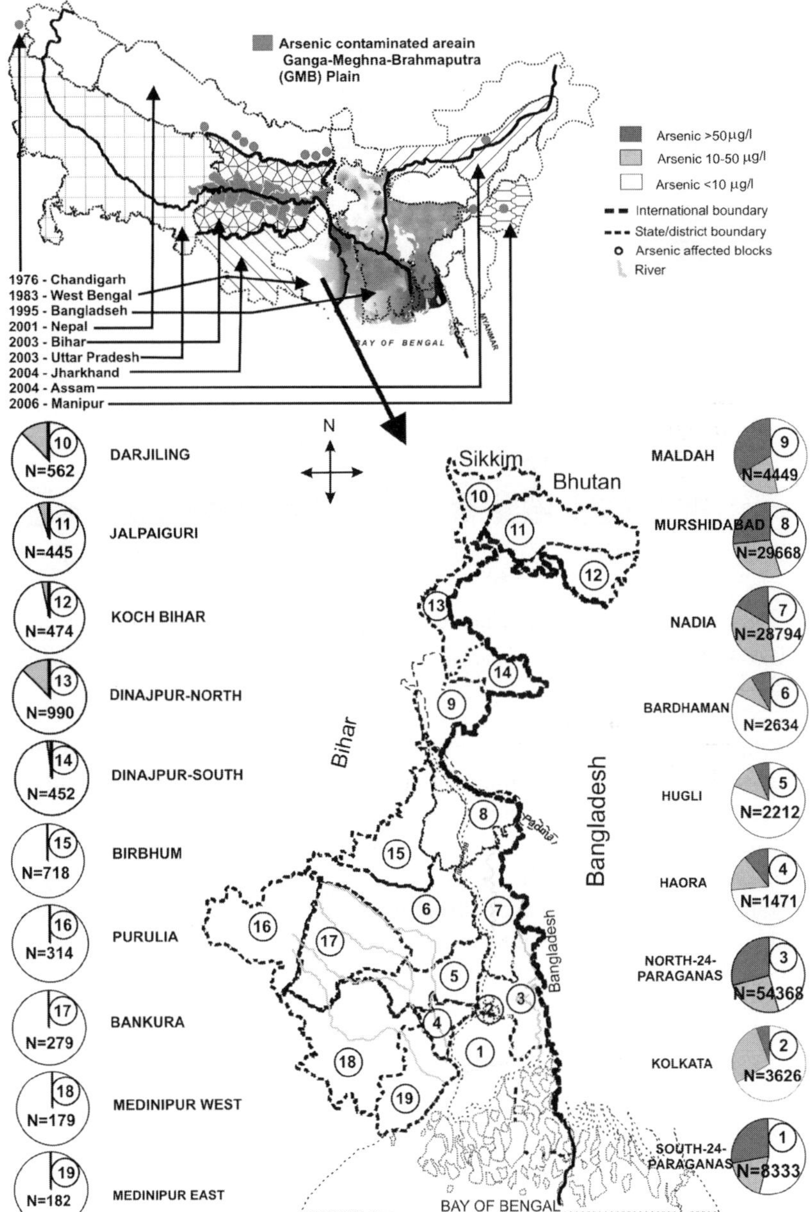

FIGURE 2 Groundwater arsenic contamination status in all 19 districts of West Bengal and in the inset, the GMB plain (Chakraborti et al., 2009).

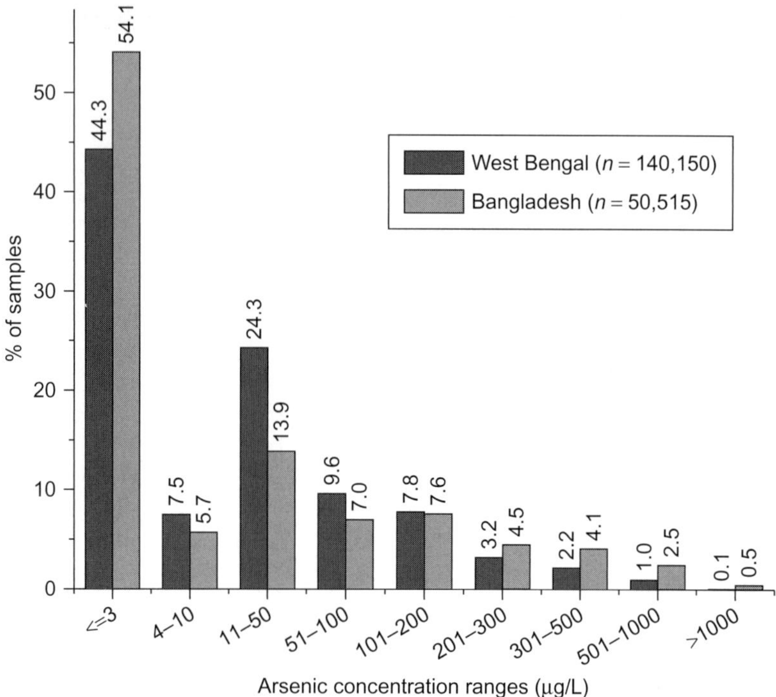

FIGURE 3   Distribution of arsenic concentration in tube-well water samples in West Bengal and Bangladesh.

From the overall water analysis results of West Bengal, the districts were categorized into three contamination areas based on the arsenic concentration in hand tube wells such as highly contaminated, less contaminated, and non-contaminated (Chakraborti et al., 2009). Nine districts (Murshidabad, Malda, North 24-Parganas, Nadia, South 24-Parganas, Bardhaman, Hoara, Hoogly, and Kolkata) of West Bengal were considered as highly contaminated (Chakraborti et al., 2009). From these nine highly contaminated districts, 135,555 hand tube-well water samples were analyzed. The analytical results showed that in the nine affected districts of West Bengal, 49.7% of hand tube wells contained arsenic above 10 µg/L and 24.7% above 50 µg/L (Chakraborti et al., 2009). About 3.4% samples had arsenic above 300 µg/L.

Kolkata city is one of the nine highly affected districts of West Bengal. Altogether 3,626 hand tube-well water samples were analyzed for arsenic from 100 out of 141 administrative wards of Kolkata. The results indicated that tube wells of 65 wards had arsenic above 10 µg/L and in 30 wards above 50 µg/L (Chakraborti et al., 2009).

So far 2,923 hand tube-well water samples were analyzed from the districts situated in the northern part of the West Bengal such as North Dinajpur, East Dinajpur, Jalpaiguri, Darjeeling, and Cooch Bihar (Chakraborti et al., 2009). The analysis results of water samples from northern part of West Bengal showed

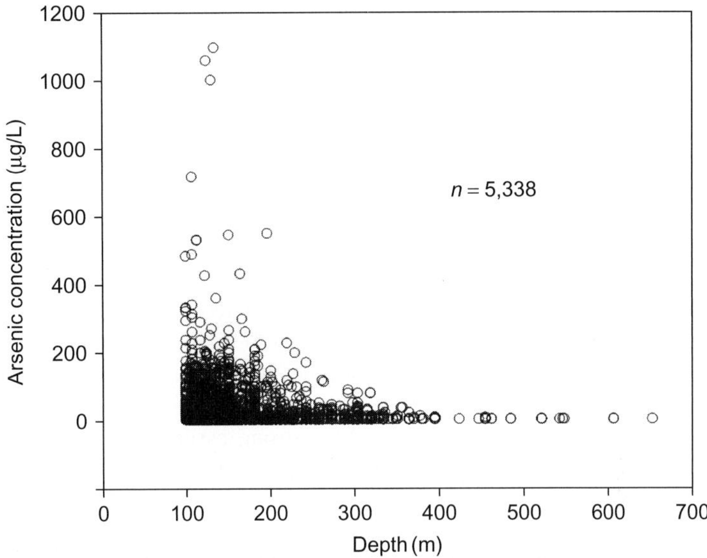

FIGURE 4 Distribution of arsenic in 5,338 tube wells (depth exceeding 100 m) of West Bengal.

that only six samples had arsenic above 50 μg/L and 157 samples were in the range of 11–50 μg/L. These districts are considered as less contaminated areas of West Bengal.

Water samples from 1,672 hand tube wells were also analyzed from the districts situated in the western and southwestern parts of West Bengal (Birbhum, Bakura, Purulia, Mednipur East, and Medinipur West districts). The results do not show arsenic contamination in groundwater above 3 μg/L, the minimum determination level of the FI-HG-AAS instrument (Chakraborti et al., 2009). These districts are considered as non-contaminated areas of West Bengal (Chakraborti et al., 2009).

Arsenic Contamination Situation of Deep Tube Wells (Exceeding 100 m) in West Bengal

So far 5,338 hand tube wells at depth range 100–651 m have been analyzed for arsenic from four highly affected districts (North 24-Parganas, Nadia, Murshidabad, and South 24-Parganas) of West Bengal (Chakraborti et al., 2009). Figure 4 shows arsenic concentrations against depth of the 5,338 tube wells from West Bengal. From this figure, it appears that at depths greater than 350 m, arsenic is not present in hand tube-well water above 50 μg/L. Although in West Bengal it was considered that tube wells exceeding 100 m deep would be arsenic safe and a large number of tube wells were installed in villages to get arsenic safe water, we cannot say, from our result, that all tube wells greater than 100 m deep would be arsenic safe. Out of 5,338 tube wells of all depth, 54% tube wells are within 100–149 m and 80.8% within 100–200 m. Only 19.2% tube wells are greater than 200 m deep (Chakraborti et al., 2009). However, if a deep aquifer is tapped under a thick clay barrier, arsenic safe water is expected in the arsenic-contaminated areas of West Bengal and Bangladesh (Chakraborti et al., 2009).

Relation Between Arsenic, Iron, and Depth of Tube Wells in West Bengal

We collected depth information of 107,253 hand tube wells from the West Bengal. Arsenic concentration decreased with increasing depth and we could not find arsenic concentration above $50 \mu g/L$ in depth greater than 350 m (Chakraborti et al., 2009). We analyzed 17,050 water samples for both arsenic and iron from West Bengal. Average concentration of iron was detected as $3,756 \mu g/L$ with the range of 40–$77,000 \mu g/L$ (Chakraborti et al., 2009). Bivariate analysis showed a poor relationship between iron and arsenic in water. A negative correlation ($r = -0.137$, n = 15,611) has been observed between depth of the tube well and iron concentration. Chi-square test ensures strong association between arsenic concentration ranges and depth segments (Chakraborti et al., 2009).

Arsenic Contamination Status in Groundwater of Two Highly Contaminated Districts (Murshidabad and North 24-Parganas) of West Bengal

Murshidabad

In 2000 we concentrated our study on Murshidabad, one of the nine arsenic-affected districts of West Bengal. The Murshidabad district was selected as we had the maximum preliminary information from the district as well as some of our field workers are from the same district. Although we worked on this district sporadically since 1991, we surveyed this district very systematically with our entire effort from June 2000 to July 2003 to understand its arsenic contamination situation in details (Rahman, 2004; Rahman et al., 2005a).

Murshidabad district lies between the latitudes of 23°43′30″ to 24°50′20″ N and longitudes of 87°49′17″ to 88°44′ E. The River Ganga forms its northern and eastern boundaries and separates it from Bangladesh. The administrative structure of West Bengal consists of 19 districts and Murshidabad is one among them. Each district of West Bengal has several blocks. In Murshidabad, there are 26 blocks. Each block has several Gram Panchayets (GP) and each GP contains numerous villages. There are 2,414 villages and municipal areas (known as wards) in this district. There are 262 GPs including municipal areas in the Murshidabad district. Extending over an area of 5324 km², this district has 5,396,351 inhabitants. The River Bhagirathi flows across the district and divides it into two parts—eastern and western.

Hand tube-well water samples were collected at random from all 26 blocks of Murshidabad. Around 5,800 man-hours (4 persons $\times$ 8 h $\times$ 182 days) were spent for collecting water samples from Murshidabad (Rahman et al., 2005a). In this district, 29,612 hand tube-well water samples were analyzed from 1,833 villages/wards covering the entire area of the villages/wards (Rahman et al.,

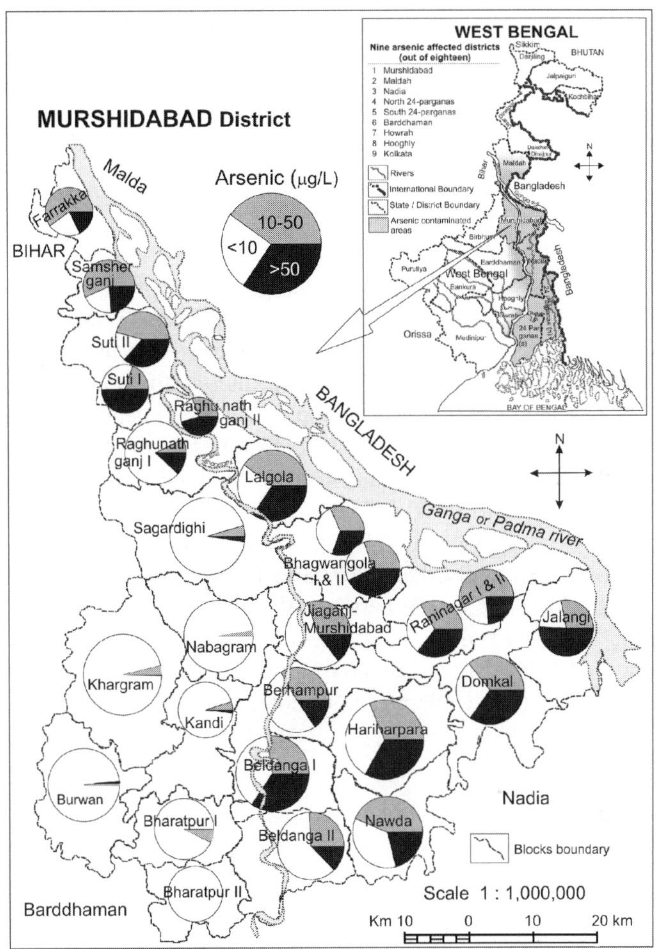

FIGURE 5 Arsenic contamination status in different blocks of Murshidabad district of West Bengal (Rahman et al., 2005a).

2005a). Water samples were collected from 250 GPs/municipal areas out of total 262. Based on the analysis results of 29,612 water samples from the district, arsenic concentration above 10µg/L was detected in 1,380 villages and wards and above 50µg/L in 994 villages and wards (Rahman et al., 2005a). Figure 5 shows the groundwater arsenic contamination status in all 26 blocks of Murshidabad. It appears that the blocks/thanas situated in the western side of River Bhagirathi were less affected ($n = 8,303$, 30.1% above 10µg/ L, 11.7% above 50µg/L) than those located on the eastern part ($n = 21,309$, 64.7% above 10µg/L, 32.5% above 50µg/L). The overall results showed that 53.8% of the hand tube wells had arsenic above 10µg/L and 26% had above 50µg/L (Rahman et al., 2005a). The results showed that only the groundwater of Bharatpur-II block was safe to drink according to the WHO guideline value of arsenic at 10 μg/L. Although the blocks situated in the western part of River Bagirathi were less arsenic contaminated, some of the blocks such as Suti-I, Suti-II, and Raghunathganj-I were identified with high degree of contamination (Rahman et al., 2005a).

A Study on Variation of Arsenic Concentration with Depth in Murshidabad

The objective of this study is to find out the possibility of water supply option from arsenic safe (arsenic < 10 μg/L) aquifer and to know the depths where arsenic is more abundant. To fulfill the above objective, an in-depth study was performed to assess arsenic content in groundwater at different depths. For this purpose, 29,612 hand tube-well water samples of varying depths were randomly collected covering all 26 blocks of Murshidabad. But during our field survey in 26 blocks of Murshidabad, we could collect the depth information of 25,629 out of 29,612 hand tube wells. Out of 25,629 hand tube wells, 1.90% are <9.4 m deep; 14.98%, between 9.4 and 15.2 m; 37.73%, between 15.5 and 23.0 m; 21.96%, between 23.2 and 30.5 m; 10.08%, between 30.8 and 38 m; 5.05%, between 38.4 and 45.7 m; 3.35%, between 46 and 53.3 m; 3.59%, between 53.6 and 61 m; 1.13%, between 61.3 and 76.2 m; 0.19%, between 76.5 and 91.4; 0.03%, between 91.7 and 122 m; and 0.01%, exceeding 122 m (Rahman, 2004). Thus shallow depth tube wells are more dominating in Murshidabad district. Figure 6 shows the change of arsenic concentration in tube well water with

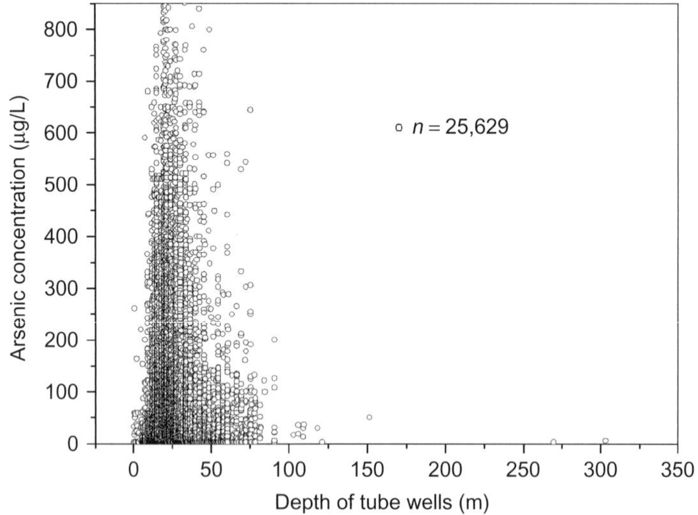

FIGURE 6 The change of arsenic concentration in tube-well water with depth in Murshidabad district of West Bengal.

depth in Murshidabad district. The results showed that arsenic concentration from lower depth to higher depth gradually increases up to the depth range of 23.2–30.5 m, and then it decreases except for depth ranges 61.3–76.2 and 76.5–91.4 m (Rahman, 2004). It also appears that there is no significant amount of arsenic present at depths greater than 91.7 m. However, the numbers of samples from these depths are not quite high.

North 24-Parganas

North 24-Parganas is one of the nine highly arsenic-affected districts in West Bengal. It is in southeast part of the state and lies in the subbasin of the Bhagirathi–Hooghly rivers and has bordering areas with Satkhira and Jessore districts of Bangladesh (both of these two districts of Bangladesh are highly arsenic-affected). In North 24-Parganas, there are 22 blocks. Total area and population of North 24-Parganas are 4,093 km^2 and 7.3 million, respectively (1991 census). More than 95% of the population use hand tube well water for drinking and around 70% for cooking (Rahman et al., 2003). Most of the tube wells are of shallow depth (15–50 m).

Extensive work has been conducted on North 24-Parganas district to know the actual magnitude of the arsenic calamity in West Bengal. The district was chosen for detail survey because of certain reasons, and these are (i) from our preliminary survey results up to 1994, we found North 24-Parganas to be of intermediate magnitude in its severity of arsenic problem compared to the other affected districts; (ii) communication with North 24-Parganas is not difficult; (iii) we have a group of local youths in this district who are working in our group and doing our preliminary field survey. Most of them are arsenic victims and have mild arsenical skin lesions. Although sporadically we surveyed North 24-Parganas from 1988, a systematic approach for the study was adopted in September 1994. Until 2003, about 4600 h were spent for our study in North 24-Paraganas district (Rahman et al., 2003).

Until 2002, 48,030 hand tube-well water samples were collected and analyzed from 22 blocks of North 24-Parganas (Rahman et al., 2003). Figure 7 shows the groundwater arsenic contamination situation in all 22 blocks of North 24-Parganas. It appears from this figure that out of 22 blocks in North 24-Parganas only two blocks, i.e., Sandeshkhali-I and Sandeshkhali-II are at present safe with respect to maximum permissible limit of arsenic (50 μg/L). In Sandeshkhali-I and Sandeshkhali-II, most of the tube wells are deep (exceeding 150 m). Shallow tube wells are saline, so people do not construct them. Due to the same reason, we have not found arsenic in tube-well water of southern part of Hingalganj block as all tube wells are of higher depth. But we got arsenic in the northern part of Hingalganj area close to Hasnabad block, where shallow tube wells with sweet water are abundant (Rahman et al., 2003). The detail arsenic contamination status of North 24-Parganas has been reported in our earlier publication (Rahman et al., 2003). Figure 8 presents the comparative bar diagram distribution of total

NORTH 24-PARGANAS District

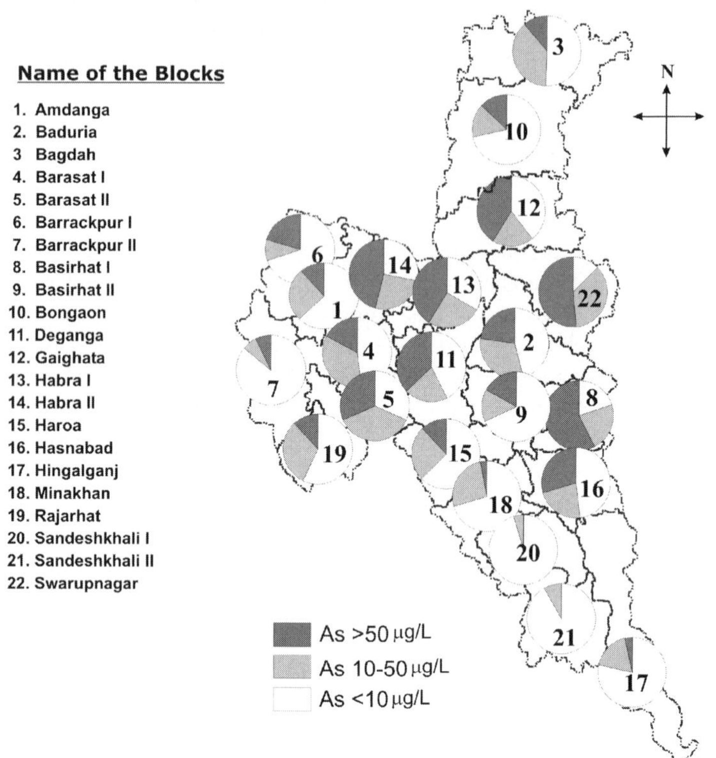

FIGURE 7   Groundwater arsenic contamination status in all 22 blocks of North 24-Parganas district of West Bengal (Rahman et al., 2003).

samples analyzed in different arsenic concentration ranges from two highly contaminated districts (Murshidabad and North 24-Parganas) of West Bengal with a highly arsenic-contaminated district Noakhali of Bangladesh. In Noakhali, 10.7% of the analyzed samples had arsenic above 1000 μg/L, whereas only 0.2% and 0.1% of the samples had arsenic above 1000 μg/L in Murshidabad and North 24-Parganas, respectively. From the results, it appears that higher As concentration is more in hand tube wells of Noakhali district compared to Murshidabad and North 24-Parganas districts.

Arsenic Contamination in Groundwater of the Jalangi—One of the Highly Arsenic-Affected Blocks in Murshidabad, West Bengal

Jalangi is one of the highly arsenic-affected blocks in Murshidabad district of West Bengal. The Jalangi block has 10 GPs and 117 villages. The area and population of the Jalangi block are 122 km^2 and 215,538, respectively. The detail

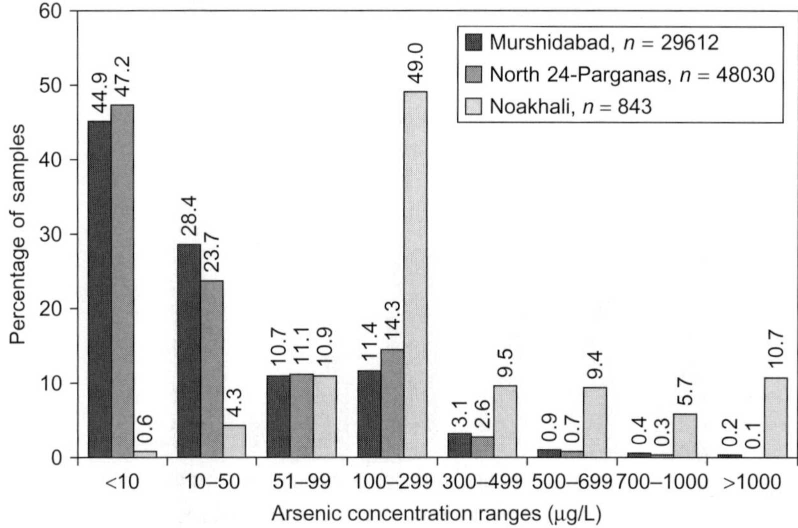

FIGURE 8 Comparative bar diagram distribution of total hand tube-well water samples in different arsenic concentration ranges from Murshidabad and North 24-Parganas districts of West Bengal with Noakhali district of Bangladesh.

results of water analyses and the impact on human have been reported in our previous publication (Rahman et al., 2005b). The groundwater of 102 villages in Jalangi contained arsenic above 10 µg/L and 95 villages had arsenic above 50 µg/L (Rahman et al., 2005b). From this block, 1,916 hand tube-well water samples were analyzed for arsenic from 104 surveyed villages. The results showed that 77.8% of water samples contained arsenic at concentration above 10 µg/L, 51% contained above 50 µg/L, and 17.2% above 300 µg/L (Rahman et al., 2005b). This is the only arsenic-affected block of West Bengal where 17.2% samples contained arsenic above 300 µg/L. Out of 1,916 water samples, 38 (2%) had arsenic above 1000 µg/L (Rahman, 2004). Figure 9 shows the comparative bar diagram distribution of total water samples in different arsenic concentration ranges from Jalangi block of West Bengal with a highly arsenic-contaminated thana (Sharsa, Jessore district) of Bangladesh.

Arsenic Contamination in Groundwater of the Sagarpara—One of the Highly Arsenic-Affected GP in Murshidabad, West Bengal

To understand better about arsenic contamination in groundwater, we present here hand tube-well water analysis data for arsenic of Sagarpara, one of the affected GPs in West Bengal. In the Sagarpara GP, there are 21 villages. The total area of Sagarpara is 20 km² and the population is about 24,419 (Rahman et al., 2005c). Our field survey information shows that almost 100% of the villagers of Sagarpara use hand tube-well water for drinking (Rahman et al., 2005c). Based

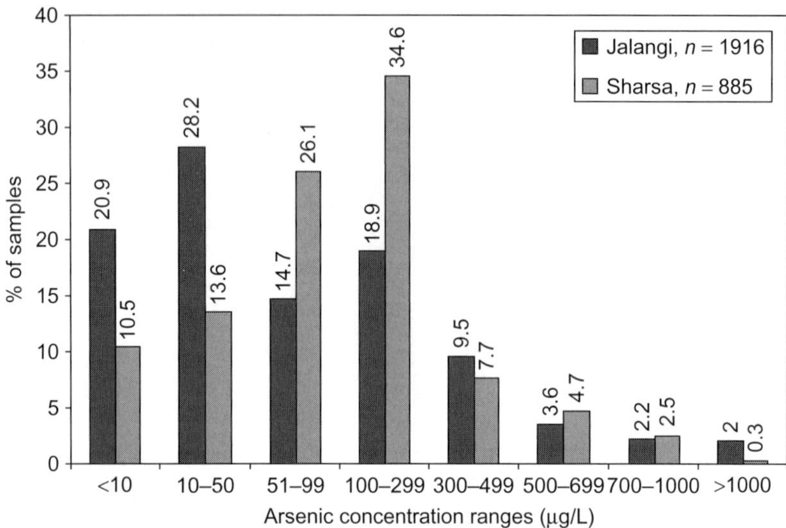

FIGURE 9   Comparative bar diagram distribution of total hand tube-well water samples in different arsenic concentration ranges from Jalangi block of West Bengal with Sharsa thana of Bangladesh.

on our survey, we estimated that 43 people use one hand tube well in Sagarpara (Rahman et al., 2005c). From the water analysis data of 565 hand tube wells, it appears that 86.2% of the water samples contained arsenic at concentration above 10 μg/L and 58.8% contained above 50 μg/L (Rahman et al., 2005c). Only 13.8% of hand tube wells were safe to drink from based on the WHO guideline value for arsenic in drinking water. Most notably, 26.5% of the analyzed samples had arsenic above 300 μg/L and 4.2% had arsenic above 1000 μg/L. Arsenic above 50 μg/L was detected in the groundwater of all 21 villages in Sagarpara. Out of 21 villages of Sagarpara, there are some villages where 80–90% hand tube wells were contaminated with arsenic at above 50 μg/L (Rahman et al., 2005c). Overall, the results indicated that the magnitude of arsenic contamination in Sagarpara GP is severe. Figure 10 shows the arsenic groundwater status of Sagarpara GP.

Arsenic in Hand Tube Wells in All 64 Districts of Bangladesh

During our field survey in Bangladesh from February 1996 to December 2002, water samples from 50,515 hand tube wells were collected and analyzed from all the 64 districts of Bangladesh by FI-HG-AAS (Rahman, 2004). Overall 3,600 villages in 331 thanas were surveyed out of the total of 490 thanas in Bangladesh. The analytical results of water samples showed that arsenic was found above the WHO's recommended level of arsenic in drinking water (10 μg/L) in 60 districts covering an area of 131,783 km^2 and population 119.3 million, and in

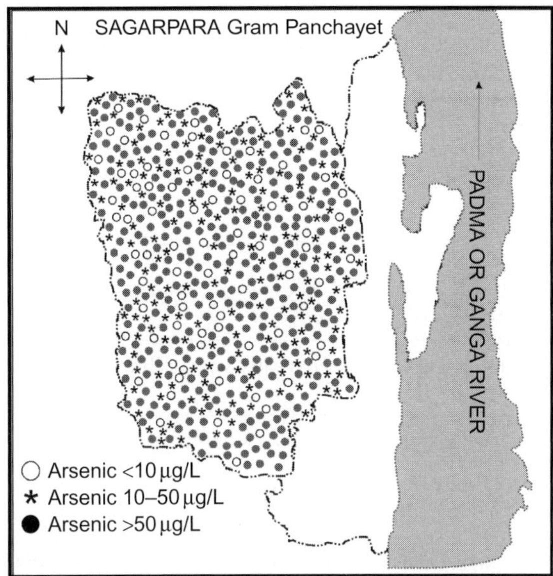

FIGURE 10   Groundwater arsenic contamination status of Sagarpara GP, one of the highly arsenic-contaminated Block Jalangi in Murshidabad district, West Bengal (Rahman et al., 2005c).

50 districts above maximum permissible limit (50 µg/L). The area and population of these 50 districts are 118,849 km^2 and 104.9 million, respectively. Our survey indicated that in 2,000 villages arsenic above 50 µg/L was detected, and in 2,500 villages groundwater contained arsenic above 10 µg/L (Rahman, 2004). Table 3 shows the arsenic contamination situation in 64 districts of Bangladesh at a glance. Figure 11 shows four geomorphological regions and the present groundwater arsenic contamination status in all 64 districts of Bangladesh. Figure 3 shows the bar diagram of concentration ranges of arsenic (µg/L) against the percentage of samples for 50,515 hand tube-well water samples from 64 districts of Bangladesh. From the analytical results, it has been revealed that 56.7% of hand tube-well water samples were safe to drink (arsenic <10 µg/L). From this figure, it is also evident that higher concentration of arsenic in samples is more prevalent in Bangladesh than West Bengal. About 40.1% and 26.2% of the samples contained arsenic above 10 and 50 µg/L, respectively. From 50 affected districts of Bangladesh, we analyzed 44,696 hand tube-well water samples, and the overall result showed that water from 51.5% of tube wells was safe to drink (<10 µg/L), whereas 48.5% and 31% of the tube wells contained arsenic above 10 and 50 µg/L, respectively. Although in 50 arsenic-affected districts of Bangladesh, only 31% of hand tube wells contained arsenic levels above 50 µg/L, these are overall results covering all affected districts. Although 50,515 hand tube wells were analyzed from all over Bangladesh, this number is very small compared to 8 million–10 million tube wells that exist in Bangladesh.

TABLE 3 Arsenic Contamination Situation at a Glance in 64 Districts of Bangladesh (Rahman, 2004)

Physical Parameters	Bangladesh
Area (km^2)	147,620
Population in millions	122
Total number of districts	64
Number of districts we have surveyed	64
Number of arsenic-affected districts (groundwater arsenic above $10\,\mu g/L$)	60
Number of arsenic-affected districts (groundwater arsenic above $50\,\mu g/L$)	50
Area of arsenic-affected districts (km^2)	118,849
Population of arsenic affected in millions	104.9
Total number of hand tube-well water samples analyzed	50,515
Percentage of samples having arsenic $>10\,\mu g/L$	40.1
Percentage of samples having arsenic $>50\,\mu g/L$	26.2
Total number of hand tube-well water samples analyzed from affected districts	44,696
Percentage of samples having arsenic $>10\,\mu g/L$ in affected districts	48.5
Percentage of samples having arsenic $>50\,\mu g/L$ in affected districts	31.0
Total number of thanas	490
Number of thanas we have surveyed	331
Number of arsenic-affected thanas with arsenic above $50\,\mu g/L$	189
Number of arsenic-affected villages (approx.) with groundwater arsenic above $50\,\mu g/L$	2,000
People drinking arsenic-contaminated water $>10\,\mu g/L$ (in million)	52
People drinking arsenic-contaminated water $>50\,\mu g/L$ (in million)	32
Highest arsenic concentration found ($\mu g/L$)	4,730
Districts surveyed for arsenic patients	33
Number of districts where we have identified people with arsenical skin lesions	31

(*Continued*)

TABLE 3 (Continued)

Physical Parameters	Bangladesh
People screened as arsenic patients from affected villages (preliminary survey)	18,991
Number of registered patients with clinical manifestations, including children	3,762 (19.8%)
Percentage of children having arsenical skin lesions based on number of total patients	6.1
Population drinking arsenic-contaminated water above $10\,\mu g/L$ (in million)	52
Population drinking arsenic-contaminated water above $50\,\mu g/L$ (in million)	32
Number of total deep tube wells (>100 m depth) analyzed	1,217
Percentage of deep tube wells having arsenic $>10\,\mu g/L$	26.9
Percentage of deep tube wells having arsenic $>50\,\mu g/L$	8.7

Distribution of Arsenic in Groundwater at Four Geomorphological Regions of Bangladesh

There are four geomorphological regions in Bangladesh (Figure 11). These are hill tract, tableland, floodplain, and deltaic regions (including coastal region) (Chakraborti et al., 1999). The analytical results of hand tube-well water samples in the four geomorphological regions are described in the following subsections.

Tableland Region

So far 9,693 hand tube-well water samples were analyzed from the tableland region of Bangladesh comprising 66 thanas of 17 districts (Rahman, 2004). The results showed that 2.3% samples contained arsenic above $10\,\mu g/L$ and 0.16% above $50\,\mu g/L$. The maximum arsenic concentration in the tableland areas has been detected as $134\,\mu g/L$ (Rahman, 2004). So, 9,677 (99.8%) water samples analyzed from this region appear safe to drink according to the standard level of arsenic in drinking water of Bangladesh ($50\,\mu g/L$), and 9,472 (97.7%) samples are safe to drink according to the WHO's guideline value ($10\,\mu g/L$) of arsenic in drinking water (Rahman, 2004). The reason why these 22 (2.3%) samples have arsenic above $10\,\mu g/L$ can be either when we had collected the samples without filtering through membrane filter, small invisible particles containing some arsenic compounds had dissolved on addition of preservative or the area

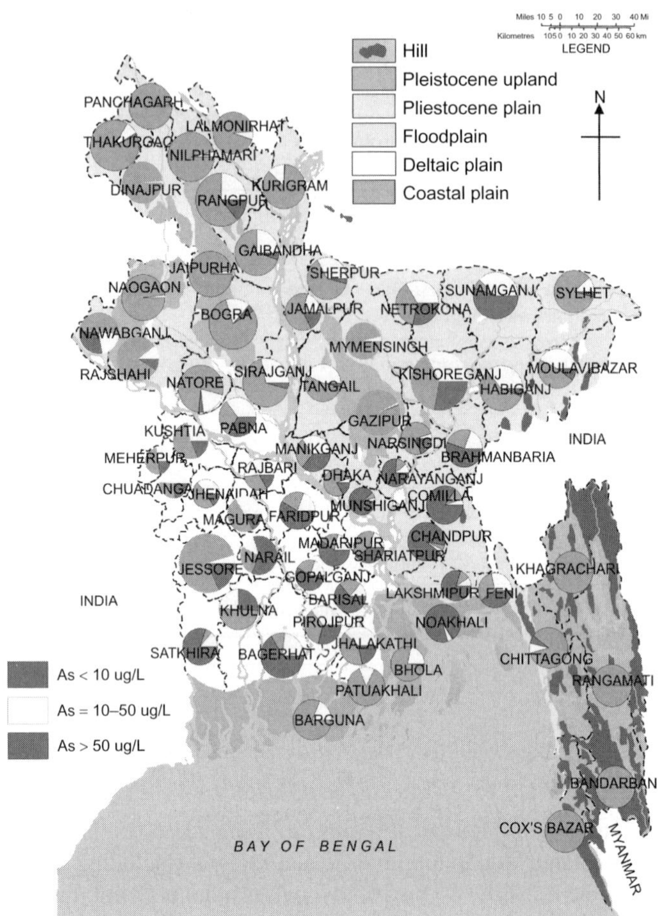

FIGURE 11   Four geomorphological regions of Bangladesh and the status of arsenic contamination in groundwater of 64 districts in Bangladesh (see Plate 2 of Color Plate section).

of these thanas is slightly contaminated (maybe these areas are in the fringe area of the tableland with arsenic-contaminated floodplain).

Floodplain Region

Hand tube-well water samples were collected from 147 surveyed thanas out of 276 in the floodplain region. From this region, 18,760 hand tube-well water samples were analyzed for arsenic. Of these, 51.9% and 35.8% samples contained arsenic above 10 and 50 µg/L, respectively, and 48.1% samples were safe to drink (arsenic <10 µg/L) (Rahman, 2004). About 11.3% samples contained arsenic above 300 µg/L. The percentage of samples contained >300 µg/L indicated the presence of arsenicosis patients in this region. The maximum concentration of arsenic was detected as 4,730 µg/L in a tube well of this region.

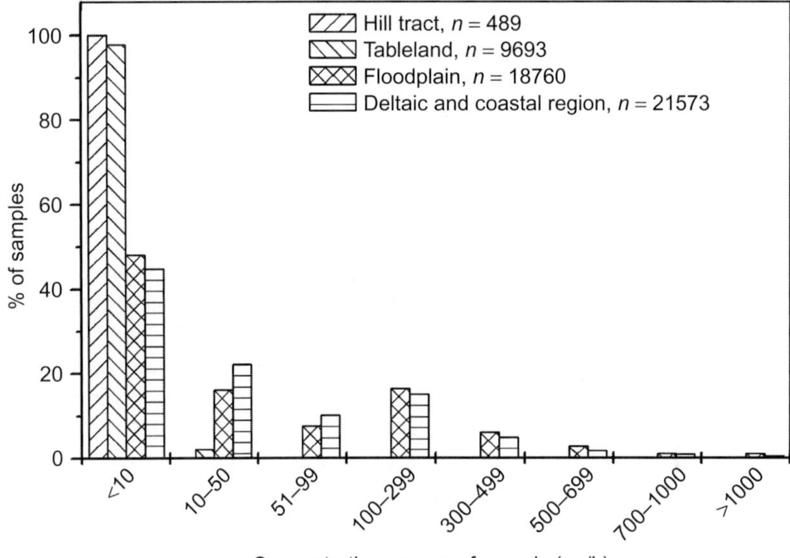

FIGURE 12 Distribution of arsenic in tube-well water samples in four geomorphological regions of Bangladesh.

Deltaic Region Including Coastal Belt

In total, 96 thanas were surveyed out of 137 for sampling. Water samples from 21,573 hand tube wells were analyzed for arsenic from deltaic region and coastal region. Out of these, 55.2%, 33.1%, and 7.8% samples contained arsenic above 10, 50, and 300 μg/L, respectively. In fact 44.8% samples were safe to drink (arsenic <10 μg/L) (Rahman, 2004). The results showed that arsenic concentration above 50 μg/L was detected in 83 thanas (Rahman, 2004). The maximum arsenic concentration was found in a tube well as 3,143 μg/L.

Hill Tract Region

So far, 489 hand tube-well water samples have been analyzed for arsenic from hill tract region of Bangladesh. All analyzed tube-well water samples contained arsenic <50 μg/L. Only 20 samples contained arsenic between 10 and 50 μg/L (Rahman, 2004).

Figure 12 shows a comparative study of groundwater arsenic contamination status for all geomorphological regions of Bangladesh. It appears that there is more groundwater arsenic contamination of higher concentration (100 μg/L and above) in floodplain compared to deltaic plain with coastal belt. The tableland and hill tract regions are almost arsenic contamination free. Out of 50 arsenic-affected districts of Bangladesh, we observed that two districts (Noakhali and Lakshmipur) of floodplain area are the highest arsenic-contaminated districts. We found 274 hand tube wells with arsenic above 1000 μg/L from all 64 districts of Bangladesh

(Rahman, 2004). Out of 274 samples, 178 samples are from these two districts. The highest arsenic concentration in groundwater (4,730 µg/L) has been detected from Chiladi village of Senbagh thana in Noakhali district of Bangladesh.

From the overall water analysis results, we noticed that in some parts of Bangladesh, arsenic contamination in groundwater is less (between 10 and 50 µg/L), some parts are arsenic safe (<10 µg/L), and some are highly contaminated (between 50 and 4730 µg/L). While trying to find out the reason, we have noticed that out of four principal geomorphological regions of Bangladesh, the tableland and hill tract regions are usually arsenic safe, but the area of floodplain and deltaic region are highly arsenic-contaminated. We also noticed that some contaminations are there in the fringe areas of the tableland with floodplain and hill tract with floodplain, and if rivers of floodplain have eroded tableland and hill tract areas (Chakraborti et al., 1999). The probable reason of contamination may be heavy deposition of Holocene sediments to floodplain and deltaic region. However, the areas partly in floodplain, partly in hill tract; partly in floodplain–partly in tableland are less contaminated.

Results on Arsenic in Tube-Well Water Samples: Comparison Between Our Data and Other International Data

We have compared the data of our water analysis from districts of four geomorphological regions of Bangladesh with the available hand tube-well water data of other organizations such as BGS with Department of Public Health Engineering (DPHE), Bangladesh; the nongovernmental organization (NGO) Forum for Drinking Water Supply and Sanitation, Bangladesh; the Bangladesh Rural Advancement Committee (BRAC), Bangladesh; the CARE, Bangladesh; the Gono Shasthya Kendra (GSK). The comparative study of our data and others' data from the deltaic, floodplain, tableland, and hill tract regions are presented in Figure 13 based on the percentage of samples contained arsenic above 50 µg/L. From the deltaic region, 21,573 hand tube-well water samples were analyzed covering 96 thanas of the total of 137. Of these, 33.1% samples contained arsenic above 50 µg/L. BGS–DPHE (1999) analyzed 12,145 water samples from 126 thanas of this region, and 27.5% samples had arsenic above 50 µg/L, which is lower than that of our study. NGO Forum analyzed 3,768 samples covering 83 thanas of this region and reported that 33.1% samples contained arsenic above 50 µg/L, which is exactly similar to our study (NGO Forum report, 2002). BRAC–Bangladesh conducted a detailed study in Jikargachha thana of Jessore district to know the status of arsenic contamination in groundwater. They collected and analyzed 26,637 water samples from this thana and found that 48.1% samples had arsenic concentration above 50 µg/L (BRAC Research Monograph, 2000). GSK analyzed 3,133 water samples from Kashinathpur union of Santhia thana in Pabna district and reported that 32.6% samples had arsenic over 50 µg/L (Disaster Forum Publication, Fact Sheet 12, Dhaka, Bangladesh).

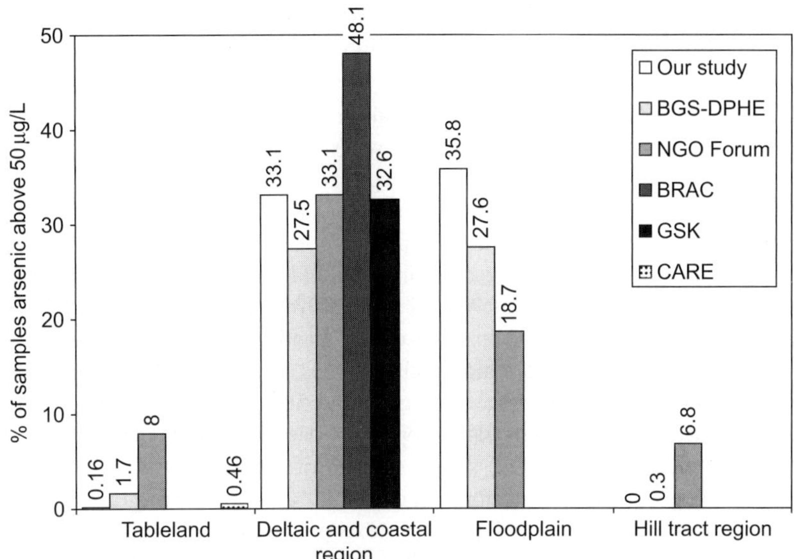

FIGURE 13 Comparative study of our data and data from other organizations from four geomorphological regions of Bangladesh (based on the percentage of the samples contained arsenic above 50 μg/L).

Water samples from 18,760 hand tube well were analyzed covering 147 thanas of floodplain region. Our results showed that 35.8% samples contained arsenic above 50 μg/L. BGS–DPHE (1999) analyzed 16,533 water samples and 27.6% had arsenic above 50 μg/L. NGO Forum study on this region showed that 18.7% of the 4,104 water samples contained arsenic above 50 μg/L (NGO Forum report, 2002).

From the tableland region, the results of water analysis revealed that only 0.16% samples contained arsenic above 50 μg/L, whereas the BGS–DPHE (1999) study showed 1.7% samples had arsenic above 50 μg/L. NGO Forum survey indicated that 8% samples out of 664 contained arsenic above 50 μg/L (NGO Forum report, 2002). CARE, Bangladesh, surveyed Tanore, Godagari thanas of Rajshahi district, and Nachole thana of Chapai Nawabganj district. They analyzed 215 water samples and only one sample had arsenic above 50 μg/L (Disaster Forum Publication, Fact Sheet 12, Dhaka, Bangladesh). Our study indicated that the groundwater of this region is mostly safe with respect to arsenic.

The results of our water analysis from hill tract region showed (Rahman, 2004) that none of the samples contained arsenic above 50 μg/L, which is almost similar to the findings of BGS–DPHE (1999) study. BGS-DPHE (1999) data shows that only one sample has arsenic above 50 μg/L from this region. Surprisingly, NGO Forum reported that 6.8% of the 295 samples contained arsenic above 50 μg/L (NGO Forum report, 2002).

Arsenic Concentration in Deep Tube Wells at Four Geomorphological Regions of Bangladesh

So far 1,217 hand tube wells have been analyzed from 100 to 415 m depth from four geomorphological regions of Bangladesh (Chakraborti et al., 1999). Table 4 shows the arsenic concentration in tube wells at different depths for each geomorphological region. It appears that deep tube wells of tableland and hill tract regions are free from arsenic contamination. Deep tube wells of both deltaic region including coastal belt and floodplain are to some extent arsenic contaminated. Figure 14 shows arsenic concentration against depth of 1,217 hand tube wells from Bangladesh. Our analysis of samples from 1,217 hand tube wells from 100 to 415 m depth indicated that out of 931 hand tube wells exceeding 200 m from floodplain and deltaic region including coastal belt, 185 samples (19.8%) contained arsenic between 10 and 49 μg/L, and 58 (6.2%) tube wells contained arsenic above 50 μg/L. The available report for deep tube wells more than 200 m deep from BGS–DPHE (1999) shows that out of 909 deep tube wells from all over Bangladesh, arsenic concentration of 34 (3.7%) is above 50 μg/L. Our results indicated that at depths exceeding 300 m (except one tube well in floodplain and three in deltaic region), all samples had arsenic concentration below 50 μg/L. Highest depth where arsenic found above 50 μg/L is 315 m and the concentration of arsenic was 225 μg/L. We analyzed 183 tube wells more than 350 m deep, and all tube wells contained arsenic below 50 μg/L, but 19 samples contained arsenic between 10 and 50 μg/L. Therefore we do not expect arsenic above 50 μg/L in groundwater in depth of exceeding 350 m in floodplain and deltaic regions of Bangladesh. But we do not know what would happen in long run.

SOURCE AND MECHANISM OF ARSENIC CONTAMINATION

From the arsenic contamination scenario in Asia, it appears that the floodplains of many rivers originating from the Himalayan Mountains and the Tibetan Plateau are affected (Chakraborti et al., 2008b). On this basis, we noticed arsenic contamination in West Bengal, Bihar, Jharkhand, UP in the Gangetic plain, Brahmaputra plain in Assam, and Padma–Meghna–Brahmaputra (PMB) plain in Bangladesh. The source is geologic. Various theories have been postulated on the sources of arsenic and the mechanism of mobilization from the source (Das et al., 1996; Bhattacharya et al., 1997; Nickson et al., 1998; Chowdhury et al., 1999; Harvey et al., 2002; Akai et al., 2004; Islam et al., 2004). The exact nature of mobilization process is still unknown.

CONCLUSIONS AND RECOMMENDATIONS

Elimination of the arsenic crisis in the Ganges basin requires (Rahman et al., 2001) concerted action that includes the following:

1. A moratorium on the installation of more tube wells in contaminated areas until all the installed tube wells are checked for arsenic contamination. The

TABLE 4 Distribution of Deep Tube Wells in Different Arsenic Concentration Range (μg/L) with Depth Collected from Four Geomorphological Regions of Bangladesh[*] (Chakraborti et al., 1999)

Geomorphological Regions	Depth Range (m)	Number of Deep Tube Wells Analyzed	Distribution of Samples in Different Arsenic Concentration (μg/L) Range				
			<10	10–49	50–99	100–199	200–299
Hill tract	205–250	82	82 (100%)	–	–	–	–
Tableland	102–204	25	25 (100%)	–	–	–	–
	205–250	47	47 (100%)	–	–	–	–
Total	72	72 (100%)	–	–	–	–	–
Floodplain	102–204	106	59 (55.66%)	8 (7.55%)	17 (16.04%)	18 (16.98%)	4 (3.77%)
	205–250	96	48 (50%)	28 (29.17%)	8 (8.33%)	11 (11.46%)	1 (1.04%)
	253–300	82	41 (50%)	33 (40.24%)	2 (2.44%)	5 (6.10%)	1 (1.22%)
	305–350	74	46 (62.16%)	27 (36.48%)	–	–	1 (1.36%)
	360–415	70	62 (88.57%)	8 (11.43%)	–	–	–
Total	428	256 (59.81%)	104 (24.30%)	27 (6.31%)	34 (7.94%)	7 (1.64%)	

(*Continued*)

TABLE 4 (Continued)

Geomorphological Regions	Depth Range (m)	Number of Deep Tube Wells Analyzed	Distribution of Samples in Different Arsenic Concentration (µg/L) Range				
			<10	10–49	50–99	100–199	200–299
Deltaic region including coastal	102–204	155	118 (76.13%)	28 (18.06%)	1 (0.65%)	4 (2.58%)	4 (2.58%)
	205–250	140	105 (75%)	22 (15.71%)	3 (2.14%)	9 (6.43%)	1 (0.72%)
	253–300	113	84 (74.34%)	16 (14.16%)	5 (4.42%)	8 (7.08%)	–
	305–350	114	71 (62.28%)	40 (35.09%)	3 (2.63%)	–	–
	360–415	113	102 (90.26%)	11 (9.74%)	–	–	–
Total	635	480 (75.59%)	117 (18.42%)	12 (1.89%)	21 (3.31%)	5 (0.79%)	
Grand total	1,217	890 (73.13%)	221 (18.16%)	39 (3.20%)	55 (4.52%)	12 (0.99%)	

Depth information received from local people/tube well owner.

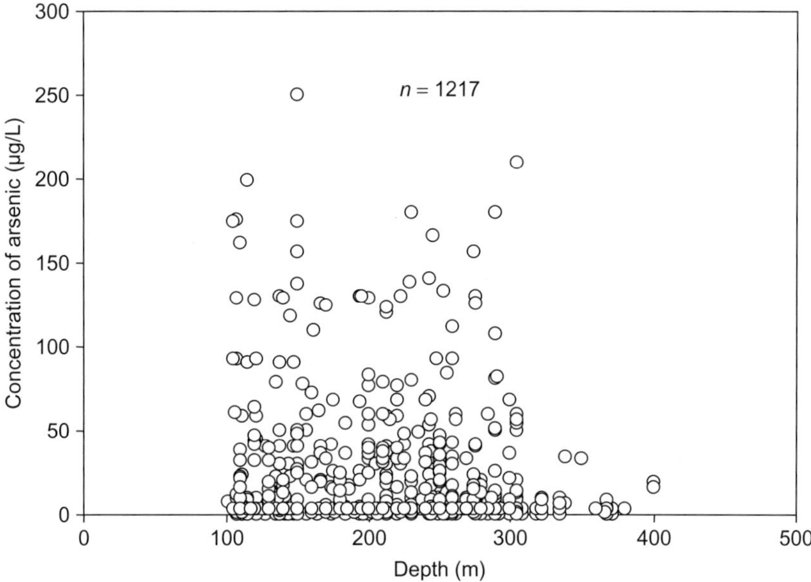

FIGURE 14 Distribution of arsenic concentration against depth (m) of tube wells in Bangladesh.

local and national governments should frame and implement regulation of new tube wells. Around 95% of the people in Bangladesh and West Bengal, India, depend on tube wells for drinking water. If the mouths of all safe tube wells are colored green, and unsafe wells are colored red, villagers can use green tube wells for drinking and cooking purposes, and the red tube wells for bathing, washing, toilet, etc. We have disturbing evidence from West Bengal, India, that previously safe tube wells now show arsenic contamination (Chakraborti et al., 2001). The currently safe tube wells require monitoring every 3–6 months to track this new development.

2. Proper watershed management.
3. Traditional water management such as dug well, three Kalsi system, and rainwater harvesting with controls of bacterial and other chemical contamination.
4. Public awareness of the arsenic calamity and assurance that it is not a curse of God.
5. Recognition that, so far, there is no effective therapy. Safe water and optimal nutrition are the only proven measures.
6. A worldwide effort by the scientific community addressing the problem that has put more than 100 million people in Bangladesh and West Bengal, India, at risk for cancer, vascular disease, and other complications.

Although tube wells provide drinking water free of microbial contamination, the merciless exploitation of groundwater for irrigation without effective watershed management to harness huge surface water resources and rainwater is seen as a gross miscalculation. In Bangladesh and West Bengal, there are huge surface resources of sweet water in the rivers, wetlands, flooded river basins, and oxbow lakes. *Per capita* available surface water in Bangladesh is about $11,000\,m^3$. These two delta areas, known as the land of rivers, have approximately 2,000 mm annual rainfall. Watershed management and villager participation are needed to assure the appropriate utilization of these huge water resources.

ACKNOWLEDGMENTS

The authors thank the field workers of the School of Environmental Studies, Jadavpur University, for their extensive help in the field sampling in arsenic-affected villages of West Bengal in India. We also thank the field workers and the Management of the Dhaka Community Hospital, Bangladesh, for their active participation in the field survey in Bangladesh. Financial support from SOES is greatly acknowledged.

REFERENCES

Agrawal, O., Sunita, G., Gupta, V.K., 1999. A sensitive colorimetric method for the determination of arsenic in environmental and biological samples. J. Chin. Chem. Soc. 46, 641–645.

Ahamed, S., Sengupta, M.K., Mukherjee, A., Hossain, M.A., Das, B., Nayak, B., et al., 2006. Arsenic groundwater contamination and its health effects in the state of Uttar Pradesh (UP) in upper and middle Ganga plain, India: a severe danger. Sci. Total Environ. 370, 310–322.

Akai, J., Izumi, K., Fukuhara, H., Masuda, H., Nakano, S., Yoshimura, T., et al., 2004. Mineralogical and geomicrobiological investigations on ground water arsenic enrichment in Bangladesh. Appl. Geochem. 19, 215–230.

Berg, M., Tran, H.C., Nguyen, T.C., Pham, M.V., Schertenleib, R., Giger, W., 2001. Arsenic contamination of groundwater and drinking water in Vietnam: a human health threat. Environ. Sci. Technol. 35 (13), 2621–2626.

Berg, M., Stengel, C., Trang, P.T., Viet, P.H., Sampson, M.L., Leng, M., et al., 2007. Magnitude of arsenic pollution in the Mekong and Red River Deltas—Cambodia and Vietnam. Sci. Total Environ. 372, 413–425.

BGS–DPHE, 1999. Groundwater studies for arsenic contamination in Bangladesh. Final Report: London, U.K., Mott MacDonald Ltd., U.K.

BGS–DPHE, 2001. Arsenic contamination of groundwater in Bangladesh. BGS Technical Report WC/00/19. British Geological Survey, Keyworth, U.K.

Bhattacharya, P., Chatterjee, D., Jacks, G., 1997. Occurrence of arsenic-contaminated groundwater in alluvial aquifers from the Delta plain, Eastern India: options for a safe drinking water supply. Int. J. Water Resour. Dev. 13, 79–92.

BRAC Research Monograph, 2000. Combating a deadly menace: early experiences with a community-based arsenic mitigation project in Bangladesh, June 1999–June 2000. BRAC Research Monograph, Series No. 16, August.

Chakraborti, D., Biswas, B.K., Basu, G.K., Chowdhury, U.K., Chowdhury, T.R., Lodh, D., et al., 1999. Possible arsenic contamination free groundwater source in Bangladesh. J. Surf. Sci. Technol. 15, 180–188.

Chakraborti, D., Basu, G.K., Biswas, B.K., Chowdhury, U.K., Rahman, M.M, Paul, K., et al., 2001. Characterization of arsenic bearing sediments in Gangetic delta of West Bengal-India. In: Chappell, W.R., Abernathy, C.O., Calderon R.L. (Ed.), Arsenic Exposure and Health Effects, Elsevier, 27–52.

Chakraborti, D., Rahman, M.M., Chowdhury, U.K., Sengupta, M.K., Lodh, D., Chanda, C.R., et al., 2002. Arsenic calamity in the Indian sub-continent. What lesions have been learned? Talanta 58, 3–22.

Chakraborti, D., Mukherjee, S.C., Pati, S., Sengupta, M.K., Rahman, M.M., Chowdhury, U.K., et al., 2003. Arsenic groundwater contamination in middle Ganga Plain, Bihar, India: a future danger. Environ. Health Perspect. 111, 1194–1201.

Chakraborti, D., Sengupta, M.K., Rahman, M.M., Ahamed, S., Chowdhury, U.K., Chowdhury, U.K., et al., 2004. Groundwater arsenic contamination and its health effects in the Ganga–Meghna–Brahmaputra plain. J. Environ. Monit. 6, 74–83.

Chakraborti, D., Singh, E.J., Das, B., Shah, B.A., Hossain, M.A., et al., 2008a. Groundwater arsenic contamination in Manipur, one of the seven North-Eastern Hill states of India: a future danger. Environ. Geol. 56, 381–390.

Chakraborti, D., Das, B., Nayak, B., Pal, A., Rahman, M.M, Sengupta, M.K., et al., 2008b. Groundwater arsenic contamination and its adverse health effects in the Ganga-Meghna-Brahmaputra plain. In: Roy, K. (Ed.), Arsenic Calamity of Groundwater in Bangladesh: Contamination in Water, Soil and Plants., Nihon University, Japan, 13–52.

Chakraborti, D., Das, B., Rahman, M.M., Chowdhury, U.K., Biswas, B., Goswami, A.B., 2009. Status of groundwater arsenic contamination in the state of West Bengal, India: a 20 years study report. Mol. Nutr. Food Res. DOI 10.1002/mnfr200700517.

Chatterjee, A., Das, D., Mandal, B.K., Chowdhury, T.R., Samanta, G., Chakraborti, D., 1995. Arsenic in ground water in six districts of West Bengal, India: the biggest arsenic calamity in the world, Part I. Arsenic species in drinking water and urine of the affected people. Analyst 20, 643–650.

Cherukurii, J., Anjaneyulu, Y., 2005. Design and development of low-cost, simple, rapid and safe, modified field kits for the visual detection and determination of arsenic in drinking water samples. Int. J. Environ. Res. Public Health 2, 322–327.

Chowdhury, T.R., Basu, G.K., Mandal, B.K., Biswas, B.K., Chowdhury, U.K., Chanda, C.R., et al., 1999. Arsenic poisoning in the Ganges delta. Nature 401, 545–546.

Das, D., Samanta, G., Mandal, B.K., Chowdhury, R.T., Chanda, C.R., Chowdhury, P.P., et al., 1996. Arsenic in groundwater in six districts of West Bengal, India. Environ. Geochem. Health 18, 5–15.

Deshpande, L.S., Pande, S.P., 2005. Development of arsenic testing field kit—a tool for rapid on-site screening of arsenic contaminated water sources. Environ. Monit. Assess. 101, 93–101.

Dhar, R.K., Biswas, B.K., Samanta, G., Mandal, B.K., Chakraborti, D., Roy, S., et al., 1997. Groundwater arsenic calamity in Bangladesh. Curr. Sci. 73, 48–59.

Dhar, R.K., Zheng, Y., Rubenstone, J., van Geen, A., 2004. A rapid colorimetric method for measuring arsenic concentrations in groundwater. Anal. Chim. Acta 526, 203–209.

Disaster Forum Publication. Fact Sheet 12 on Arsenic: A Disaster Forum Publication, Dhaka, Bangladesh (5/8 Sir Syed Ahmed Road, Mohammadpur, Block A, Dhaka 1207, Bangladesh, e-mail: mailto:df@bangla.net).

Erickson, B.E., 2003. Field kits fail to provide accurate measure of arsenic in groundwater. Environ. Sci. Technol. 37, 35A–38A.

Gallagher, P.A., Schwegel, C.A., Wei, X., Creed, J.T., 2001. Speciation and preservation of inorganic arsenic in drinking water sources using EDTA with IC separation and ICP-MS detection. J. Environ. Monit. 3, 371–376.

Garai, R., Chakraborty, A.K., Dey, S.B., Saha, K.C., 1984. Chronic arsenic poisoning from tubewell water. J. Indian Med. Assoc. 82, 34–35.

Goessler, W., Kuehnelt, D., 2002. Analytical methods for the determination of arsenic and arsenic compounds in the environment. In: Frankenberger Jr. W.T. (Ed.), Environmental Chemistry of Arsenic. Marcel Dekker, New York, pp. 27–50.

Harvey, C.F., Swartz, C.H., Badruzzaman, A.B.M., Keon-Blute, N., Yu, W., Ali, M.A., et al., 2002. Arsenic mobility and groundwater extraction in Bangladesh. Science 298, 1602–1606.

International Agency for Research on Cancer (IARC), 2004. IARC Monograph 84: Some Drinking Water Disinfectants and Contaminants Including Arsenic. World Health Organization, IARC, Lyon, France.

International Conference on Arsenic, 1995. International conference on arsenic in groundwater: cause, effect and remedy. School of Environmental Studies, Jadavpur University, Kolkata, India, 6–8 February.

International Conference on Arsenic, 1998. International conference on arsenic pollution of groundwater in Bangladesh: cause, effects and remedies. School of Environmental Studies, Jadavpur University, India and Dhaka Community Hospital, Bangladesh, Dhaka, February 8–12.

Islam, F.S., Gault, A.G., Boothman, C., Polya, D.A., Charnok, J.M., Chatterjee, D., et al., 2004. Role of metal reducing bacteria in arsenic release from Bengal delta sediments. Nature 430, 68–71.

Jakariya, M., Vahter, M., Rahman, M., Wahed, M.A., Hore, S.K., Bhattacherya, P., et al., 2007. Screening of arsenic in tubewell water with field test kits: evaluation of the method from public health perspective. Sci. Total Environ. 379, 167–175.

Karagas, M.R., Le, C.X., Morris, S., Blum, J., Lu, X., Spate, V., et al., 2001. Markers of low level arsenic exposure for evaluating human cancer risks in a US population. Int. J. Occup. Med. Environ. Health 14, 171–175.

Karagas, M.R., Stukel, T.A., Tosteson, T.D., 2002. Assessment of cancer risk and environmental levels of arsenic in New Hampshire. Int. J. Hyg. Environ. Health 205, 85–94.

Khandaker, N.R., 2004. Limited accuracy of arsenic field test kit. Environ. Sci. Technol. 38, 479A.

Mosaferi, M., Yunesion, M., Mesdaghinia, A., Naidu, A., Nasseri, S., Mahvi, A.H., 2003. Arsenic occurrence in drinking water of I.R. of Iran: the case of Kurdistan province. In: Ahmed, M.F., Ali, M.A., Adeel, Z. (Eds.), Fate of Arsenic in the Environment. Bangladesh University of Engineering and Technology and United Nation University International Symposium, International Training Network Centre, Bangladesh University of Engineering and Technology, Dhaka, Bangladesh, pp. 1–6.

Mukherjee, A., Sengupta, M.K., Hossain, M.A., Ahamed, S., Das, B., Nayak, B., et al., 2006. Arsenic contamination in groundwater: a global perspective with emphasis on the Asian scenario. J. Health Popul. Nut. 24, 142–163.

NGO Forum Report, 2002. Half Yearly Report. NGO Forum for Drinking Water Supply & Sanitation. Dhaka, Bangladesh, January.

Nickson, R., MacArthur, J.M., Burgess, W.G., Ahmed, K.M., Ravenscroft, P., Rahman, M., 1998. Arsenic poisoning in Bangladesh groundwater. Nature 395, 338.

Nickson, R., Sengupta, C.S., Mitra, P., Dave, S.N., Banerjee, A.K., Bhattacharya, A., et al., 2008. Current knowledge on the distribution of arsenic in groundwater in five states of India. J. Environ. Sci. Health 42, 1707–1718.

Pillai, A., Sunita, G., Gupta, V.K., 2000. A new system for the spectrophotometric determination of arsenic in environmental and biological samples. Anal. Chim. Acta 408, 111–115.

Rahman, M.M., 2004. Ph.D. thesis. Present status of groundwater arsenic contamination in Bangladesh and detailed study of Murshidabad, one of the affected neighboring districts in West Bengal, India. Jadavpur University, Kolkata, India.

Rahman, M.M., Chowdhury, U.K., Mukherjee, S.C., Mandal, B.K., Paul, K., Lodh, D., et al., 2001. Chronic arsenic toxicity in Bangladesh and West Bengal, India—a review and commentary. Clin. Toxicol. 39, 683–700.

Rahman, M.M., Mukherjee, D., Sengupta, M.K., Chowdhury, U.K., Lodh, D., Chanda, C.R., et al., 2002. Effectiveness and reliability of arsenic field testing kits: are the million dollar screening projects effective or not? Environ. Sci. Technol. 36, 5385–5394.

Rahman, M.M., Mandal, B.K., Roychowdhury, T., Sengupta, M.K., Chowdhury, U.K., Lodh, D., et al., 2003. Arsenic groundwater contamination and sufferings of people in North 24-Parganas, one of the nine arsenic affected districts of West Bengal, India: the seven years study report. J. Environ. Sci. Health 38, 25–59.

Rahman, M.M., Sengupta, M.K., Mukherjee, S.C., Pati, S., Ahamed, S., Chowdhury, U.K., et al., 2005a. Murshidabad—one of the nine groundwater arsenic affected districts of West Bengal, India. Part I: Magnitude of contamination and population at risk. J. Toxicol. Clin. Toxi. 43, 823–834.

Rahman, M.M., Sengupta, M.K., Ahamed, S., Chowdhury, U.K., Das, B., Hossain, M.A., et al., 2005b. Magnitude of arsenic groundwater contamination and human suffering in Jalangi— one of the 78 arsenic affected block/thanas in West Bengal—India. Sci. Total Environ. 338, 189–200.

Rahman, M.M., Sengupta, M.K., Ahamed, S., Chowdhury, U.K., Lodh, D., Hossain, M.A., et al., 2005c. Status of groundwater arsenic contamination and human suffering in a Gram Panchayet (cluster of villages) in Murshidabad, one of the nine arsenic affected districts in West Bengal— India: a semi-microlevel study. J. Water Health 3, 283–296.

Roychowdhury, T., 2008. Influence of several factors during collection and preservation prior to analysis of arsenic in groundwater: a case study from West Bengal, India. J. Int. Environ. Appl. Sci. 3, 1–20.

Saha, K.C., 1984. Melanokeratosis from arsenic contaminated tubewell water. Indian J. Derm. 29 (4), 37–46.

Samanta, G., Chakraborti, D., 1997. Flow injection atomic absorption spectrometry for the standardization of arsenic, lead and mercury in environmental and biological standard reference materials. Frenius J. Anal. Chem. 357, 827–832.

Samanta, G., Roy Chowdhury, T., Mandal, B.K., Biswas, B.K., Chowdhury, U.K., et al., 1999. Flow injection hydride generation atomic absorption spectrometry for determination of arsenic in water and biological samples from arsenic affected districts of West Bengal, India and Bangladesh. Microchem. J. 62, 174–191.

Shibata, Y., Morita, M., 1989. Speciation of arsenic by reversed-phase high performance liquid chromatography-inductively coupled plasma mass spectrometry. Anal. Sci. 5, 107–109.

Shraim, A., Chiswell, B., Olszowy, H., 1999. Speciation of arsenic by hydride generation atomic absorption spectrometry in hydrochloric acid reaction medium. Talanta 50, 1109–1127.

Shraim, A., Chiswell, B., Olszowy, H., 2000. Use of perchloric acid as a reaction medium for speciation of arsenic by hydride generation atomic absorption spectrometry. Analyst 125, 949–953.

SM 3114. Standard methods for the examination of water and wastewater. Arsenic and selenium hydride by hydride generation atomic absorption spectrometry, American Water Works Association, 1999.

SM 3120. Standard methods for the examination of water and wastewater. Metals by plasma emission spectrometry, American Water Works Association, 1999.

Steinmaus, C.M., George, C.M., Kalman, D.A., Smith, A.H., 2006. Evaluation of two new arsenic field test kits capable of detecting arsenic water concentrations close to 10 µg/L. Environ. Sci. Technol. 40, 3362–3366.

USEPA, 1999. USEPA Office of Water. Analytical methods support documents for arsenic in drinking water. December, p. 36.

van Geen, A., Cheng, Z., Seddique, A.A., Hoque, M.A., Gelman, A., Graziano, J.H., 2005. Reliability of a commercial kit to test groundwater for arsenic in Bangladesh. Environ. Sci. Technol. 39, 299–303.

Whitnack, G.C., Brophy, R.G., 1969. A rapid and highly sensitive single sweep polarographic method of analysis for arsenic (III) in drinking water. Anal. Chim. Acta 48, 123–127.

Whitnack, G.C., Brophy, R.G., Martens, H.H., 1972. Rapid and highly sensitive polarographic analysis for arsenic (III) in drinking water; application to determination of arsenic in potable desert groundwater uncovers an analysis problem with these waters. http://www.stinet.dtic.mil/oai/oai?verb=getRecord&metadataPrefix=html&identifier=AD0894934

World Health Organization (WHO), , 2001. Environmental Health Criteria 224 Arsenic and Arsenic Compounds. International Programme on Chemical Safety, World Health Organization, Geneva.

Chapter 6

Forensic Water Quality Investigations: Identifying Pollution Sources and Polluters

Lawrence B. Cahoon
Department of Biology and Marine Biology and Center for Marine Science, University of North Carolina Wilmington, Wilmington, NC 28403, USA

Robert H. Cutting
Department of Environmental Studies, UNC Wilmington, Wilmington, NC 28403-5949, USA

INTRODUCTION

The aim of water quality analysis is to determine the presence of pollutants of concern and estimate their concentrations within acceptable levels of precision. Water pollution can then be defined as concentrations of harmful materials or their indicators at or above certain levels that have been established by epidemiological or other methods, or set by regulation.[1] Remediation or mitigation of water pollution then requires the identification and quantification of the

---

1. For example, the Clean Water Act defines "pollutant" as "…dredged spoil, solid waste, incinerator residue, sewage, garbage, sewage sludge, munitions, chemical wastes, biological materials, radioactive materials, heat, wrecked or discarded equipment, rock, sand, cellar dirt and industrial, municipal, and agricultural waste discharged into water.…" 33 U.S. Code Sec. 1362(6)

Handbook of Water Purity and Quality
Copyright © 2009 by Elsevier Inc. All rights of reproduction in any form reserved.

sources. In the case of pollution caused by human actions, source identification also entails determination of responsibility, which may engender civil actions or even criminal charges, and burden polluters with penalties and remediation costs. Consequently, polluters may challenge the methods, results, and interpretations of water quality investigations, as well as the skill and veracity of investigators. The term "forensic" is used to describe situations such as trials or administrative hearings in which adversarial argumentation is used to establish facts, eliminate incorrect observations and interpretations, and test propositions. Clearly, skillful analytical work is a requisite for effective environmental forensic investigation, but a larger set of skills and methods must be employed to yield satisfactory outcomes from good field and laboratory work in a forensic context. The skills and techniques that support positive outcomes in a forensic setting are broadly useful in many other settings as well.

Regulatory agencies frequently perform the investigative functions described here, as they should, but often lack the resources to monitor water quality as thoroughly as they might. Anyone involved in water quality analyses may detect evidence of a problem; regulators admit that many of their investigations are prompted by information given by other water quality professionals, other members of the regulated community, and even untrained citizens who observe something that appears problematic. Professional investigators have frequently obtained their skills through *ad hoc* experience rather than formal training. Consequently, the approaches described here may be helpful to many in the practice of water quality analysis.

Sometimes evidence is uncovered as a result of routine sampling programs. In other cases, a complaint about a fish kill or discolored water prompts a more specific inquiry. We counsel our students to treat any case as if it were criminal in nature so that the highest standards of evidence and persuasion would apply, thus the results should be sufficient for any other application.

We first discuss general considerations for any water quality investigation and then address specific issues related to evidence and presentation of your findings. Most cases are won or lost at the investigative stages, hence the efforts are worthwhile. Note that although shows like "CSI" focus on field and lab analysis, real-life investigations often require more emphasis on careful review of documentation and questioning live witnesses.

QUALITY ASSURANCE/QUALITY CONTROL

The correct approach to any forensic investigation is to consider the question, "How do you know the things you claim to know?" as if asked by an adversary. Many good forensic practices are of common sense, but others require more training and experience to apply effectively. The key is preparation, as "fortune favors those who are prepared." Effective preparation begins with planning, more specifically a quality assurance/quality control, or "QA/QC," plan. The American Public Health Association Standard Methods manual (APHA, 2001) provides excellent descriptions of the elements of QA/QC plans for standard laboratory method validation. Extension of the same principles and approaches to the entire investigative process is a logical and wise practice in a forensic context.

The elements of a QA/QC plan include delineation of responsibility for each task, sample control (including chain of custody procedures), specification of methods and techniques (including standardization, calibration, and equipment maintenance procedures), data assessment and reduction, and reporting procedures. Quality control procedures *per se* can be divided into internal and external components. Internal components address sample analysis quality including calibration and standards, analysis of blanks, analysis of duplicates, true replication (as opposed to subsamples), determination of precision, and recovery of known additions (when sample matrix effects may be a confounding factor). External quality control procedures include training and certification procedures, competence testing, analysis of unknowns and external standards, and external data review. These measures are obviously best considered prior to commencement of field sampling efforts, *Ex post facto* attempts to implement quality control measures may be better than nothing, but can be used against investigators in forensic situations, and beg the question of how reliable initial results can be. Documentation is necessary and is best handled from the beginning with forethought given to the nature of the information to be collected, creation of routine sample data entry procedures, and appropriate metadata (investigator identity, place, date/time, environmental conditions, and other relevant observations) along with written notes on any unusual observations. Many labs use standardized data sheets for field sampling efforts, with one sheet per sampling location, coupled with data forms for laboratory analyses and procedures (see Appendix A for chain of custody samples). Data security should be also considered; accidental losses can be embarrassing, but deliberate tampering can be devastating.

BASIC SITE RESEARCH

To set the scene both for the investigation and for any prospective audience, it is usually helpful to obtain maps from your agency, the local tax office, and now sources such as Google Earth, NASA World Wind, and Microsoft's Terraserver and state GIS repositories.[2] U.S. Environmental Protection Agency (EPA) and many local agencies also offer interactive mapping that will reveal potential pollution sources. Maps and on-site inspections are essential to develop your own

---

2. Maps and aerial photos
- Street map from Google Maps, or Yahoo! Maps, or Windows Live Local
- Satellite image from Google Maps, Windows Live Local, WikiMapia
- Google Earth
- Topographic map from TopoZone
- Topographic map from U.S. Geological Survey Topographic Maps
- Aerial image or topographic map from TerraServer-USA
- An extensive listing of maps and imagery is at Wikipedia

narrative of what happened, where, when, etc. A wealth of information on ownership, sales and prior ownership, building information, wastewater and other utility information, as well as aerials and parcel maps is usually available at the local real property tax office. Much of the documentation on transfers is online at recorders' or register of deeds offices. Information on corporate forms and other business entities is available often from the Secretary of State or corporate regulatory agency. Shareholder information, however, is seldom available. The Web site of the entity of interest can also provide not only company organization and details but also names of key personnel. Other agencies such as a state coastal or resources agency may also have records, but much of this information is unavailable even to other state agencies.[3] An umbrella organization for agencies, R@IN, The Regional Environmental Enforcement Associations, can be found at http://www.regionalassociations.org/ (last visited April 20, 2009) but even member states do not always utilize the resource. Some information on compliance and enforcement history is available through limited portals like EPA's ECHO (http://www.epa.gov/compliance/data/systems/multimedia/echo.html), but it is not particularly user friendly if information on a particular company or product class is sought. Nor is it frequently updated. ECHO does allow searches by Standard Industrial Classification (SIC) and its successor standard, but even the new network established by EPA (The Environmental Exchange Network http://www.exchangenetwork.net/basics/how_it_works.htm) currently has no planned public access.

Often personal inspection of the locations of investigated incidents can reveal bits of relevant information or even major insights that are valuable to subsequent case development, so it is a rare case where a site visit is unnecessary. Case presenters should *always* be familiar with the scene(s). If your agency has the resources and the case is significant, there is a class of scene re-creation devices, for example, the total station hardware (surveying tools) and software for 3D modeling.[4] Law enforcement agencies such as the FBI use them to do crime scene re-creation and just did so in the Blackwater/Iraq case.

---

2. (Continued)
- EPA EnviroMapper®: http://www.epa.gov/enviro/html/em/
- EPA Window to My Environment: http://www.epa.gov/enviro/wme/
- NASA World Wind: http://www.worldwind.arc.nasa.gov/
  Be sure also to contact local, regional, and state governments, which you can locate here. Contact the local government list of state environmental agencies: http://www.epa.gov/epa-home/state.htm
3. Cutting, Cahoon & Leggette, Enforcement Data: A Tool for Environmental Management, 36 environmental law reporter 10060-10072 (January 2006), discussing how inter- and intra-state and sharing ins primitive compared with criminal resources like NCIC (National Crime Information Center).
4. Total Station hardware and software: http://totalstation.org/ e.g., http://www.leica-geosystems.com/corporate/en/lgs_8276.htm FBI's Blackwater investigation: NPR Nov 1, 2007—http://www.npr.org/templates/story/story.php?storyId=15835037

SAMPLING

Some water quality investigations focus on recorded information and procedures; however, the majority involves sampling. Here, the QA/QC process is essential and should be specifically documented. The equipment, calibration, methods, handling, and analysis should all be carefully documented through reporting procedures that comport with your QA/QC methodology. We recommend that the required "chain of custody" be documented to criminal law standards whenever possible, which means recording every event and having each person who handles the sample indicate their participation on a straightforward form (see discussion and samples in Appendix A). Naturally, collection and handling should occur according to the recommended procedures (such as the APHA Standard Methods) and documented in writing and through photos if possible. For example, fecal coliform samples should be refrigerated, kept out of the light and tested within a specified time period, all of which can easily be documented. In addition, actual documentation such as calibration records and equipment manuals must be readily available. Although water quality investigations typically focus on the water column, be aware that significant clues can be uncovered in the benthic materials, where everything from fecal coliform to pharmaceuticals can exist for some time and may resurface when the material is disturbed.

WORKING WITH THE LABORATORY

Laboratory work is often required to identify and quantify pollutants in water quality cases. Whether this work is done "in-house" or by a certified external laboratory, the entire process must be carefully documented, including of course the chain of custody. In all cases, it is wise to retain control (untested) samples, if possible. The lab staff will likely prepare a report that can itself become evidence, or the technician may also testify. Again, the object is to ensure that the credentials of the lab are adequate and available, the results are clear, and the results are presented in the context of the written legal standard. As noted in the discussion of quantitative analysis above, it is usually helpful if the technician or another witness can articulate the reason that the test result is significant. Audiences are skeptical of violations within what might appear to them to be small percentages of a standard they might not understand. Some agencies, for example, will not pursue enforcement unless the standard is exceeded by 40%. The audience must see why the numbers mean something, especially if there is a public health component to the standard.

TYPES OF EVIDENCE

There are three general classifications of evidence: (1) real evidence including most samples, other tangible items, maps, chart models, and even on-site viewing of the incident scene; (2) documentary evidence such as documents (permit applications, manifests, discharge records, memos, and other books and records), photographs, and electronic documents such as the increasingly important e-mail and computer hard-drive evidence; and (3) testimonial evidence, whether in the form of percipient (eyewitness) testimony or opinion testimony by either lay or expert witnesses (most readers may qualify in the latter category).[5]

Evidence is also classified according to whether it is direct or circumstantial. Direct evidence such as eyewitness testimony establishes a fact. Circumstantial evidence requires someone (e.g., judge or jury) to make an inference about what happened. An example might be evidence of a visual plume in a water body, pollutants in a ditch or soil, and a pipe (particularly with residue of the pollutant) upgradient from the soil and water body. In contrast, a stream of the pollutant from the pipe flowing across the ground into the water would be a direct evidence. Circumstantial evidence may actually be more reliable, as eyewitnesses often have difficulty with details. Forensic cases often involve piecing together various types of circumstantial evidence. There is also a distinction commonly made between physical evidence, such as equipment or chemical results, and biological evidence, such as dead fish or evidence that associates a perpetrator, such as hair. Finally, reconstructive evidence, such as pieces of shattered equipment found at a spill site, can assist in constructing inferences as to what happened (e.g., an explosion) and even when, where, and how it happened. Different rules apply to each category, although frequently all types are involved in any given case.

A common pattern might involve a witness who reports (testimonial) a fish kill or plume (real and perhaps documentary). The sampling should reveal the nature of the pollutant involved (real evidence, although the lab report is documentary evidence). If a discharge permit for that material (NPDES permit) is held by an upstream entity, the public reports and permits (documentary) can be reviewed for compliance. However, some of these records are generated by the potential defendant, such as the daily monitoring reports (DMR) required of NPDES permit holders. Hence, whenever possible, employees (and former employees) should be interviewed to verify accuracy and company practice (testimonial evidence, leading perhaps to more documentary evidence, such as records and procedure manuals), and/or appropriate agency staff should be interviewed (testimonial evidence for compliance histories, enforcement cases, and/or other useful information). A review of online EPA and other federal and NGO materials and release data for the area could demonstrate other potential sources of the material in question.[6]

---

5. See, for example, Federal Rules of Evidence, copy available at http://www.law.cornell.edu/rules/fre/index.html

6. See the citations at note 3 above, and

   Centers for Disease Control and Prevention (CDC): Environmental issues: http://www.cdc.gov/Environmental/

   Environmental defense, scorecard: http://www.scorecard.org/

   Federal Emergency Management Agency (FEMA) hazard discussions: main page: http://www.fema.gov/hazard/index.shtm

Information on the type of business may also reveal sources of chemicals for which permits to discharge were not held (e.g., a photo lab on premises with no discharge permit for silver), or hazardous materials for which adequate disposal records are not available, both of which may indicate a possible malfunction or even intentional release (such as burying the classic 55-gallon drums) (Clifford, 1998). Moreover, documentary evidence in the form of (1) computer data on production, upsets, and breakdowns; (2) internal company memos and e-mails can either substantiate or refute compliance data; and (3) agency compliance documents, enforcement documents, or permits. More than a few cases involve falsification of or errors in self-reporting data. Company procedure manuals may indicate strict QA/QC policies, but recently, in an air pollution case, a major oil company was found to have failed habitually to follow its own safety and compliance plans, resulting in significant environmental impacts and injuries that culminated in criminal prosecution and multimillion dollar penalties (http://yosemite.epa.gov/opa/admpress. nsf/o/70AFE4F098BEB51F85257562006C2581). Thus initial witness reports and sampling may only be the beginning of the inquiry. In "paper cases," however, care must be taken to organize documents, authenticate them, and connect them so that an audience (particularly a jury) will not doze off when the file boxes are opened.

Another common fact pattern is usually more difficult: nonpoint sources such as storm water or agricultural runoff. A witness may report sedimentation or discoloration in receiving waters. Once the constituent pollutants are understood, there still must be a connection forged between the result and the cause. Often, this means review of ownership records along possible drainage routes (topographic maps and the hall of records—now frequently online, as noted above), as well as pollutants likely to be found (EPA or first-responder records). In a leaking underground storage tank (UST) scenario, markers in petroleum products may provide necessary clues as sophisticated field and laboratory analyses. In the case of animal wastes, it may be difficult even to distinguish among types of animals, let alone ownership of the particular source, although waste characteristics can provide some clues (Cutting et al., 2006). In both cases, reviews of construction and operating permits (building records) and a careful tracing of the drainage area via mapping involve amassing documentary evidence. Still, it may be testimonial evidence that connects the dots. In one

---

6. (Continued)

Fish advisories: from EPA, listed by state: http://www.epa.gov/waterscience/fish/states.htm

U.S. Environmental Protection Agency home page, with quick index for hazards such as lead, indoor air pollution, water pollution http://www.epa.gov/ as well as the "Where you Live Section" http://www.epa.gov/epahome/whereyoulive.htm with information such as EnviroMapper and state environmental agencies and also the "Window to My Environment mapping tool" http://www. epa.gov/enviro/wme/ and surf your Watershed http://www.cfpub.epa.gov/surf/locate/index.cfm

US Geological Survey Office of Groundwater: http://www.water.usgs.gov/ogw/

US Geological Survey Watershed Information: http://www.water.usgs.gov/wsc/

National Institute of Health, National Medical Library, Tox Town: information about chemical hazards in your area http://www.toxtown.nlm.nih.gov/

recent North Carolina criminal case, an owner pled to intentional draining of a waste lagoon into a creek after interviews with employees revealed that lagoon levels mysteriously lowered overnight while employees were away. Multiple sources present different problems, and they may yield more than one defendant. In storm water, for example, soil analysis from adjacent parcels revealed by building records may be identical, but interviews with neighbors, building inspectors, employees, and subcontractors (even though they may not be friendly witnesses) as well as on-site inspections may reveal which actor caused the sedimentation—or that both contributed and are therefore culpable. The question then becomes how to apportion liability if there are damages, as well as to determine relative fault for enforcement purposes.

LEGAL TOOLS TO OBTAIN EVIDENCE

As noted, agency personnel may have administrative and search warrant powers, and some agencies also have administrative subpoena powers. Agencies and private parties always have informal devices such as informational requests, which often surprisingly yield useful documentary or testimonial evidence. But if informal avenues fail, consult counsel about the techniques available in your jurisdiction. In private civil litigation, for example, the powers to examine any potential witness and to require production of documentary and real evidence, including the right to inspect facilities, are routinely available through the subpoena power and depositions. Records and physical evidence can be identified and located through written questions, called interrogatories, and then examined through requests for production and inspection. We recommend consulting counsel early on if you believe there may be difficulty obtaining evidence you consider essential. Of course, counsel must be consulted for the more precise requirements of inspection and search warrants. Administrative search warrants require that (1) the affiant work for an agency that has extensive regulatory powers, such as a water control agency and (2) pursuant to the regulatory plan, access is needed to determine compliance at a particular site. A full search warrant typically requires clearly presented facts that (1) a crime has been committed and (2) there is reasonable cause to believe that evidence of the crime exists at the location to be searched. Typically, the criminal prosecutor should be consulted if a serious crime is suspected. In any case, if documents are the target, persistence is the key. More than one "smoking gun" has been found at the bottom of the file box or "misfiled." The federal Freedom of Information Act (FOIA) provides a mechanism for private parties to obtain public documents. The procedures are generally straightforward but may involve some cost. See discussion and samples at Appendix B.

ACCESS AND ENTRY

Any time you need to view a scene, you must ask whether you have a legal right to be there. If the entity is subject to permit conditions that require entry, regulatory officials may still need either an inspection warrant or a search warrant if entry is refused, so counsel should be consulted if in doubt. Evidence may be excluded if it is unlawfully obtained, particularly in criminal cases. Deliberate trespassing must be avoided both to ensure that the evidence can be used and to protect the investigator from prosecution, so it is imperative to know where you are when walking that stream course. Use photographic tools (see Section Photography) and your own GPS or other locational devices so that you know and can prove where you are, and that you have express permission to be on that property (even if it is public land). See discussion of sources of ownership information in Section Quality Assurance/Quality Control.

WITNESSES

Most scientists and environmental professionals are trained in sampling techniques, but not in interview skills. Yet reporting parties and even neighbors can often provide key facts or opinions that can help you focus the inquiry and save time and resources. A reporting party may observe discoloration once, but a neighbor can simplify the investigation by volunteering that it occurs at intervals and provide other conditions, such as the local factory working overtime or an occasional dump truck in the area. Be alert to employees and former employees and ask who knows about personnel and where the records are kept. Many owe allegiance to the potential defendant, but some find it difficult to lie, especially if they have been treated poorly (e.g., the movies *A Civil Action* and *Erin Brockovich*[7]). Of course, with these and all witnesses, you must also be aware of the potential biases and credibility issues. Employees and former employees, or individuals who have had a relationship of some type with the potential defendant, are often called "turncoats." However, many white-collar crime prosecutors relate that without such witnesses, connecting the pollution to any particular defendant can be difficult.

Even suspected polluters are frequently happy to talk, for example, if the investigator presents a relaxed, friendly approach. Be aware, though, that if the case may become criminal and the witness is a target of the investigation, counsel should be consulted to avoid constitutional issues, such as the Fifth Amendment limitation on self-incrimination. We recommend open-ended questions such as, "What do you think happened?" rather than cross-examining a witness, at least at the outset. Sometimes that is not possible, and techniques such as "good cop/bad cop" can work wonders, but listening is the key skill in any event. Recording testimony, electronically if permission is secured (or not

---

7. See *A Civil Action* (Buena Vista Studios, 1998) (depicting the infamous toxic pollution case in Woburn, MA); Erin Brockovich (Universal Pictures, 2000) (depicting the Pacific gas and electric water pollution case in California).

necessary in your jurisdiction) or with notes, is essential. If the witness will provide a written statement, it is usually desirable.

A word on expert witnesses, whether you are one or you need to consult an expert, prosecutors counsel, that you must be (or find) experts whose *time* is for sale, rather than those whose *opinions* are for sale. The credentials (and limitations) of an expert should be readily at hand and the expert's track record understood (for example, does he or she almost always testify for the defense?). The "expert" need not be a professional expert, either, as line employees can be experts in company procedures or operation of particular equipment.[8] Expert witnesses are generally permitted to offer an opinion (often qualified as their best professional judgment) as to some relevant issue in the case, such as what pollutant might be involved or where the source of the pollutant might be. As with all science, there will be conflicting opinions as to methodology, concepts, interpretation of data, and conclusions. In jurisdictions that follow the federal Rules of Evidence, the judge generally determines whether the expert testimony is sufficiently reliable (the "gatekeeper" approach). Some forums may require that the expert view represents the prevailing view in a given field (the *Frye* test), whereas others, such as North Carolina, allow nearly any expert testimony subject to cross-examination and rebuttal and allow the jury (or judge if no jury) to weigh conflicting views. It is important to understand what rules your forum will follow. It is equally important to understand the nature of the expert testimony. Remember to pace yourself when you are a witness, and to counsel witnesses that you suggest to testify simply and directly. One practical pointer to remember is to take a brief pause before answering when you are being cross-examined to allow your counsel time to object.

DOCUMENTARY EVIDENCE

Documents can be of direct evidence (of ownership, for example) or circumstantial evidence (that an entity knew or should have known of the standard to which they are held, or knew of a condition that was not reported, for example). Documents must be properly categorized and marked from the outset to ensure ready access (more than one agency has lost key documents). Ownership, permitted activity and standards, internal procedures, maintenance and repair records, and internal memos can establish individual facts or patterns and practices of conduct. The hall of records in your jurisdiction (often available electronically, above), local and state GIS offices, secretaries of state (e.g., interlocking business entities), and other agencies can provide key documents for all these issues; in spite of this be aware that even agencies are often

---

8. There is a great scene in *My Cousin Vinny*, where the attorney's girlfriend (Marisa Tomei) takes the stand as an expert because she grew up working for her family's car repair shop and the issue had to do with mechanical characteristics of a particular vehicle. (Palo Vista Productions 1992).

unable to access records easily in other jurisdictions, or even their own agency. Electronic records are not yet the norm, but are a critical link in the documentary chain. Discovery orders (in civil cases) and search warrants (in criminal cases) are typically required to secure the actual hard drive of a suspected defendant's electronic records, although sometimes servers, networks, and even your own e-mails can provide key clues as to who was involved when and what they knew. Documentary evidence must usually be authenticated, so make sure your sources are known and that you can demonstrate the source of all documentation, such as certification by the public office that prepared them.

PHOTOGRAPHY

Photography may be the single best tool available in a forensic situation. Photography can show third parties the answers to the critical questions of any investigation: "What happened? Where did it happen? How did it happen? Who is or might be responsible?" Photography can document the presence of witnesses, corroborating evidence, location, sequences of events, the presence of the investigators themselves, and their sampling and measurement work. Photography at a developing scene can document unexpected events or features, so it is wise to take a lot of pictures or run a video camera constantly; editing can take out unneeded imagery but can never restore images not photographed in the first place. Digital imagery may lend itself to doctoring more easily than chemical film, but some digital media, such as CD-ROM disks, are as secure as chemical film against tampering. Recorded imagery frequently preserves details that escaped the observers on the scene, as many crime scene surveillance tapes, crash recordings, and other publicly displayed forms of forensic photography have demonstrated. Imagery can be extraordinarily more persuasive than verbal testimony. Pictures of a fish kill, for example, can convey the magnitude of the deaths far better than tabulations of numbers of individuals. Cell phones with cameras allow instant transmission of visual evidence, so forensic investigators can transmit imagery to regulators, home base, or the news media, however, be aware of the caveats listed below.

Photography poses several inherent challenges to environmental investigators. First, the human eye captures imagery quite differently than mechanical lenses and chemical or digital media. The human eye is a very sophisticated imaging organ, and its interconnectedness with the optical processing centers of the brain endows the whole visual sensory apparatus with remarkable ability to recognize patterns, detect motion, and react to unusual signals. Cameras, on the contrary, have to be used as tools without the intimate sensor–brain connection. Photographers, therefore, must give conscious thought to recording imagery that establishes the temporal and spatial contexts of a scene. The human eye can adapt to and see usefully across over six orders of magnitude of light intensity, making it somewhat paradoxically a poor light meter. Cameras are usually much more limited in their useful range of lighting, so consideration must be given to that through familiarity with one's camera and the employment of other light sources, avoidance of shadows, etc. Humans tend to conserve film or disk space, but this reflex works against the necessity of investigative work. One must be prepared to shoot prolifically; a bad shot is better than no shot at all. However, it is possible to present imagery that is too "persuasive," even so graphic as to revolt third parties, so one must be careful to show only those images one is prepared to defend as necessary to the case. One must also be cautious to avoid doing things before the camera (or otherwise) that would hurt a case—evidence of improper technique, evidence tampering, trespass, etc. Defense counsel will likely be entitled to all images, not just those you want to use. In summary, the images ultimately must help tell the story unfolded by investigation. This requires the presence of mind in using cameras and preparation beforehand. The obvious mistakes—dead batteries, no film, dropped camera—can happen to anyone, but preparation is again a key. Murphy's Law is a useful guide here: bring extra batteries, an extra camera, and someone else who knows how to use it. In some court cases, only camera-dated evidence may be allowed as evidence, so make sure this feature is activated on your camera if you can. Note that a camera often is the best security device field personnel can have to guard against landowner rage or any other hazard.

QUANTITATIVE REASONING

Quantitative analysts are trained to work with numbers and can generally interpret and understand quantitative results with the ease of long experience. Many target audiences, however, lack such training and experience. Effective communication of the significance of results, such as violations of water quality standards, must employ effective modes of presentation more intuitively. Humans receive information in several basic ways. We are primarily visual creatures taking in more than two-thirds of our information through visual imagery. The old phrase, "a picture is worth a thousand words," is quite true. For the analyst, converting numerical information into pictures, such as graphs, is therefore an effective communication mode. Edward Tufte has developed a very sophisticated set of insights into the visual communication of quantitative information. Verbal descriptions lie somewhere between abstract numerical symbols and imagery in their ease of understanding. Effective forensic presentations combine all three modes of communication, so that a numerical datum representing a violation of a given water quality standard, for example, can be illustrated graphically as a point lying well above a line corresponding to the standard and clearly distinguishable from other data points in the set. Verbal descriptions can then explain what the standard represents and how the unusual datum might most likely be explained as a violation and not operator error or some other trivial result. It is more important that the presenter explains the significance of the departure of the violation from the standard in understandable terms, that is, "what is the effect (on fish, human health, etc.) of a turbidity reading that is 100 times the level of the standard?"

PRESENTATION OF FINDINGS

The ultimate test is whether your findings are persuasive. We have several suggestions, but the bottom line is to "Keep It Simple, Stupid (KISS)". The reality is that while scholarly presentations should be at a high level, most newspapers are keyed to fifth grade reading level and juries are much the same. We suggest that you avoid patronizing—and anesthetizing—your audience unless you know that a highly technical presentation is required. That does not mean that you do not cover all the bases: it simply means that once your credentials are established, your methodology and findings must be understandable as well as technically (and legally) sufficient. You must be able to persuade your target audience (1) what you know, (2) how you know what you know, and (3) why it is significant.

Preparation is once again the key. Practice your presentation. Have your counsel or colleagues question your presentation and practice response convincingly. This will help you anticipate adversity and bolster your presentation. "Walk a mile" in your adversaries' shoes to understand their position and objections. To the extent possible, review their presentation and positions. Review your own after testing by your colleagues.

Organize your presentation. Have your documentation and your evidence ready. Often it can be useful to have official-looking samples and equipment out and visible even when not used so the target audience is curious. Case management software, or even free, open source programs (such as Free Mind) can provide excellent outlining tools.

Models, diagrams, charts, and other real evidence can be immensely helpful along with photographs. If, for example, models, charts, or written materials such as articles or treatises are to be utilized ensure that they are clearly understood (and understandable to your intended audience) and that they are available both in advance for inspection and at any subsequent hearing. The FBI and other high-end operations have total station hardware and software available to recreate scenes in 3-D such as the Blackwater investigation in Iraq. For most investigations, it is still useful to have large-scale presentation boards with the scene presented clearly. If samples are to be utilized, it is often useful to have them available for the audience to examine. One successful trial attorney even placed sample jars that clearly had two substances of different viscosity on the counsel table, but never used them during the entire hearing.

If you elect PowerPoint or some other electronic methods to present evidence, make sure your equipment is reliable and your presentation straightforward (and that it is allowed by the rules of your forum). The current debate about the efficacy of these devices suggests that for charts, graphs, and photos/videos they may be useful, but for words and number charts much less so for the reasons stated above.

CONCLUSIONS

Successful forensic presentation of water quality issues is grounded on good science, but depends in part on the art of persuasion. Appearances can be significant, so anticipate the ultimate use of your evidence and findings at all stages, and keep asking yourself how your presentation will be received.

REFERENCES

APHA, 2001. In: Clesceri, L.S., Greenberg, A.E., Eaton, A.D. (Eds.), Standard Methods for the Examination of Water and Waste Water, Quality Assurance, twenty first ed. American Public Health Association, Washington, D.C.

APHA, 2005. In: Clesceri, L.S., Greenberg, A.E., Eaton, A.D. (Eds.), Standard Methods for the Examination of Water and Waste Water, Quality Assurance/Quality Control, twenty second ed. American Public Health Association, Washington, D.C.

Centers for Disease Control and Prevention (CDC). Environmental Issues. http://www.cdc.gov/Environmental/

Clifford, M., 1998. Environmental Crime: Enforcement, Policy and Social Responsibility. Aspen Publishers, Gaithersburg, MD.

Cutting, R., Cahoon, L.B., 2006. Forensic Environmental Science: Ecological Society of America. In: Ecology 101. http://www.esapubs.org/bulletin/current/pdfweb87_2/ecol101_forensic.pdf

Cutting, R., Cahoon, L., Leggette, 2006. Enforcement Data: A Tool for Environmental Management, 36 Environmental Law Reporter 10060-10072.

Drielak, S.C., 1998. Environmental Crime: Evidence Gathering and Investigative Techniques Book. C.C. Thomas, Springfield, IL 2007.

Faigman, D.L., 1999. Legal Alchemy: The use and Misuse of Science in the Law. W.H. Freeman, New York.

Federal Emergency Management Agency (FEMA). http://www.fema.gov/hazard/index.shtm

Free Mind. http://www.freemind.sourceforge.net/wiki/index.php/Main_Page

Kubasek, N.K., Silverman, G.S., 2008. Environmental Law, sixth ed. Prentice Hall, Englewood Cliffs, NJ.

Mallin, M.A., Cahoon, L.B., Toothman, B.R., Parsons, D.C., McIver, M.R., Ortwine, M.L., et al., 2007. Impacts of a raw sewage spill on water and sediment quality in an urbanized estuary. Mar. Pollut. Bull. 54, 81–88.

National Institute of Health, National Medical Library. Tox Town. http://toxtown.nlm.nih.gov/

Powell, F.M., 1998. Law and the Environment. Thomson/West, Eagan, MN.

Salzman, J., Thompson, B.H., 2003. Environmental Law and Policy. Foundation Press, New York.

Total station software. http://www.leica-geosystems.com/corporate/en/lgs_8276.htm

Tufte, E.R., 1983. The Visual Display of Quantitative Information. Graphics Press, Cheshire, CT.

Tufte, E.R., 1990. Envisioning Information. Graphics Press, Cheshire, CT.

Tufte, E.R., 1997. Visual Explanations: Images and Quantities, Evidence and Narrative. Graphics Press, Cheshire, CT.

Tufte, E.R., 2003. The Cognitive Style of PowerPoint. Graphics Press, Cheshire, CT.

U.S. Environmental Protection Agency home page. http://www.epa.gov/
US Geological Survey Office of Groundwater. http://www.water.usgs.gov/ogw/
US Geological Survey Watershed Information. http://www.water.usgs.gov/wsc/
Watershed. http://www.cfpub.epa.gov/surf/locate/index.cfm

CASES

Air Pollution Variance Board v. Western Alfalfa Corp, 416 U.S. 861, 1974
Daubert v. Merrell Dow Pharmaceuticals, Inc, 509 U.S. 579 (1993)

Appendix A: Chain of Custody

Forms and additional discussion can be found at

- EPA: EPA Chain of Custody: http://www.epa.gov/apti/coc/
- City of San Diego: http://www.sandiego.gov/mwwd/pdf/baf/14_appendixd _9.pdf
- State of Virginia Division of Consolidated Laboratory Services: http://www.dgs.virginia.gov/LinkClick.aspx?fileticket=2PJopG%2F7t40%3D &tabid=507
- Johns Manville Company: http://www.jm.com/corporate/labs_services/ Chain_of_Custody.pdf
- State of New Jersey Department of Environmental Protection: http://www. state.nj.us/dep/srp/guidance/fspm/pdf/chapter10.pdf

Appendix B: Freedom of Information Acts

- Sample FOIA Request from MMS: http://www.mrm.mms.gov/foia/sam-plelettert.htm
- Your right to federal records *Questions and Answers on the FOIA and Privacy Act May 2006* general information from GSA: http://www.pueblo. gsa.gov/cic_text/fed_prog/foia/foia.htm; FOIA Reference Guide, Revised May 2006, http://www.usdoj.gov/04foia/04_3.html; Principal FOIA Contacts at Federal Agencies, http://www.usdoj.gov/04foia/foiacontacts.htm; U.S. Government Manual, 2006-07, http://www.gpoaccess.gov/gmanual/; A Citizens Guide on Using the FOIA, http://www.fas.org/sgp/foia/citizen.html
- *Reporters committee for freedom of the press*
- Letter generator for federal and state government agencies: http://www.rcfp. org/foialetter/index.php
- *FOIA center, MISSOURI school of journalism:* http://www.nfoic.org/ states; and http://www.firstamendmentcenter.org//press/information/topic. aspx?topic = how_to_FOIA&SearchString = sample#request
- The first Amendment Center, offers FOIA resources are available at: http://www.firstamendmentcenter.org/Press/information/topic.aspx? topic = how_to_FOIA#request

Chapter 7

Regulatory Considerations to Ensure Clean and Safe Drinking Water

Craig L. Patterson and Roy C. Haught

U.S. Environmental Protection Agency, National Risk Management Research Laboratory, Water Supply and Water Resources Division, Cincinnati, OH, USA

INTRODUCTION

Federal drinking water regulations are based on risk assessment of human health effects and research conducted on source water, treatment technologies, residuals, and distribution systems. This chapter focuses on the role that U.S. Environmental Protection Agency (EPA) research plays in ensuring pure drinking water in the United States and throughout the world. The first part of this chapter explains EPA's strategic goals for drinking water, the rule-making process, and applicable drinking water regulations. The second part of this chapter highlights EPA human health and drinking water research.

EPA STRATEGIC GOALS FOR "CLEAN AND SAFE WATER"

EPA updates and creates a strategic plan every five years. The current strategic plan encompasses the 5-year period from 2006 to 2011. EPA's strategic goals for "clean and safe water" have evolved from focusing on contaminants in water to protecting source water and water infrastructure. Objectives include protecting human health, protecting water quality, and enhancing science and research. EPA's first objective protects human health by reducing exposure to contaminants in drinking water (including protecting source waters), fish and shellfish, and recreational areas. The second objective protects the quality of rivers, lakes, and streams on a watershed basis and protects coastal and ocean waters. The third objective enhances science and research by conducting leading-edge, sound scientific studies to support the protection of human health through the reduction of human exposure to contaminants in drinking water, fish and shellfish, and recreational waters and to support the protection of aquatic ecosystems, specifically the quality of rivers, lakes, and streams, and coastal and ocean waters (USEPA, 2006a).

The strategic plan targets the improvement of drinking water quality in community water systems (CWSs) serving 6% of the U.S. population (2002) that do not meet all applicable health-based drinking water standards. EPA plans to accomplish these goals through effective treatment and source water protection and improvements in regulatory monitoring and reporting.

EPA's Human Health Research Strategy improves the scientific foundation of human health risk assessment and enables evaluation of public health outcomes from risk management decisions. This strategy develops a multidisciplinary program with linkages among exposure, dose, effect, and risk assessment methods. Human health effect decisions are based on harmonizing risk assessment approaches to predict aggregate and cumulative risk (exposure to mixture of pollutants from multiple sources) and to protect susceptible subpopulations (USEPA, 2003a). More information on implementation of this strategy is provided in the section on Human Health Research.

EPA's National Risk Management Research Laboratory (NRMRL) completed a strategic plan (USEPA, 2006b) in September 2006 that focuses on the following research areas for clean and safe water:

- Integrated watershed management
- Water quality restoration and protection
- Nutrient trading
- Source water protection
- Water infrastructure
- Regulated and unregulated drinking water contaminants
- Water distribution

More details on these issues are provided in the section on Drinking Water Research.

Since the terrorist attacks of September 11, 2001, the United States has been faced with potential threats to its water systems including drinking water distribution systems. In September 2002, the EPA prepared a Strategic Plan for Homeland Security (USEPA, 2002a). The goals include four key areas:

1. Critical infrastructure protection
2. Preparedness, response, and recovery
3. Communication and information
4. Protection of EPA personnel and infrastructure

EPA's strategic plan also lists emerging issues that can affect EPA goals for clean and safe water, decaying water infrastructure and population growth, water scarcity, nanotechnology, remote sensing technologies, climate change, pharmaceuticals in wastewater, and renewable energy (USEPA, 2006a).

DRINKING WATER REGULATIONS

History

EPA was created in 1970 to protect human health and the environment (air, water, and land). A variety of Federal research, monitoring, standard-setting, and enforcement activities were consolidated to form the U.S. EPA. For more than 35 years, EPA has been working to improve water quality and ensure pure drinking water for the American people.

After the creation of EPA, two major regulations were enacted that have had a great impact on drinking water in the United States. The Clean Water Act (CWA) of 1972 prohibits releases of toxic amounts of pollutants into U.S. waters and provides protection and propagation for fish, shellfish, and wildlife and recreation in and on the water. The Safe Drinking Water Act (SDWA) of 1974 protects public health by regulating the nation's drinking water supplies. Figure 1 provides a timeline of drinking water acts and rules created after the enactment of the CWA and SDWA.

The Rule-making Process

A description of the rule-making process provides insight into why EPA regulations are necessary and how they lead to the protection of human health and the environment. The rule-making process requires EPA to follow these nine steps:

1. Identify potential contaminants

EPA identifies a potential contaminant that adversely affects public health and occurs with a frequency at levels that pose a threat. It then monitors certain water systems for the presence of the contaminant.

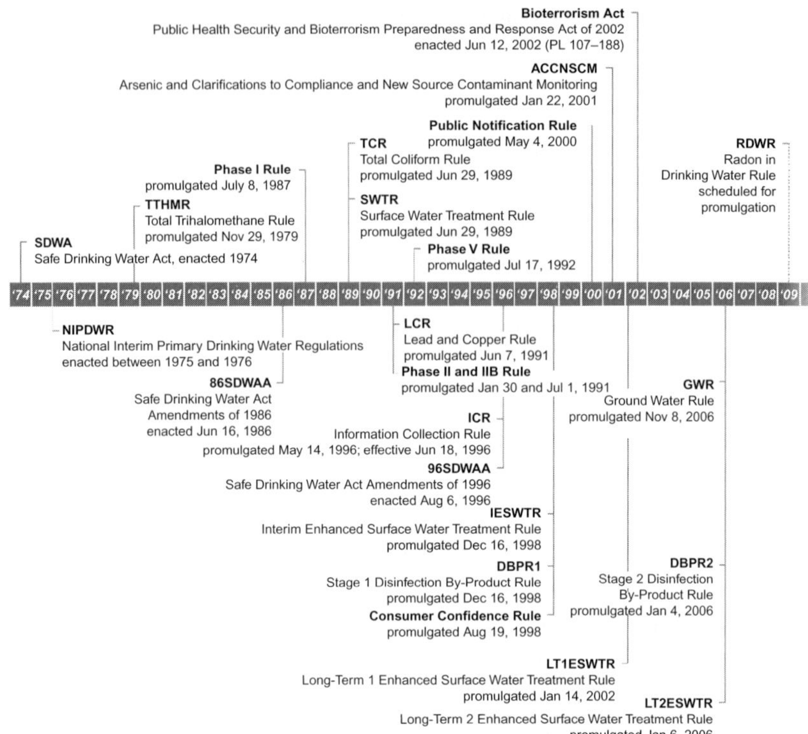

FIGURE 1 Timeline of federal drinking water acts and rules (USEPA, 2005b).

2. Perform a cost–benefit analysis

EPA performs a cost–benefit analysis to determine the economic impact of regulating the contaminant based on its occurrence and the costs associated with applicable treatment technologies and analytical methods.

3. Obtain input from interested parties

EPA obtains input from interested parties on all aspects and repercussions of regulating the contaminant.

4. Determine a maximum contaminant level goal

EPA determines a maximum contaminant level goal (MCLG) that reflects no known or expected health risk associated with the contaminant and allows for a margin of safety. The MCLG is based on risk to the most sensitive people (infants, children, pregnant women, elderly, and immunocompromised individuals). The MCLG is often a nonenforceable level set at levels that cannot be met with best available technologies (BATs).

5. Specify a maximum contaminant level

EPA specifies a maximum contaminant level (MCL) or an enforceable standard. The MCL is the maximum permissible level of the contaminant delivered to the consumer from a public water system (PWS).

6. Set a required treatment technique

EPA sets a required treatment technique that specifies a way to treat water to remove the contaminant in the absence of an economically or technically feasible treatment technology, or when there is no reliable or economic method to detect contaminants at very low levels. For instance, the Surface Water Treatment Rule (SWTR) specifies filtration and disinfection as a treatment technique.

7. Provide grants to implement individual state drinking water programs

States are authorized to grant variances from standards for a PWS serving up to 3,300 people. However, the PWS must install EPA-approved variance technologies. EPA can approve variances to systems serving 3,301 to 10,000 people. However, no variances are allowed for microbial contaminants and all variances must ensure no unreasonable risk to public health. Exemptions may be granted to allow extra time to seek compliance options or financial assistance. However, exempt water systems must comply after the exemption period expires.

EPA provides grants to implement state drinking water programs. The Drinking Water State Revolving Fund (DWSRF) serves as a financial mechanism to assist communities with implementation of regulatory standards. The DWSRF provides capital for infrastructure and management improvements, source water assessment and protection, and training. The DWSRF provides assistance to drinking water systems including financial assistance for purchase of alternative low-cost water treatment systems protective of public health.

8. Enforce water system safety standards with state regulators

EPA and state primacy agencies enforce water system safety standards, issue administrative orders, take legal actions, and fine utilities that do not comply with regulations. These actions are designed to increase water system compliance and water system operator certification.

9. Require disclosure of public information

Finally, EPA requires disclosure of public information. Water suppliers must notify consumers quickly when there is a serious problem with water quality. Annual consumer confidence reports must be provided to water customers describing the source and quality of their tap water to document compliance with drinking water safety standards.

Clean Water Act

The CWA eliminates releases of toxic amounts of pollutants into waters of the United States for protection and propagation of fish, shellfish, and wildlife, and recreation in and on the water. The CWA reduces direct pollutant discharges into waterways, finances municipal wastewater treatment facilities, manages polluted runoff, and restores and maintains the integrity of the nation's waters.

The CWA regulates water pollution from 56 industrial categories and enforces pretreatment requirements for industrial users contributing wastewater to publicly owned treatment works. Some of the major programs resulting from the CWA include the national pollutant discharge elimination system (NPDES) program, the total maximum daily load (TMDL) program, and the state water pollution control revolving fund.

Safe Drinking Water Act

The SDWA protects public health by regulating the nation's drinking water supplies. The SDWA provides a framework for developing drinking water standards that include health-based goals for the implementation of technically achievable and enforceable standards. In cases where PWSs cannot comply with MCL standards, EPA allows the use of alternative treatment techniques. Enforceable requirements under the SDWA include the National Primary Drinking Water Regulations (USEPA, 2003b). These regulations apply to PWSs with at least 15 connections or 25 persons at least 60 days per year. Ensuring water quality in private wells is typically the responsibility of the homeowner. However, some states do set standards for private wells. The National Secondary Drinking Water Regulations are nonenforceable guidelines for contaminants under the SDWA. These secondary contaminants may cause cosmetic (such as skin and tooth discoloration) or aesthetic effects (such as taste or odor). States may adopt and enforce these standards as state regulations.

Contaminant Candidate List

The SDWA Amendments of 1996 establishes a list of known or anticipated contaminants that occur in PWSs, which is known as the contaminant candidate list (CCL). The CCL is revisited periodically to prioritize contaminants for additional research and data gathering. The National Contaminant Occurrence Database assists EPA with the development of regulations for monitoring certain unregulated contaminants. With significant input from the scientific community and other interested parties, EPA determines whether or not a regulation for a CCL contaminant is appropriate based on projected adverse health effects, the extent of occurrence in drinking water, and whether the regulation of a CCL contaminant reduces risks to health.

Drinking Water Compliance Issues

Many rules have been written and are now being enforced since the enactment of the SDWA in 1974. Several of the regulations that directly impact water quality and purity are summarized in the following paragraphs.

Arsenic Rule

The Arsenic Rule changed the arsenic MCL from 50 to 10 μg/L. All CWSs and nontransient, noncommunity water systems (NTNCWSs) were required to reduce exposure to arsenic in drinking water by January 23, 2006. The Arsenic Rule is designed to improve public health by reducing the number of fatal and nonfatal bladder and lung cancers. The Arsenic Rule also requires monitoring of new systems and new drinking water sources and clarifies the procedures for determining compliance with MCLs for inorganic chemicals (IOCs), synthetic organic compounds (SOCs), and volatile organic compounds (VOCs) (USEPA, 2001a).

Disinfectants/Disinfection By-product Rules

The Stage 1 and Stage 2 Disinfectants/Disinfection By-product (D/DBP) Rules reduce the total trihalomethane (TTHM) MCL to 0.080 mg/L and the haloacetic acids 5 (HAA5) MCL to 0.060 mg/L. The Stage 1 D/DBP Rule established an MCL of 0.010 mg/L for bromate for plants that use ozone and an MCL of 1.0 mg/L for chlorite for plants that use chlorine dioxide. The Stage 1 D/DBP Rule also regulates maximum residual disinfectant levels (MRDLs) for chlorine and chloramine (4.0 mg/L) and chlorine dioxide (0.8 mg/L). Surface water and groundwater systems under the direct influence of surface water were required to comply with the Stage 1 D/DBP Rule by January 1, 2004.

The Stage 2 D/DBP Rule reduces the potential risk of adverse health effects by monitoring for DBPs throughout the distribution systems. The rule applies to CWSs and NTNCWSs that either add a primary or residual disinfectant other than ultraviolet (UV) light or deliver water that has been treated with a primary or residual disinfectant other than UV light. Owners must begin complying with the rule requirement to determine compliance with the operational evaluation levels for TTHMs and HAA5s between January 2013 and July 2014 depending on the population served by their system (USEPA, 2001b).

Groundwater Rule

The Groundwater Rule (GWR) specifies the appropriate use of disinfection in groundwater systems. The GWR reduces the risk of exposure to fecal contamination in groundwater sources used by PWSs. The GWR is expected to reduce the average number of viral illnesses by 23%. The GWR requires periodic sanitary surveys; source water monitoring for *Escherichia coli* (*E. coli*), enterococci, or coliphage; corrective actions for any system with significant deficiencies or source water fecal contamination; and compliance monitoring to ensure that the treatment technology reliably achieves 99.99% (4-log) inactivation or removal of viruses. The GWR was promulgated in October 2006 (USEPA, 2006c).

Lead and Copper Rule

The Lead and Copper Rule (LCR) establishes action levels of 0.015 mg/L for lead and 1.3 mg/L for copper based on 90th percentile level of tap water samples.

The LCR is designed to reduce damage to brain, red blood cells, and kidneys caused by exposure to lead and to reduce stomach and intestinal distress, liver or kidney damage, and complications of Wilson's disease caused by exposure to copper. EPA estimates that approximately 20% of human exposure to lead is linked to drinking water. If action levels are exceeded, other requirements may be triggered including water quality parameter monitoring, corrosion control treatment, source water monitoring/treatment, public education, and lead service line replacement. All CWSs and NTNCWSs are required to comply with the LCR (USEPA, 2004a).

Surface Water Treatment Rules

The Long-Term 1 (LT1ESWTR) and Long-Term 2 (LT2ESWTR) Enhanced Surface Water Treatment Rules strengthen filtration requirements and provide protection against disinfection-resistant microbial pathogens such as *Cryptosporidium* in drinking water. The LT1ESWTR requires a minimum 2-log removal (99%) of *Cryptosporidium* for PWSs serving less than 10,000 people using surface water or groundwater under the influence of surface water. It also establishes individual filter turbidity monitoring to minimize poor performance.

The LT2ESWTR controls microbial contaminants by targeting systems with elevated *Cryptosporidium* risk. Systems must monitor their source water and calculate an average *Cryptosporidium* concentration to determine if their system requires additional treatment. The LT2ESWTR sets log removal or inactivation requirements for *Cryptosporidium* based on the treatment technologies employed by the PWS. Owners must install and operate additional treatment technologies based on the removal capabilities of their existing system and their average *Cryptosporidium* concentration between March 31, 2012, and September 30, 2014, depending on the population served by their system (USEPA, 2006d).

Radionuclides Rule

EPA has updated its standards for radionuclides in drinking water. The Radionuclides Rule retains existing MCLs for beta–photon emitters (4 mrem/year), gross alpha particles (15 pCi/L), and combined radium 226/228 (5 pCi/L) and reduces radioactive exposure by establishing an MCL for uranium of 30 μg/L. The Radionuclides Rule is designed to improve public health by reducing toxic kidney effects and the risk of cancer. All PWSs must have completed initial monitoring for regulated radionuclides excluding the beta particle and photon emitters by December 31, 2007 (USEPA, 2001c).

Total Coliform Rule

The Total Coliform Rule (TCR) controls total coliform bacteria including fecal coliforms and *E. coli* in all PWSs. The TCR minimizes fecal pathogens in drinking water by establishing an MCL based on the presence or absence of total coliforms. The TCR reduces the risk of illness from disease causing organisms associated with animal waste or sewage by testing for fecal coliforms or *E. coli*. It requires collection of samples at sites that are representative of water quality throughout the distribution system. It provides a schedule for routine monitoring frequencies based on the population served by the PWS and requires sanitary surveys for systems collecting less than five samples per month (USEPA, 2001d).

HUMAN HEALTH RESEARCH

Human health research drives human health risk assessment by providing methods, models, tools, and data that enable evaluation and protection of public health. EPA focuses on problem-driven health risk issues made public while managing source water and drinking water systems throughout the United States.

EPA establishes human health criterion for pollutants in water by determining the highest concentration that is not expected to pose a significant risk to human health. EPA evaluates the potential hazard of a pollutant to human health using toxicological studies based on scientifically sound test designs, exposure conditions, and measured endpoints.

EPA's Office of Water provides information to the public on health effects of specific chemicals in drinking water through Criteria and Health Advisory (HA) documents. Criteria documents provide a comprehensive summary of toxicological and exposure data for the regulation of a specific contaminant. HA documents provide information on health effects of drinking water contaminants that are not regulated and are likely to be without adverse effects on health and aesthetics. In addition, the HA summarizes information on available analytical methods and treatment techniques for the contaminant. HA documents for contaminants are subject to change as more research information becomes available (USEPA, 2001e). More information on specific contaminants is available at the EPA's Water Science home page (USEPA, 2006e).

Toxicokinetics

EPA published Health Effects Test Guidelines (1996) designed to minimize variations in toxicokinetic testing procedures (USEPA, 1996). Toxicokinetic tests determine the bioavailability of pollutants (test substances) after dermal or oral treatment, ascertain whether the metabolites of the pollutant are similar after dermal and oral administration, and examine the effects of a multiple-dosing regimen on the metabolism of the pollutant after oral administration. Absorption toxicokinetics refers to the rate and extent of absorption of the pollutant, metabolism, and excretion rates of the test substance after absorption. The tests document the distribution of toxicokinetic effects in the body and if there are signs of bioaccumulation.

Health Effects Data

Human Studies

Human exposure studies provide valuable health effects data on specific contaminants. Data on human health effects of a pollutant are found in available literature on epidemiology studies and case reports. Short-term exposure data and human symptoms are found in records on accidental exposure to toxic pollutants. Long-term exposure data are derived from exposure to workers during the production or long-term use of a chemical. Human reproductive and development effects are obtained from studies of neighborhoods in close proximity to contaminated groundwater. Data on carcinogenicity can be uncovered by searching hospital records and cancer registration files. However, exposure–dose information is often hard to document in these cases (USEPA, 2004b).

Animal Studies

Contaminants are evaluated in short-term and long-term exposure studies with mammals (rats, mice, guinea pigs, rabbits, and dogs). Studies on health effects in several species of animals assess the acute, chronic, and subchronic toxicity of chemicals. Acute studies are conducted by introducing a chemical via mouth (oral) or skin (dermal), or by injection into blood veins, muscles, or abdominal cavity (intraperitoneal). Studies on the acute effects of a chemical on lungs (inhalation), eyes (ocular toxicity), and the nervous system (neurotoxicity) are also common.

Subchronic and chronic toxicity and oncogenicity studies provide data defining the toxicity and carcinogenic potential of a contaminant. Single- and multigenerational studies assess the potential effects on reproductive and developmental indices. Genetic toxicology studies determine whether or not a chemical induces chromosomal aberrations or DNA damage and repair. Chronic studies determine if a chemical affects mortality, food consumption, food efficiency, hematology, clinical chemistry, or urinalysis parameters. In cancer studies, animals are treated with various dose levels of a contaminant to determine treatment-related increases in tumor incidence (USEPA, 2004b).

Quantification of Toxicological Effects

HA documents provide guidance values based on noncancer health effects for 1-day, 10-day, longer-term (up to 7 years), and lifetime exposures. However, lifetime exposure values are not recommended for known or probable human carcinogens (USEPA, 2001e).

Evaluation of Carcinogenic Potential

Determination of the carcinogenic potential of a contaminant focuses on epidemiological studies on humans and animals found in available literature. In March 2005, EPA issued "Final Guidelines for Carcinogen Risk Assessment"

(USEPA, 2005a) that uses a narrative approach to describe the potential of a contaminant to cause cancer in humans using weight of evidence descriptors:

- Carcinogenic to humans (H)
- Likely to be carcinogenic to humans (L)
- Likely to be carcinogenic above a specified dose but not likely to be carcinogenic below that dose because a key event in tumor formation does not occur below that dose (L/N)
- Suggestive evidence of carcinogenic potential (S)
- Inadequate information to assess carcinogenic potential (I)
- Not likely to be carcinogenic to humans (N)

Information on human health guidelines for risk assessment other than cancer (neurotoxicity, reproductive toxicity, exposure assessment, developmental toxicity, mutagenicity, and chemical mixtures) is available at EPA's National Center for Environmental Assessment (NCEA) Web site. NCEA has prepared and maintains the Integrated Risk Information System (IRIS) that can be reached from the IRIS home page (USEPA, 2005a). IRIS is an electronic database that provides descriptive and quantitative information on both cancer and noncancer human health effects that may result from exposure to various contaminants in the environment.

Risk Assessment

Risk assessment combines estimates of environmental exposure with known adverse effects of exposure to determine an overall estimate of the potential public health risk.

Risk assessment provides a process of determining how best to protect public health (e.g., determine allowable levels of contamination in drinking water). The risk assessment process follows four basic steps:

1. Hazard identification
2. Dose–response assessment
3. Exposure assessment including sensitive populations
4. Risk characterization

In hazard identification, available toxicokinetic and health effects data are examined to identify health problems associated with a contaminant. Exposure studies are reviewed to determine the amount and length of contaminant exposure that will cause harm (CA EPA, 2001). Human and animal studies are evaluated to determine the likelihood that the toxic effects of a contaminant will cause health injury or disease in humans.

In dose–response assessment, hazards are evaluated to determine the amount of contaminant required to cause varying degrees of toxic injury or disease. Dose–response relationships vary with the conditions of exposure and the concentration of the contaminant. Quantification of dose–response relationships of toxic contaminants is critical to estimating risk to human health.

In exposure assessment, scientists consider the nature and size of the exposed population to determine the amount, duration, and pattern of exposure to a contaminant. Humans may be exposed differently to contaminant doses because of differences in human behavior, physiology, and metabolism that are accounted for when estimating actual human exposure. The evaluation may examine past, present, and future exposures to a wide range of contaminant concentrations in the general population as well as sensitive populations (infants, children, pregnant women, elderly, and immunocompromised individuals).

Risk characterization estimates the risk of health effects in an exposed population by describing risks to humans in terms of the extent and severity of probable harm. By weighing the uncertainty and scientific judgments proposed in the first three steps, scientists determine the likelihood that individuals or populations will experience any toxic health effects associated with a contaminant. The risk assessment process is designed to come up with the most accurate estimate of risk to human health possible.

A staff paper entitled "An Examination of EPA Risk Assessment Principles and Practices" describes how risk assessment is conducted at EPA (USEPA, 2004c). EPA continues to focus on particular risk assessment issues in discussions with EPA's Science Advisory Board, consultative groups, professional societies, states, nongovernmental organizations, tribal groups, and other interested parties. More emphasis is being placed on cumulative risk assessment to evaluate and estimate the potential human risks associated with multichemical and multipathway exposures to contaminants (FR, 2002).

DRINKING WATER RESEARCH

EPA's National Risk Management Research Laboratory (NRMRL) is responsible for managing drinking water research within the Water Supply and Water Resources Division (WSWRD) located in Cincinnati, Ohio, and Edison, New Jersey. Drinking water research within WSWRD serves several purposes:

- Provides relevant risk management research for the control of contaminants in drinking water that pose a threat to human health
- Develops approaches and tools for characterization and restoration of impaired water bodies primarily in urban areas
- Supports EPA rule-making and policy development for clean and safe water (USEPA, 2006a)
- Lends technical support to the water industry in the United States and around the globe on issues related to waterborne disease and water quality
- Works cooperatively with governmental and nongovernmental organizations to respond to waterborne disease outbreaks

- Protects PWSs from intentional hazards
- Delivers research product results to public water utilities and their consumers
- Collaborates on multidisciplinary inter/intraagency research projects in support of the design, development, testing, and evaluation of environmental protection technologies

EPA collaborates with multidisciplinary research strategy groups comprised of a wide variety of governmental and nongovernmental constituents to assist with the planning and implementation of drinking water research. Drinking water research results strengthen risk assessment resulting in directed improvement of drinking water quality and human health.

Risk Management

To provide relevant risk management research, EPA utilizes the "Drinking Water Contaminant Management Framework" (Figure 2) to identify research needs. Analysis determines risk management options that can be incorporated to control a contaminant or suite of contaminants in drinking water that pose a threat to human health. EPA focuses on four broad research areas (USEPA, 2005b) that include source water management, treatment, distribution systems (USEPA, 2003c), and residuals management.

Source Water Management

Source water protection and management can be divided into four categories: water quality criteria, source water assessments, preventative measures to address sources of contamination, and contingency planning. EPA source water protection and management research focuses on urban watershed management and source water assessment and protection programs.

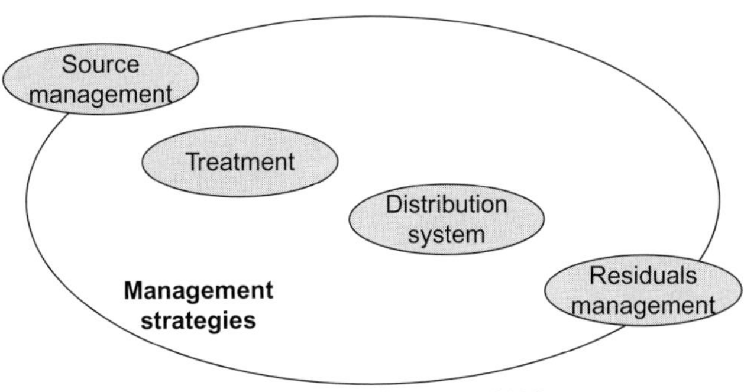

FIGURE 2 Drinking water contaminant management framework.

Research Questions

A range of scientific issues exist within each of these categories. Some of the most important questions include (USEPA, 2003d):

- How adequately do the Ambient Water Quality Criteria that address the major drinking water contaminants protect public health?
- What improved techniques are needed to better define source water characteristics and sources of contamination (e.g., natural, anthropogenic)?
- What are the fate and transport characteristics of certain types of contaminants in surface water and groundwater?
- How effective are candidate protection measures (i.e., best management practices [BMPs]) on improving the quality of the source water?
- Can contaminants be eliminated through pollution prevention or alternative approaches (e.g., isolation of zones in the subsurface, specific siting choices)?
- What are the impacts of sudden increases in source water contaminant concentrations on drinking water treatment performance?
- What early warning and monitoring systems should be developed to alert utility operators of contaminant incursions at the source so that corrective actions might be employed?

Urban Watershed Management

The CWA requires states to make a list of surface water bodies that are polluted and identify major causes of impairment such as fecal contamination, sediments, nutrients, toxics, and flow. States must also prioritize the water bodies on the list and develop TMDLs to improve the water quality.

Research studies on microbial source tracking (MST) assist in water quality evaluations and TMDL-related activities (USEPA, 2005c). MST has been used to help identify nonpoint sources responsible for fecal pollution in watersheds, and MST tools are being applied in the development of TMDLs.

Research studies on the characterization and management of stormwater, combined sewer overflows (CSO), and sanitary sewer overflows (SSO) provide states with information on which to base development of TMDL's and restoration actions. Urban watershed management research is transitioning from CSO/SSO treatment technology to urban/urban fringe watershed management and will eventually incorporate urban watersheds in mixed-use watersheds.

Research studies on the design and effectiveness of urban BMPs allow states to evaluate and recommend solutions on structural BMP placement and effectiveness and integration of nonstructural BMPs. These studies are looking at what BMPs to deploy, how many are required, and where they are needed to determine how these decisions will impact water quality.

Developing models of urban watershed management practices (e.g., wet weather flow models) provide regulators with intelligent decision-making tools. The outputs from these models determine the effect of watershed management actions on water quality. The outcome from implementation of these research study findings will improve water quality in urban and mixed-use watersheds.

Source Water Assessment and Protection Programs

The 1996 Amendments to the SDWA (Section 1453) reinforces source water protection, requiring each state to develop and implement a source water assessment program (SWAP). A source water assessment delineates the land area that most directly contributes raw water to the drinking water supply and assesses potential sources of contamination. The information gathered is used to determine how susceptible the PWS is to contamination threats. Water treatment costs are reduced by prioritizing ways to protect and minimize contamination of ground and surface water supplies. States and PWSs are required to make a summary of assessment results available to the public.

Once a SWAP is completed, the results are used to develop and implement a source water protection (SWP) program. Sanitary surveys are also performed to inspect all components of a PWS from source to tap. "State Source Water Assessment and Protection Programs Guidance" (USEPA, 1997) provides implementation guidance for state SWAP and SWP programs.

Treatment Technologies

Drinking water research at EPA focuses on a wide variety of treatment technologies that safely and economically treat drinking water. Research studies test and evaluate filtration and disinfection technologies for the removal of protozoa, bacteria, and viruses for compliance with the SDWA and subsequent regulations (e.g., LT2ESWTR, Stage 2 D/DBP Rule).

Research Questions

Can existing processes be modified to meet new treatment goals?
How can research address treatment of multiple contaminants?
What emerging and innovative technologies need to be developed?
What are the costs associated with water treatment processes?
Should research focus on low-cost treatment technologies rather than "cutting-edge" technologies, which tend to be more expensive?

Best Available Technologies

By definition, the MCL is set as close to MCLG as feasible using the best available treatment technology, treatment technique, or other means (e.g., evaluating treatment efficiency under field conditions) by taking cost into consideration.

State primacy agencies are responsible for applying BATs to comply with EPA's primary drinking water MCL standards. For instance, the director of Ohio EPA has identified BATs for the removal of organic, inorganic, and microbial contaminants from drinking water under Chapter 3745-81 of the

Ohio Administrative Code. Chapter 3745-81 constitutes the primary drinking water rules for Ohio PWSs (OH EPA, 2008).

Using Ohio as an example, BATs for primary drinking water MCLs vary by inorganic contaminant and encompass 15 filtration, corrosion control, adsorption, ion exchange (IX), and disinfection technologies. The MCLs apply to all community and nontransient noncommunity PWSs. BATs for organic contaminant MCLs also vary, but rely on the removal capabilities of granular activated carbon and packed tower aeration technologies. For microbial contaminants, the BAT for achieving compliance with the MCL for total coliforms is based on development and implementation of source water assessment and protection programs, disinfection for groundwater systems, filtration and disinfection for surface water systems, maintenance of a disinfectant residual, and proper maintenance of water distribution systems (USEPA, 2002b). The BATs for contaminants evolve through research and development of treatment technologies and as emerging and innovative treatment technologies become commercially available.

Conventional Treatment

The primary goal of conventional filtration is to remove particulates from drinking water. Conventional treatment is typically used in municipal treatment plants and consists of the addition of coagulant chemicals, flash mixing, coagulation–flocculation, sedimentation, and filtration. After conventional treatment, drinking water is disinfected prior to the water distribution system. A majority of the drinking water in the United States follows this treatment regimen. Direct filtration includes coagulation and filtration without sedimentation. Advances in membrane filtration technology and the disinfection by-product regulations are altering the traditional treatment processes that make up conventional treatment, but the treatment concepts remain the same.

Alternative Treatment Technologies

Many of the PWSs cannot afford the infrastructure and costs associated with conventional treatment. Alternative treatment technologies can serve as viable options to produce safe drinking water and provide protection to surface water and groundwater supplies. With limited budgets and without the resources required to deal with complex equipment or everchanging regulatory requirements, factory built, skid mounted, remotely controlled treatment plants become potential alternatives. Pre-engineered package plants offer low construction and operating costs, simple operation, low maintenance, and adaptability to part-time operation. Package plants can be purchased and retrofitted as pretreatment, treatment, and post-treatment processes.

Treatability of Water Supplies

Surface water is typically more difficult to purify than groundwater. The ability to treat surface water varies with its type (spring water, snowmelt, glacial water, lake water, river water), the time of year (seasonal, temperature effects), and many other factors that impact water quality characteristics. Groundwater is typically considered clean because of natural filtration underground or may be relatively clean when it is under the influence of surface water. However, naturally occurring inorganic contaminants and runoff containing animal wastes can negatively impact the water quality characteristics of both groundwater and surface water.

Types of Treatment Processes and Removal Capabilities

Many drinking water treatment technologies are designed and built to be contaminant specific for the removal of a recalcitrant compound such as arsenic or perchlorate. To simplify this discussion, treatment technologies are categorized as physical (filtration), biological (disinfection), and chemical (sorption).

Physical Treatment (Filtration)

With technological advances in membrane materials and cleaning processes, membranes are capable of producing drinking water with higher purity at higher flux rates and lower pressures than they were in the past. Research on reverse osmosis, nano-, ultra-, and microfiltration units continues to focus on membrane integrity and bacteria and virus removal capability. Rapid sand, radial flow garnet, and multimedia filters provide pretreatment options for particulate and turbidity removal. Low-cost filtration alternatives include disposable bag and cartridge microfilters for use in small communities. Reductions in filter pore size to 0.01 μm have made ceramic filters effective for the removal of virus-sized particles without the addition of chemicals.

Biological Treatment (Disinfection)

Disinfection is the process used to purify water by destroying or inactivating microbial pathogens. Many forms of chlorine are used as drinking water disinfectants in the United States. Some treatment plants are using ozone, ultraviolet irradiation, and other disinfectants followed by the addition of chlorine as a disinfectant residual. The effectiveness of the disinfectants in killing microorganisms (i.e., biocidal efficiency) varies with the type of microorganism and the water quality conditions (such as pH).

Chlorine and Disinfection By-products

The chlorine dosage and free residual chlorine are critical performance parameters.The optimum amount of disinfecting agent is required to achieve appropriate disinfection and minimize DBP formation. The MCL for total trihalomethanes is 0.080 ppb and for haloacetic acids it is 0.060 ppb. In some cases, this might result in a change to an alternative preoxidant or disinfectant, use of membranes, or elimination of the use of free chlorine (Pollack et al., 1999).

On-Site Chlorine Generators

On-site chlorine generators remove the transport and operating hazards associated with chlorine gas or liquid (sodium hypochlorite or calcium hypochlorite) systems. Salt is used to produce sodium hypochlorous acid for the disinfection of microorganisms in drinking water. EPA conducts studies that evaluate on-site brine-based chlorine generators and compares them to each other and to liquid bleach. Brine concentration levels are critical for proper operation. The accumulation of salt residue requires maintenance of system tanks and piping.

Ozonation and Disinfection By-products

Ozone is an extremely effective oxidizing agent. However, it dissipates quickly in drinking water leaving treated water unprotected in the distribution system. Some form of chlorine is typically added after ozone treatment to provide a disinfectant residual. Ozone also reacts with bromide (if present in the source water), leading to the formation of carcinogenic DBPs that include bromate, bromoform, and dibromoacetic acid.

UV Irradiation

The germicidal effects of UV light are well known. The use of UV disinfection has increased dramatically following research on inactivation of *Cryptosporidium*. UV disinfection processes do not use chemicals but are most effective in waters with low turbidity and color. Some form of chlorine is typically added after UV disinfection to provide a disinfectant residual and protect against viruses (high UV doses are needed to kill viruses).

Advanced Oxidation Process for Organic Removal

EPA continues to evaluate advanced oxidation processes for the reduction of organics from drinking water. Several combinations of oxidants (UV, ozone, hydrogen peroxide) have been studied for the destruction of Methyl Tertiary Butyl Ether (MTBE) in groundwater. The combination of UV and ozone oxidizes the MTBE instantaneously. Several by-products are generated as a result of MTBE treatment, including tertiary butyl formate, butane, methyl acetate, acetone, and acetaldehyde (Graham et al., 2004).

Sorption Technologies (Chemical Treatment)

Sorption is the common term used for both absorption and adsorption. Absorption is the process of integration of one substance into another substance. Adsorption is the process of physical adherence or bonding of one substance to the surface of another substance. Sorption mechanisms are generally categorized as physical adsorption, chemisorption, or electrostatic adsorption. Common sorption technologies include ion exchange (IX), activated alumina (AA), iron-based media, and granular activated carbon (GAC) (USEPA, 2000).

Ion Exchange

The IX is a physical/chemical process in which ions held electrostatically on the surface of a solid phase are exchanged for ions of similar charge in a solution (i.e., drinking water). The solid is typically a synthetic IX resin, which is used to preferentially remove particular contaminants of concern. IX is commonly used in drinking water treatment for softening (i.e., removal of calcium, magnesium, and other cations in exchange of sodium), as well as removing nitrate, arsenate, chromate, and selenate from municipal water. Because of its higher treatment cost compared to conventional treatment technologies, IX application is limited.

Activated Alumina

The AA adsorption is a physical/chemical process by which ions in solution are removed by the available adsorption sites on an oxide surface. AA is porous and highly adsorptive. It filters various contaminants including fluoride, arsenic, and selenium. The alumina can be regenerated. When all available adsorption sites are occupied, the AA media is typically regenerated with a strong base such as sodium hydroxide. Competition for adsorption sites by other ions such as phosphate, silicate, sulfate and fluoride may limit the use of AA.

Powdered Activated Carbon/Granular Activated Carbon

Activated carbon is produced by exposing carbon to very high temperatures increasing the surface area of internal pores. Powdered activated carbon (PAC) and GAC remove organic contaminants through adsorption, primarily a physical process in which dissolved contaminants adhere to the porous surface of the carbon particles. PAC is good for taste and odor control and is added directly to raw water and removed by settling in sedimentation basins. After saturation, activated carbon is typically regenerated for reuse (NDWC, 1997).

Lime Softening

Lime softening is suitable for treating groundwater. Either hydrated lime ($Ca(OH)_2$) or quicklime (CaO) may be used in the softening process. Hydrated lime is generally used more in smaller plants because it stores better and does not require slaking (producing a chemical change in lime by combining it with water) equipment. On the contrary, quicklime costs less per ton of available calcium oxide and is thus more economical for use in large plants. Softened water has high causticity and scale-formation potential. Recarbonation with carbon dioxide is employed to reduce pH and mitigate scaling of downstream processes and pipelines (NDWC, 1998b).

Point-of-Use/Point-of-Entry Applications

As a final barrier before consumption, properly designed and operated point-of-use (POU)/point-of-entry (POE) treatment systems are capable of providing clean and safe drinking water at the consumer's tap. Since the tragic events of September 11, 2001, POU/POE treatment devices have gained importance for protection against intentional or accidental contamination of drinking water distribution systems.

EPA is testing and evaluating the capabilities of a wide variety of devices incorporating adsorption, filtration, and disinfection technologies. POU/POE research studies are being conducted on reverse osmosis, nano-, ultra-, and microfiltration units for the removal of particulates and parasitic cysts (*Giardia* and *Cryptosporidium*), adsorptive media cartridges for the removal of IOCs such as arsenic, and brominated and chlorinated resins for the disinfection of microbial pathogens.

Water Reuse

Water supplies from conventional surface water and groundwater resources are struggling to meet future drinking water demands. Increases in population, water usage, droughts, and water shortages are causing many communities in arid and semi-arid locations in the United States to turn to water reuse as a water supply alternative. There are numerous opportunities for water reuse applications including urban, industrial, agricultural, environmental and recreational reuse, groundwater recharge, and augmentation of potable supplies. More information on these applications is available in EPA's Guidelines for Water Reuse (USEPA, 2004d).

Advances in water reuse technologies have improved treated water quality. Membrane bioreactors, biological nutrient removal processes, microfiltration, ozonation, and UV irradiation have enhanced removal and inactivation of microbial pathogens. EPA is currently conducting studies on endocrine-disrupting compounds and pharmaceuticals and personal care products to identify new technologies for the treatment of recalcitrant compounds. The key to ensuring safe water from reuse applications is to set high standards of reliability at water reclamation plants. Many states have incorporated procedures and practices into their reuse rules and guidelines to enhance reliability.

Water reuse treatment technologies for small communities can be categorized as systems with chemical and physical agents, mechanical, aquatic, and terrestrial treatment systems. Various types of chemical and physical agents (chlorine, ultraviolet irradiation, ozone, peroxide, and permanganate) have been used to oxidize wastewater. Mechanical treatment systems include oxidation ditches, extended aeration systems, sequencing batch reactors, trickling filters, and membrane processes. Aquatic treatment systems include constructed wetlands, aquacultural treatment systems, and sand filters. Terrestrial treatment systems use rapid infiltration and slow-rate overland flow/subsurface infiltration methods.

Residuals Management

Research Questions

- What quantities of waste residuals are being generated?
- What are the transport/disposal options?

- Are there methods to eliminate/reduce the waste?
- Is the material mobile in the environment? Can we immobilize it?
- What can be done with radioactive waste residuals?
- What are the characteristics of waste produced from new treatment processes?
- What are the regulatory issues facing management of waste residuals in the future?

There are currently no regulations or standards from the EPA that specifically cover water treatment plant residuals. Depending on the residuals' composition and the method of disposal, general regulations governing the disposal of solid and liquid wastes determine the fate of these materials.

Residuals' transport, treatment, and disposal can be a significant cost to communities. Developing new techniques for the disposal of waste residuals, including on-site land application, can minimize transport and disposal costs.

Types of Waste Residuals

Liquid residuals from water treatment operations include brines, caustics, filter backwash, sedimentation basin wash water, and solutions used for recharging solid media (NDWC, 1998a). Solid residuals can include sludge, schmutzdecke (biological surface layer in slow sand filtration units), and spent treatment media. The majority of liquid waste residuals generated by PWSs are most likely disposed of on-site (land application) or by sanitary sewer. Solid waste residuals are disposed of on-site (land application) or are transported for disposal in municipal landfills. Radioactive residuals create disposal issues in some states. Some PWSs generate technologically enhanced naturally occurring radioactive materials (TENORM). TENORM residuals are both solid and liquid and may contain nonexempt levels of radioactive material.

Liquid Residuals

A significant source of liquid residuals is filter backwash. In 2001, EPA published the final version of the Filter Backwash Recycling Rule (FBRR). The primary goal of the FBRR is to minimize consumer's exposure to microbial contaminants (e.g., *Cryptosporidium*) during cleaning and backwashing operations. The FBRR, implemented on June 8, 2004, requires backwash water, thickener supernatant, or dewatering liquids to be processed through PWS conventional or direct filtration units or through an alternate recycle location as approved by the state and/or local agencies.

Liquid waste residuals may be disposed of by direct/indirect discharge, underground injection, and land disposal.

Direct Discharge of Liquids

PWSs opting to discharge liquid residuals such as filter backwash and sedimentation basin wash water to U.S. waters must obtain a National Pollutant

Discharge Elimination System (NPDES) permit (40 CFR Section 122). Currently, EPA does not have technology-based effluent limits for water treatment plants. In this situation, discharge permits are usually based on best professional judgment and water quality-based effluent limits. Individual states conduct discharge permitting.

TMDL rules apply to PWS liquid residuals. TMDL requirements are developed and approved for the receiving water body. If a TMDL is not developed, the PWS must certify that the treatment and control methods employed are most appropriate for the reduction of pollutants generated by the PWS.

Indirect Discharge of Liquids

PWSs also discharge liquid residuals to sanitary sewers (i.e., "down the drain"). Indirect discharge does not require an NPDES permit, but a pretreatment program may have to be implemented by the operator. EPA has developed pretreatment guidance and regulations for industrial discharges to water treatment plants (see Effluent Guidelines cited in 40 CFR Section 403). Several states (e.g., Ohio) require that significant industrial users discharging to a Publicly Owned Treatment Works (POTW) obtain a permit for discharge.

Land Disposal of Liquids

Liquid residuals generated by PWSs that are reused through land application and are not classified as hazardous wastes are typically regulated by the state. Liquid residuals classified as hazardous wastes are subject to comprehensive generator, transport, storage, treatment, and land disposal restrictions defined in Resource Conservation and Recovery Act (RCRA) regulations. Some PWS liquid residuals are discharged to lagoons or evaporation ponds. In these cases, the SDWA and RCRA impose requirements for nonhazardous wastes to protect surface water and groundwater supplies.

Solid Residuals

PWS solid residuals (e.g., sludge, schmutzdecke, and spent treatment media) are classified as RCRA hazardous or nonhazardous wastes (40 CFR Sections 261.21 to 261.24) based on their ignitability, corrosivity, reactivity, and/or toxicity. Toxicity is assessed by the toxicity characteristic leaching procedure (TCLP). If contaminant concentrations in the TCLP leachate are in excess of those listed in the Land Disposal Restrictions (RCRA 40 CFR 268.40), the solid waste residual is classified as hazardous and must be disposed in an RCRA Subtitle C class landfill. Transport and disposal costs are considerably higher for hazardous waste residuals than for nonhazardous waste residuals sent to municipal solid waste landfills.

Radioactive Residuals

Low-level radioactive waste (LLRW) landfills are an option for PWS residuals with radionuclide concentrations deemed to be unacceptable for disposal at a solid or hazardous waste landfill. LLRW landfills are licensed by the Nuclear Regulatory Commission (NRC) or by a state under agreement with NRC, and guidelines for disposal of radioactive sludge and solids are more stringent than those in a solid waste landfill. More information is available in an EPA guidance document entitled "A Regulators' Guide to Management of Radioactive Residuals from Drinking Water Treatment Technologies" (USEPA, 2005d).

Distribution System Integrity

Research Questions

- What is the fate of specific contaminants in the distribution system?
- What are appropriate methods for controlling the contaminant?
- Does control of this contaminant affect existing distribution system chemistry/microbiology?
- How do changes in water chemistry affect sorption/desorption of contaminants?
- What models need to be developed and maintained for water distribution systems?
- What are the interrelations between biofilms and contaminants?

EPA research indicates that there is a different level of risk associated with the various distribution system infrastructure components. The relative risk of pathogens entering a distribution system can be summarized as follows (USEPA, 2002b):

- High risk—treatment breakthrough, intrusion, cross-connections, main repair/break
- Medium risk—uncovered water storage facilities
- Low risk—new main installation, covered water storage facilities, growth and resuspension, purposeful contamination

Proper operation and maintenance (O&M) of distribution systems plays a key role in ensuring that safe drinking water is provided to the consumers. More information is available in an EPA document entitled "Water Distribution Systems Analysis: Field Studies, Modeling and Management, A Reference Guide for Utilities" (USEPA, 2005e). PWS operators need to adequately understand and address the following three categorical issues facing the distribution system infrastructure components:

1. Infrastructure issues (repair and rehabilitation)
2. Operational issues (e.g., biofilm growth/disinfectant by-product formation, nitrification, and finished water aging)
3. Contamination events (e.g., cross-connections, permeation/leaching, and intrusion/infiltration)

Infrastructure Issues

The American Society of Civil Engineers (ASCE) rates the nation's drinking water infrastructure at D− (A through F scale). The report card states that the nation's 54,000 drinking water systems face an annual shortfall of $11 billion needed to replace facilities that are nearing the end of their useful life and to comply with federal water regulations (ASCE, 2005). The American Water Works Association (AWWA) white paper "New or Repaired Water Mains" (USEPA, 2002c) indicates that the installation and/or repair of water mains provides a potential route for direct contamination of the distribution system. Contamination can occur before, during, or after construction/repair activities.

Operational Issues

PWS operators must operate their distribution system to minimize the deterioration of water quality delivered to the consumer after it leaves the treatment plant. The water quality in the distribution systems can deteriorate substantially because of excessive growth of biofilm, DBP formation, nitrification, and improper storage of finished water. PWSs must be aware of these issues and optimally operate their system to control both biofilms and DBPs, prevent nitrification, and minimize detention times that result in excessive water age.

Corrosion, Scaling, and Metal Mobility

Inorganic contaminants such as lead, arsenic, antimony, and radium may exist in public water supplies in trace amounts at the entry points to the water distribution system. The accumulation and rerelease of these contaminants in concentrated amounts, which may not be detected by current monitoring practices, may result in elevated levels at our taps. Recent research has confirmed that corrosion deposits and scale on the inside of the distribution system piping and storage facilities can serve as reservoirs where trace contaminants accumulate. Concentrated amounts of these contaminants can be rereleased into the water supply because of changes in water chemistry or mixing of waters at concentrations exceeding their MCL.

Contamination Events

Distribution systems are vulnerable to external contamination events such as cross-connections, permeation/leaching, and intrusions/infiltrations. They contain locations where nonpotable water can be accidentally cross-connected to potable sources. These cross-connections can provide a pathway for backflow of nonpotable water into potable sources.

The infrastructure and appurtenances of distribution system, including piping, linings, fixtures, and solders, can react with the water supply as well as the external environment. Permeation of piping materials and nonmetallic joints and leaching (dissolution of metals, solids, and chemicals) can result in the degradation of the distributed water. Leaching from cement linings can occur in soft, aggressive, poorly buffered waters (USEPA, 2002d).

Pressure surges in water distribution systems damage pipes, fittings, and valves, causing leaks and shortening the life of the system. Mitigation techniques include the maintenance of an effective disinfectant residual throughout the distribution system, leak control, redesign of air relief venting, installation of hydropneumatic tanks, and more rigorous application of existing engineering standards. It was found that fecal indicators and culturable human viruses were present in the soil and water exterior to the distribution system pipes making it possible for these microorganisms to infiltrate/intrude into the distribution system (LeChevallier et al., 2003).

Leak Detection

EPA leak detection studies provide valuable data on how to characterize a water system's performance and condition. A pipeline test apparatus in Edison, New Jersey, has been used to evaluate leak detection and location devices and procedures.

Hydraulic and Water Quality Models

EPA develops hydraulic and water quality models to simulate conditions in water distribution systems. For instance, EPANET software models water distribution piping systems. EPANET performs extended-period simulation of the hydraulic and water quality behavior within pressurized pipe networks and assists in exposure assessment. Pipe networks consist of pipes, nodes (pipe junctions), pumps, valves, and storage tanks or reservoirs. EPANET tracks the flow of water in each pipe, the pressure at each node, the height of the water in each tank, and the concentration of a chemical species throughout the network to simulate and trace chemical species, water age, and the source of contamination.

Water Quality Monitoring Systems

Research Questions

- What is the current status of water quality monitor usage?
- What types of SCADA systems are most suitable to PWSs?
- What parameters are monitored and is there room for improvement?
- Is the monitoring system going to be used for security or water quality?
- What are the O&M issues for on-line monitoring systems?

SCADA Systems

Supervisory control and data acquisition (SCADA) systems consist of three key components: monitoring/control device(s) (e.g., a sensor/analyzer that measures and reports the desired parameter, a variable frequency drive pump whose speed can be controlled remotely), data transmission equipment/media (e.g., phone, wire, and radio), and data collection and processing unit (typically a central computer that analyzes the reported parameter value and programmatically decides what controls are warranted based on the reported value).

The application of SCADA to operate, monitor, and control water systems from a central location can reduce violations of MCLs as well as monitoring/reporting violations. The expected results from an appropriately designed and successfully deployed SCADA system are as follows:

- Enhanced security and control
- Improved water quality
- Regulatory compliance
- Reduced overall maintenance costs

Constant remote monitoring of the water quality has the potential to provide cost savings in time and travel for O&M. It has been determined that remote telemetry can support regulatory reporting guidelines by providing real-time continuous monitoring of the water quality and reporting the information electronically.

Long-term real-time remote monitoring provides data to significantly enhance treatment system operation and reduce system downtime. Real-time remote monitoring has the following advantages (Clark et al., 2004):

- Leads to improved customer satisfaction, improved consumer relations, and other health benefits
- Satisfies regulatory recordkeeping and reporting requirements
- Reduces labor costs (associated with time and travel) for small system operators
- Provides the capability to instantly alert operators of undesirable water quality and/or other changes in treatment system(s)
- Reduces downtime and increases repair efficiency through remote troubleshooting
- Identifies monitored parameter trends and adjust operating parameters accordingly
- Provides an attractive alternative to fixed sampling and O&M schedules

Monitoring Equipment

In general, monitors can be categorized by the types of parameters (contaminants, agents, characteristics) that the monitor measures. For establishing water quality, monitors measure one or more parameters that represent physical, chemical, and/or biological characteristics of the system.

Physical Monitors

Physical monitors measure physical characteristics of the water such as flow, velocity, water level, pressure, and other intrinsic physical characteristics of water. Examples of intrinsic physical characteristics include turbidity, color, conductivity, hardness, alkalinity, radioactivity, temperature, and oxidation–reduction potential. In general, physical monitors tend to be relatively inexpensive, quite durable, and readily available.

Chemical Monitors

Chemical monitors detect and measure inorganic or organic chemicals that may be present in the water. A specific technology or multiple technologies must be properly selected for a particular chemical or group of chemicals. Examples of chemical monitors include chlorine analyzer, nitrate sensor, total organic carbon (TOC) analyzer, and many others. Typically, the same general type of technology may be available for either automated online monitoring capability or manual grab sample analysis.

Biological Monitors

Biological monitors (biomonitors) include biosensors and biosentinels. Biosensors detect the presence of biological species of concern, such as some forms of algae or pathogens. The general operating principles of biosensors may include photometry, enzymatic, and/or some form of biochemical reaction. Biosentinels use biological organisms as sentinels to determine the likely presence of toxicity in a water sample. Most biosentinels operate by observing the behavior of selected organisms such as fish, mussels, daphnia, and algae. When the sentinel organism senses the presence of toxic contaminant(s), the organism reacts in some manner.

Data Transmission

Depending upon availability, cost, user preference, and the relative location of the sensors to the data acquisition system, the communication media for data transmission can be either wired (e.g., direct, phone line) or wireless (e.g., radio, cellular). Typically, direct wire and phone line (including cellular) communication media are the most inexpensive.

Remote Monitoring and Control Systems

Most treatment systems/technologies can be equipped with sensors and operating devices that can be monitored from remote locations. Remote monitoring and control technology improves monitoring/reporting and reduces O&M costs.

Online remote monitoring devices are fairly complex devices that are designed to automatically measure, record, and display specific physical, chemical, or biological parameters.

Contamination Warning Systems

EPA is developing monitoring systems that measure standard water quality parameters such as TOC, pH, turbidity, conductivity, chlorine, oxidation–reduction potential, and temperature. A database repository is being developed based on bench-and-pilot-scale experiments that reveal how these traditional parameters, if monitored online, can serve as triggers for contamination events.

Homeland Security/Emergency Response

Research Questions

- Are information sources adequate for PWSs?
- Can information dissemination be improved through collaboration with other governmental and nongovernmental organizations?
- Are emergency response procedures/protocol adequate?

Under Presidential Decision Directive (PDD) 63-Protecting America's Critical Infrastructures, issued in May 1998, EPA was designated as the lead agency for the water supply sector. The Public Health Security and Bioterrorism Preparedness and Response Act (Bioterrorism Act), passed in June 2002 (P.L. 107–188), provided EPA the mandate to work in water security. Homeland Security Presidential Directives (HSPDs) guide the agency's research and technical support activities to protect the nation's water and wastewater as follows:

- HSPD-7—Critical Infrastructure Identification, Prioritization, and Protection
- HSPD-8—National Preparedness
- HSPD-9—Defense of United States Agriculture and Food
- HSPD-10—Biodefense for the 21st Century

EPA's Office of Research and Development (ORD) officially established the National Homeland Security Research Center (NHSRC) in February 2003. To meet the responsibilities of their directives, EPA's Office of Water established the Water Protection Task Force, which was formally organized as the Water Security Division (WSD) in August 2003. WSWRD provides technical support and complements NHSRC research by conducting bench-, pilot-, and field-scale research.

The Bioterrorism Act amended the SDWA and requires all public water suppliers serving populations >3,300 to complete vulnerability assessments (VAs) and to develop or modify emergency response plans. VAs identify potential threats, assess the critical assets of the system, evaluate the likelihood and consequences of an attack, and develop a prioritized set of system upgrades to increase security.

Threats and Risks to the Water Supply

The risk of contamination using chemical, biological, and/or radiological substances with subsequent consequences must be understood by PWS administrators and operators to provide appropriate security, employ suitable detection systems, and develop strategies to deal with contamination events.

Disinfection in Distribution Systems

Disinfection ensures that dangerous microbial contaminants are inactivated before they can enter the distribution system. Chlorine gas, hypochlorite, chlorine dioxide, and chloramines are very effective disinfectants because residual concentrations can be maintained in the water distribution system. Some European countries use ozone and chlorine dioxide as oxidizing agents for primary disinfection prior to the addition of chlorine or chlorine dioxide for residual disinfection. The Netherlands identifies ozone as the primary disinfectant, as well as common use of chlorine dioxide, but typically uses no chlorine or other disinfectant residual in the distribution system (Connell, 1998).

Alternative Drinking Water Supplies in the Event of an Incident

In the event of a contamination incident, PWSs may need to utilize an alternate source of water. This need may arise because of drought, contamination of the primary source, or failure at the source (e.g., a dam). Use of an alternate source of water can be complex and requires advance approval by the state agencies. In many cases it may be more economical and practical to contract with a neighboring water supplier and form a partnership for sharing raw and/or finished water during emergencies.

EPA's environmental technology verification (ETV) program verifies monitoring and treatment technologies relevant to U.S. drinking water supplies. ETV's Drinking Water Systems Center, operated by NSF International, conducts studies on mobile package drinking water treatment systems that could be used for emergency water supplies for short-term treatment of compromised tap water or as permanent installations in small communities.

Response Protocol Toolbox

EPA released the "Interim Final Response Protocol Toolbox: Planning for and Responding to Contamination Threats to Drinking Water Systems" in December 2003 (USEPA, 2003e). The DHS developed a document entitled "National Strategy for the Physical Protection of Critical Infrastructures and Key Assets" (DHS, 2003).

RESEARCH PRIORITIES

Current Research Priorities

EPA drinking water research is currently focused on arsenic, CCL chemicals and microbes, distribution systems, including lead–copper and homeland security issues. To improve on water quality, EPA is investigating the effectiveness and optimum placement of structural BMPs, modeling of wet weather flow, determining the effect of management actions on water quality, and MST.

Future Research Priorities

EPA is in the process of developing an approach to better evaluate new contaminant control goals in context of all control goals. Water quantity issues (e.g., sustainability, reuse) must be addressed more fully within context of agency responsibilities. To improve on water quality, EPA is incorporating urban watersheds into mixed-use watersheds, addressing issues of scale, focusing on watershed microbial communities, and addressing nonstructural BMP issues.

SUMMARY AND CONCLUSIONS

EPA has a national drinking water and water-quality research program whose mission is important to EPA policy and regulatory development. The program has a long history and is relevant as much today as it was 30 years ago. The challenges facing drinking water treatment systems are numerous as described in this chapter. Research at EPA directs resources to the most pressing issues that apply to as many PWSs as possible. The sheer number of PWSs and the degrees to which they vary make this a difficult task. Future research must be adaptable to upcoming challenges. While searching for breakthroughs in the latest technologies, future work must consider energy efficiency, ways to alleviate water shortages, and affordable technologies that meet the regulatory considerations of the CWA and SDWA.

ACKNOWLEDGMENTS

The authors would like to acknowledge the contributions of employees within the EPA's Water Supply and Water Resources Division. In 2006, WSWRD took part in a program review that prioritized drinking water research, some of which is summarized here. Background information on regulations and risk management research was provided by Shaw Environmental and Infrastructure, Inc., under contract with U.S. EPA. General information was obtained from the EPA Web site at http://www.epa.gov. Conclusions and opinions presented in this chapter are those of the authors and do not necessarily represent the positions of U.S. EPA.

REFERENCES

ASCE, 2005. Report Card for America's Infrastructure (http://www.asce.org/reportcard/2005/index.cfm).

CA EPA, 2001. A Guide to Health Risk Assessment. California EPA, Office of Environmental Health Hazard Assessment, Sacramento, CA HRSguide2001.pdf, 2001 (http://www.oehha.ca.gov/pdf/HRSguide2001.pdf).

Clark, R., Panguluri, S., Haught, R., 2004. Remote monitoring and network models: Their potential for protecting US water supplies. In: Mays, L.W. (Ed.), Water Supply Systems Security. McGraw Hill, New York, NY.

Connell, G.F., 1998. European water disinfection practices parallel U.S. treatment methods. http://www.clo2.com/reading/waternews/european.html.

DHS, 2003. National strategy for the physical protection of critical infrastructures and key assets, February 2003 (http://www.dhs.gov/xlibrary/assets/Physical_Strategy.pdf).

FR, 2002. Pesticides; guidance on cumulative risk assessment of pesticide chemicals that have a common mechanism of toxicity, federal register, January 16, 2002, vol. 67, number 11, pp. 2210–2214 (http://www.epa.gov/fedrgstr/EPA-PEST/2002/January/Day-16/p959.htm).

Graham, J., Striebach, R., Patterson, C., Krishnan, E., Haught, R., 2004. Measurement of MTBE and its oxidation products from the treatment of surface waters by ozonation and UV-ozonation. Chemosphere 54, 1011–1016 (http://www.sciencedirect.com/science?_ob=ArticleURL&_udi=B6V74-49Y98VY-2&_user=14684&_rdoc=1&_fmt=&_orig=search&_sort=d&view=c&_acct=C000001678&_version=1&_urlVersion=0&_userid=14684&md5=fc405169d73ea79cc4fd0403d9374684).

LeChevallier, M.W., Gullick, R.W., Karim, M.R., Friedman, M., Funk, J.E., 2003. The potential for health risks from intrusion of contaminants into the distribution system from pressure transients. J. Water Health, 3–14 IWA Publishing.

NDWC, 1997. National Drinking Water Clearinghouse (NDWC). NDWC Fact Sheet-Technical Brief on Organic Removal. (http://www.nesc.wvu.edu/ndwc/pdf/OT/TB/TB5_organic.pdf, 1997).

NDWC, 1998a. Tech brief: Water treatment plant residuals management National Drinking Water Clearinghouse. West Virginia University Research Corporation, Morgantown, WV 1998 (No website found).

NDWC, 1998b. NDWC Fact Sheet—Technical Brief on Lime Softening (http://www.nesc.wvu.edu/ndwc/pdf/OT/TB/TB8_lime_softening.pdf).

OH EPA, 2008. Ohio Administrative Code (OAC), Division of Drinking and Ground Waters, Primary Drinking Water Rules, Chapter 3745-81-11 (8/1/05), -12 (1/1/02), -14 (1/1/08) (http://www.codes.ohio.gov/oac/3745-81).

Pollack, A.J., Chen, A.S.C., Haught, R.C., Goodrich, J.A., 1999. Options for Remote Monitoring and Control of Small Drinking Water Facilities. Battelle Press, Columbus, Ohio (No website found).

USEPA, 1996. Health Effects Test Guidelines OPPTS 870.8500, EPA 712-C-96-257, June 1996 (http://www.epa.gov/opptsfrs/publications/OPPTS_Harmonized/870_Health_Effects_Test_Guidelines/Drafts/870-8500.pdf).

USEPA, 1997. State Source Water Assessment and Protection Programs Guidance, Final Guidance, Office of Water, EPA-816-R-97-009, August 1997 (http://www.epa.gov/safewater/sourcewater/pubs/guide_stateswpfinal_1997_toc.pdf).

USEPA, 2000. Arsenic Removal from Drinking Water by Ion Exchange and Activated Alumina Plants, EPA-600-R-00-088, October 2000 (http://www.epa.gov/nrmrl/pubs/600r00088/600r00088.pdf).

USEPA, 2001a. Arsenic and Clarifications to Compliance and New Source Monitoring Rule: A Quick Reference Guide, Office of Water, EPA 816-F-01-004, January 2001 (http://www.epa.gov/SAFEWATER/arsenic/pdfs/quickguide.pdf).

USEPA, 2001b. Stage 1 Disinfectants and Disinfection Byproducts Rule: A Quick Reference Guide, Office of Water, EPA 816-F-01-010, May 2001 (http://www.epa.gov/OGWDW/mdbp/qrg_st1.pdf).

USEPA, 2001c. Radionuclides Rule: A Quick Reference Guide, Office of Water, EPA 816-F-01-003, June 2001 (http://www.epa.gov/SAFEWATER/radionuclides/pdfs/qrg_radionuclides.pdf).

USEPA, 2001d. Total Coliform Rule: A Quick Reference Guide, Office of Water, EPA 816-F-01-035, November 2001 (http://www.epa.gov/ogwdw/disinfection/tcr/pdfs/qrg_tcr_v10.pdf).

USEPA, 2001e. Health advisory values for drinking water contaminants and the methodology for determining acute exposure values, Joyce Morrissey Donohue and John C. Lipscomb, US Environmental

Protection Agency, Mail Code 4304, 1200 Pennsylvania Avenue NW, Washington, D.C., 20460, USA Received 27 June 2001; accepted 20 November 2001. Available online 14 December 2001 (http://www.sciencedirect.com/science?_ob=ArticleURL&_udi=B6V78-44NM1YW-2&_user=14684&_rdoc=1&_fmt=&_orig=search&_sort=d&view=c&_acct=C000001678&_version=1&_urlVersion=0&_userid=14684&md5=e87903d25fc9e5bd1af94dc39a87bf8f).

USEPA, 2002a. Strategic Plan for Homeland Security, U.S. EPA, September 2002 (http://www.epa.gov/epahome/downloads/epa_homeland_security_strategic_plan.pdf).

USEPA, 2002b. Health Risks from Microbial Growth and Biofilms in Drinking Water Distribution Systems. Office of Ground Water and Drinking Water, Washington, D.C. 2002 (http://www.epa.gov/safewater/tcr/pdf/biofilms.pdf).

USEPA, 2002c. New or Repaired Water Mains, Office of Water, OGWDW, Distribution System Issue Paper, August 15, 2002 (http://www.epa.gov/safewater/disinfection/tcr/pdfs/whitepaper_tcr_watermains.pdf).

USEPA, 2002d. Permeation and Leaching, Office of Water, OGWDW, Distribution System Issue Paper, August 15, 2002 (http://www.epa.gov/safewater/disinfection/tcr/pdfs/whitepaper_tcr_permation-leaching.pdf).

USEPA, 2003a. Human Health Research Strategy, U.S. EPA, Office of Research and Development, EPA/600/R-02/050, September 2003 (http://epa.gov/nheerl/humanhealth/HHRS_final_web.pdf).

USEPA, 2003b. National Primary and Secondary Drinking Water Regulations, Office of Water, EPA 816-F-03-016, June 2003 (http://www.epa.gov/safewater/consumer/pdf/mcl.pdf).

USEPA, 2003c. Small Drinking Water Systems Handbook, A Guide to "Packaged" Filtration and Disinfection Technologies with Remote Monitoring and Control Tools, EPA/600/R-03/041, May 2003 (http://www.epa.gov/nrmrl/pubs/600r03041/600r03041.pdf).

USEPA, 2003d. Drinking Water Research Program Multi-Year Plan, (http://www.epa.gov/osp/myp/dw.pdf, 2003).

USEPA, 2003e. Interim final response protocol toolbox: Planning for and responding to contamination threats to drinking water systems. Washington, D.C., 2003 (http://cfpub.epa.gov/safewater/watersecurity/home.cfm?program_id=8#response_toolbox).

USEPA, 2004a. Lead and Copper Rule: A Quick Reference Guide, Office of Water, EPA 816-F-04-009, March 2004. (http://www.epa.gov/OGWDW/lcrmr/pdfs/qrg_lcmr_2004.pdf)

USEPA, 2004b. Drinking Water Health Advisory for Oxamyl, EPA-822-B-04-002, September 2004 (http://www.epa.gov/waterscience/criteria/drinking/oxamyl-dw-ha.pdf).

USEPA, 2004c. An Examination of EPA Risk Assessment Principles and Practices, EPA/100/B-04-001, March 2004 (http://www.epa.gov/OSA/pdfs/ratf-final.pdf).

USEPA, 2004d. Guidelines for Water Reuse, Office of Wastewater Management, Office of Water, PEA/625/R-04/108 pp 106-113 (http://www.epa.gov/nrmrl/pubs/625r04108/625r04108.pdf).

USEPA, 2005a. Guidelines for Carcinogen. Risk Assessment. Risk Assessment Forum. U.S. Environmental Protection Agency, Washington, D.C. EPA/630/P-03/001B. March 2005 (http://www.epa.gov/iris/cancer032505.pdf).

USEPA, 2005b. Impellitteri, C., Patterson, C., and Haught, R., Small Drinking Waters Systems: State of the Industry and Treatment Technologies to Meet the Safe Drinking Water Act Requirements, EPA600/X-05/021, November, 2005 (http://www.epa.gov/nrmrl/pubs/600r07110/600r07110.pdf).

USEPA, 2005c. Microbial Sources Tracking Guide Document. Office of Research and Development, Washington, D.C. EPA-600/R-05/064, 131 pp, June 2005 (http://www.epa.gov/nrmrl/pubs/600r05064/600r05064.pdf).

USEPA, 2005d. A Regulators' Guide to Management of Radioactive Residuals from Drinking Water Treatment Technologies, Washington, D.C., EPA/816/R-05/004, June 2005 (http://www.epa.gov/ogwdw/radionuclides/pdfs/guide_radionuclides_regulatorsguide.pdf).

USEPA, 2005e. Water Distribution System Analysis: Field Studies, Modeling and Management, a Reference Guide for Utilities, EPA/600/R-06/028, December, 2005 (http://www.epa.gov/nrmrl/pubs/600r06028/600r06028prelithruchap4.pdf).

USEPA, 2006a. 2006–2011 EPA Strategic Plan Charting Our Course, EPA-190-R-06-001, September 29, 2006 (http://www.epa.gov/cfo/plan/2006/entire_report.pdf).

USEPA, 2006b. NRMRL Strategic Plan, Office of Research and Development. National Risk Management Research Laboratory, Cincinnati, Ohio 45268, September, 2006 (http://www.epa.gov/nrmrl/pdf/nrmrl_strat_plan2006.pdf).

USEPA, 2006c. Final Ground Water Rule, Office of Water, EPA 815-F-06-003, October 2006 (http://www.epa.gov/OGWDW/disinfection/gwr/regulation_factsheet_final.html).

USEPA, 2006d. Long Term 2 Enhanced Surface Water Treatment Rule: A Quick Reference Guide, Office of Water, EPA 816-F-06-005, June 2006 (http://yosemite.epa.gov/r10/water.nsf/c6e3c862e806dd688825688200708c97/aabb1ca50d7acf3b882570d500695cbe/$FILE/94075184.pdf/LT2%20QRG%20Sch1%20Final.pdf).

USEPA, 2006e. 2006 Edition of the Drinking Water Standards and Health Advisories. Office of Water, Washington, D.C. EPA/822/R-06/013, 2006. (http://www.epa.gov/waterscience/criteria/drinking/dwstandards.pdf).

Chapter 8

Microbiological Threats to Water Quality

Lawrence B. Cahoon and Bongkeun Song
Department of Biology and Marine Biology and Center for Marine Science, University of North Carolina Wilmington, Wilmington, NC 28403, USA

INTRODUCTION

The United Nations has declared 2008 to be the year of sanitation, recognizing the urgency of resolving threats to public health posed by contaminated drinking water and poor sewage treatment practices. Microbiological hazards in water supplies are among the most serious hazards facing developing nations, in particular. Diarrheal diseases caused by contaminated drinking water and poor sanitation practices kill an estimated 1.6 million children each year, making this the third leading cause of death among children under 15 years in less well developed countries (Cohen, 2008). However, even developed countries with modern sanitation and water supply systems are vulnerable as outbreaks of cryptosporidiosis in the United States and infections with pathogenic *Escherichia coli* O157: H7 strains exemplify (MacKenzie et al., 1994; MMWR, 1996).

The United States Environmental Protection Agency (U.S. EPA) has developed a short list of microbiological hazards for which testing protocols are to be implemented in municipal water supply systems (EPA, 2008a–c). As of 2008, EPA has promulgated three such candidate contaminant lists (CCLs) of microbes. The original inclusive list (universe) was based on earlier surveys, including that of Taylor et al. (2001), who listed 1,415 microbial agents then known to infect humans. Others have been added and the lists include bacteria,

Handbook of Water Purity and Quality
Copyright © 2009 by Elsevier Inc. All rights of reproduction in any form reserved.

viruses, prions, rickettsias, helminths, fungi, and protozoans. Clearly, testing for all such agents is impossible and unwarranted, so a shorter list is desirable. U.S. EPA conducted two levels of screening, based on the best available scientific characteristics, and identified 29 hazardous organisms for further consideration, from which 11 have been identified for the most recent CCL: two viral agents, seven bacterial agents, and two protozoans (EPA, 2008c) (Table 1).

Viral agents can cause a wide array of diseases of which many are difficult to treat. Among them, the caliciviruses include the noroviruses, which are responsible for a large proportion of viral gastroenteritis outbreaks, particularly in confined settings, such as institutions and cruise ships (Lang, 2003). Infection is usually by the fecal–oral route or direct contact. Hepatitis A virus is one of the several hepatitis-causing viruses, but unlike the "serum hepatitis" forms (B and C), it is most frequently contracted through exposure to contaminated foods, particularly fecally contaminated shellfish and uncooked vegetables (Lappalainen et al., 2001).

Bacterial agents include species causing a wide array of diseases, many quite dangerous. *Campylobacter jejuni* is responsible for significant

TABLE 1 Characteristics of Microbial Pathogens Included in EPA's Candidate Contaminant List 3 (Data from EPA, 2008c)

Pathogen	US WBDO[1]	DW detects[2]	Mortality risk[3]	Serious sequelae[4]
Viruses				
Caliciviruses	Multiple	Yes	No	No
Hepatitis A virus	Multiple	Yes	Yes	Yes
Bacteria				
Campylobacter jejuni	Multiple	Yes	Rare	Yes
Escherichia coli O157	Multiple	Yes	Rare	Yes
Helicobacter pylori	None	Yes	>0.1%	Yes
Legionella pneumophila	Multiple	Yes	Yes	Yes
Salmonella enterica	Multiple	Yes	Rare	Yes
Shigella sonnei	Multiple	Yes	Rare	Yes
Vibrio cholerae	Multiple	Yes	Rare	Rare
Protozoans				
Entamoeba histolytica	Multiple	Yes	>0.1%	Yes
Naegleria fowleri	1	Yes	High	High

[1]*Incidence of water-borne disease outbreaks in the United States.*
[2]*Detection in drinking water supplies in the United States.*
[3]*Risk of death from primary infection.*
[4]*Risk of serious complications, including hospitalization and other conditions.*

proportions of bacterial gastroenteritis cases and is often contracted through unsanitary handling of poultry products (Peyrat et al., 2008). The O157: H7 strain of *E. coli* harbors a plasmid that codes for the synthesis of Shiga-toxin; other strains of *E. coli* are known to produce this toxin, but the O157: H7 strain is most well studied. Infections occur from consumption of contaminated beef products and, according to some reports, from contaminated drinking water as well (Mead and Griffin, 1998). *Helicobacter pylori* is a bacterium now known to account for as much as 90% of gastric ulcers and a significant portion of stomach cancers, contracted through the fecal–oral route in contaminated foods and drinking water (Kandulski et al., 2008). *Legionella pneumophila* causes "Legionnaire's disease," a serious form of pneumonia contracted through inhalation of water droplets contaminated with this bacterium, frequently in air-conditioning or other water-handling systems (Simmons et al., 2008). *Salmonella enterica* includes many strains of enteric disease-causing bacteria including *Salmonella typhi*, the causative agent of typhoid fever. Salmonellosis is contracted from animal sources and products, including poultry and hoofed animals, but recently in the United States through contaminated fresh produce as well (D'Aoust, 1994; Rabsch et al., 2001). *Shigella sonnei* is one of the group of *Shigella* spp., closely related to *E. coli*, that cause a severe form of gastro-enteritis termed shigellosis, contracted primarily through the fecal–oral route (Ekdahl and Andersson, 2005). *Vibrio cholerae* is widespread in estuarine and coastal waters and is hosted by crustacean zooplankton. Only certain strains cause cholera, perhaps the most severe and deadly form of gastroenteritis, notably the El Tor strain, responsible for the most recent (and ongoing) cholera pandemic (Sack et al., 2004). Cholera is contracted through the fecal–oral route of transmission. Cholera outbreaks are often tied to water pollution or other nutrient loading events that stimulate algal blooms and subsequent blooms of herbivorous zooplankton that naturally host the cholera vibrio.

Protozoan agents are particularly problematic in that medications effective in treating infections caused by them are rare. *Entamoeba histolytica* causes a form of amoebic dysentery, usually contracted through the fecal–oral route in contaminated drinking water (Gopal Rao and Padma, 1971). *Naegleria fowleri* is one of the several protozoans capable of infecting humans through exposure to contaminated waters while swimming, and causing infections of the central nervous system, relatively rare but usually fatal (Schuster and Visvesvara, 2004).

Most of the organisms selected for the short list infect humans via fecal contamination of drinking water or by unsanitary food handling, but other life modes, routes of exposure or infection, and mechanisms of morbidity and mortality are represented among the "universal" lists. Some opportunistic pathogens or parasites can occur naturally in the environment independently of fecal contamination, such as *Klebsiella pneumoniae* (APHA, 1998). Moreover, some types of water-borne organisms responsible for very high profile outbreaks of human disease or other harmful outcomes are not included in the most recent, short CCL, e.g., the protozoan, *Cryptosporidium parva*, helminth and nematode parasites, harmful algal bloom (HAB) species, or fungal pathogens and parasites.

QUANTIFICATION METHODS

Risk of disease from water-borne pathogens and parasites is almost always a function of received dose. Some pathogens can cause disease at very low doses, e.g., *Shigella* sp. (Bagamboula et al., 2002), but others require much higher doses to yield a thriving infection. Consequently, the concentration of water-borne disease-causing organisms in the medium is an important measure of risk. Quantification methods range from direct counts to various culture-based (filtration and/or fermentation) approaches (usually used with bacterial agents) to far more sophisticated molecular techniques, including quantitative polymerase chain reaction (q-PCR) methods, e.g., newer techniques for *C. parva*, a protozoan (Fontaine and Guillot, 2002). Although detailed descriptions of these methods are available from standard method manuals, e.g., APHA (2005) and published literature, brief description of some of the different approaches is given here.

Direct count methods typically target larger, distinctive, or manipulable pathogens. For example, the EPA-approved standard method for the quantification of *C. parva* involves filtration of large volumes of water, staining with a fluorescent-tagged antibody, and microscopy to distinguish *C. parva* from other closely related forms that cross-react in the staining procedure (EPA, 1999). Suspended material can interfere with the filtration step, the fluorescent-tagged antibody is relatively expensive to use routinely, and a skilled microscopist must still distinguish *C. parva* from other nontarget organisms. Consequently, these techniques have somewhat limited applicability and utility.

Culture-based approaches include fermentation tube, colorimetric, and filtration–incubation methods. Some bacterial pathogens or indicators of their presence produce gas by fermenting certain substrates. Gas production in conjunction with serial dilution techniques can yield quantitative estimates of specific bacterial concentrations. For example, the standard most probable number (MPN) technique for coliform bacteria utilizes their ability to ferment lactose, using the appearance of gas bubbles in a dilution series to yield estimates of original concentration. Colorimetric methods employ the differential abilities of bacteria to metabolize colored substrates or yield colored products as an identification tool and for quantification. Commercial colorimetric assays have been developed as presence/absence tests for rapid evaluations and, in combination with dilution techniques, as modified MPN assays as well.

Filtration–incubation methods are similar to the fermentation tube methods, but rely on a combination of appropriately labile substrates and metabolic inhibitors to select for the bacteria of interest on membrane filters incubated over a mix of these bacteria. For example, the MFC method for coliform bacteria allows direct counts of colony-forming units (CFU) in a volume of water passed through the filter (APHA, 1998). All culture-based approaches have the common drawbacks that some samples will exhibit colony overgrowth, making accurate counts difficult, and that many species of bacteria cannot be cultured at all (by some estimates >99% of all field organisms), and even many individual cells of "culturable" species may be damaged and unable to grow in culture. Moreover, culture-based methods for some of the culturable but highly pathogenic bacteria, e.g., *Shigella* sp., pose an elevated risk to laboratory personnel and require extra care.

Molecular methods have been developed for the identification of microbial contaminants, and they offer significant promise for circumventing the issues of detection, identification, and viability vs. culturability. Rapid DNA and RNA extraction protocols, application of polymerase chain reaction (PCR) techniques to amplify even rare genomes, and development of species- and even clone-specific primers for the identification allow far more sophisticated analysis of the microbial assemblages present in a sample than was possible using traditional microbiological methods. It must be conceded that these methods still require relatively sophisticated laboratory procedures and are not generally suited for routine monitoring. However, these approaches and the insights offered by them have allowed development of commercially available kits for the detection and quantification of some number of important pathogens; e.g., kits for *L. pneumophila*, *Salmonella*, *Shigella*, and *V. cholerae* are available from Fisher Scientific Co. However, many of these molecular methods are not yet certified as standard protocols for water-quality assessments.

MICROBIAL CONTAMINATION INDICATORS

The challenges inherent in quantifying even a short list of microbial threats to human health include difficulties with handling and culturing target organisms, and the risk of exposing field and laboratory personnel to dangerous pathogens. Consequently, indicator organisms that are not themselves pathogenic or that can be handled with little risk, and whose presence correlates with the presence of one or more pathogen types are frequently quantified. The most widely used methods for monitoring or predicting the presence of potential pathogens in aquatic environments are based on cultivation and enumeration of the fecal coliform group (including *E. coli* specifically), the fecal enterococcus group (including the fecal streptococcus subgroup), and *Clostridium perfringens* (APHA, 1998). The coliform and enterococcus groups confer the advantage for monitoring work of being unable to form resting spores, so that their presence is the evidence of recent contamination. *C. perfringens* has a resting stage, however, which makes it more useful as a historical indicator of contamination.

Use of fecal microbial indicators of microbial contamination entails several qualifying conditions. First, these indicators denote the risk of fecal contamination, the major, but not exclusive, source of microbial contamination. Some pathogens derive from other sources, i.e., *Legionella*, whereas others can occur in the environment without a direct human source of contamination, e.g.,

Klebsiella and *V. cholerae*. Second, significant positive correlation between concentrations of these indicators and actual health effects must be established. In the case of fecal coliforms and enterococcus, correlations between these and the incidences of gastroenteritis above background levels have been established by epidemiological studies. Subsequently standards for water uses have been established, e.g., geometric means of 200 and 33 CFU per 100 ml of water for fecal coliforms and enterococcus, respectively, for human body contact in salt water (EPA, 2003). Third, fecal contamination indicators can derive from multiple sources, including wild and domestic animals, so evaluation of risk to humans and remediation of contamination can be complicated. Source identification is obviously critical to remediation efforts. This consideration has led to attempts to devise microbial source tracking (MST) techniques.

MICROBIAL SOURCE TRACKING

The aim of MST is to identify sources, both species (humans vs. animals) and location. Although human sources may be most dangerous in terms of pathogen compatibility and transmissibility, animal sources can yield unusual pathogens. Some animals are natural reservoirs for certain pathogens, e.g., waterfowl that host influenza viruses (Suarez, 2000). It is important to recognize that MST methods may fail to identify a specific source of fecal microbial contamination, but may allow some sources to be ruled out, which may be almost as useful in taking preventative measures to protect public health.

MST methods may be broadly categorized as library-dependent or library-independent and molecular or nonmolecular techniques (McCorquodale et al., 1996; Sargeant, 1999; Bernstein et al., 2002) (Table 2). Library-dependent techniques require sampling and characterization of as many likely microbial contamination source types as possible to develop a "library" of sources and characteristics, against which unknowns can be compared. Library development is preliminarily necessary to unknown source tracking efforts, and studies have shown that libraries can be site- or region-specific. Larger libraries tend to be more powerful, but require more effort and time, which may be necessary in settings with many candidate sources (Wiggins et al., 1999). Library-independent methods rely on characteristics of different microbial contamination sources that are considered to be more reliable indicators, although some such distinctions, e.g., the ratios of fecal coliform and fecal streptococcus, have proven less useful than proposed (Brion and Lingireddy, 1999; Sankararamakrishnan and Guo, 2005). Nonmolecular techniques include physiological markers, such as antibiotic resistance patterns (Hagedorn et al., 1999), and host-specific indicators (Mandaville, 2002).

Molecular techniques, which have become significantly more powerful and accessible, have focused on identification of bacteria and viruses that associate with specific host organisms (Sargeant, 1999). The discovery and refinement of "genetic markers" of target organisms allow discrimination among sources of

TABLE 2 Classification of MST Methods

Non-molecular methods	Library dependent	Antibiotic resistance patterns; carbon source profiling
	Library independent	Fecal bacteria ratios; host-specific indicators; F+ coliphage serotyping; enterotoxin biomarkers
Molecular methods	Library dependent	Polymerase chain reaction; pulsed field gel electrophoresis; ribotyping, randomly amplified polymorphic DNA
	Library independent	Bacteriophage indicators; virus indicators; molecular host-specific indicators; bacterial endemism and cospeciation; terminal restriction fragment length polymorphism; amplified fragment length polymorphism

pathogens or indicators, as well as more recently quantification of these microbial contamination sources (Bernstein et al., 2002; Meays et al., 2004). Typical coliform or enterococcus counting methods are not able to identify the source of fecal contamination (Field et al., 2003; Scott et al., 2002). In addition, fecal coliforms can survive and grow after they are released into the receiving water (Desmarais et al., 2002; Solo-Gabriele et al., 2000). This raises the question of the method's accuracy (Scott et al., 2002; Simpson et al., 2002). Alternatively, molecular methods based on nucleic acid detection have been widely used for current MST protocols. Most molecular methods are based on PCR, which determines the presence or absence of target organisms by detecting specific pathogenic genes or small-subunit ribosomal RNA (rRNA) genes. Various molecular methods have been developed to detect *E. coli* (Bej et al., 1990, 1991; Franks et al., 1998), *Salmonella* spp. (Fukushima et al., 2002; Lofstrom et al., 2004; Malorny et al., 2003), *Shigella* spp. (Fukushima et al., 2002; Horman et al., 2004), *Campylobacter* spp. (Lubeck et al., 2003a,b), *Legionella* (Wellinghausen et al., 2001), and *Vibrio vulnificus* (Panicker et al., 2004) in water samples. Among these microorganisms, the *Bacteroides–Prevotella* group, which are not members of the coliform group, was proposed as an alternative fecal pollution indicator as the members of this group are highly abundant in the feces of warm-blooded animals (Daly et al., 2001; Hold et al., 2002; Leser et al., 2002). They also have host species- or group-specific distributions (Bernhard and Field 2000a,b; Simpson et al., 2004). PCR methods with primers targeting host-specific *Bacteroides–Prevotella* 16S rRNA genes have been developed to distinguish fecal pollution sources between human and other animals (Bernhard and Field, 2000a; Dick et al., 2005a,b; Dick and Field, 2004; Matsuki et al., 2002;

Seurinck et al., 2004; Simpson et al., 2002; Okabe et al., 2007). In addition, terminal restriction fragment length polymorphism (T-RFLP) and length heterogeneity PCR have been developed to track the source of *Bacteroides–Prevotella* groups as fecal contaminants (Bernhard and Field, 2000a,b; Boehm et al., 2003). T-RFLP is a molecular technique to examine microbial community structures by comparing DNA fingerprint profiles. The profiles are generated by digesting fluorescent labeled PCR-amplicons of a target gene using one or more restriction enzymes. The T-RFLP profiles of the *Bacteroides–Prevotella* 16S rRNA genes were used to identify the sources of microbial contaminations in coastal water (Bernhard and Field, 2000a). In addition, amplified fragment length polymorphism (AFLP) and enterobacterial repetitive intergenic consensus polymerase chain reaction (ERIC-PCR) were used to determine different host-specific *E. coli* strains (Guan et al., 2002; Leung et al., 2004; Parveen et al., 1999).

Real-time PCR techniques have been adopted to quantify the numbers of host-specific *Bacteroides–Prevotella* groups, which were interpreted as microbial contaminants from different fecal resources such as human, cow, and pig (Okabe et al., 2007; Santoro and Boehm, 2007; Shanks et al., 2008). The q-PCR techniques combine the power of oligonucleotide probe hybridization with PCR amplification and provide a sensitive and streamlined alternative to membrane-based probe hybridization approaches for estimating gene abundance (Heid et al., 1996). The q-PCR assays use two different methods depending on how they determine the increase of PCR products. In the SYBR green assay, PCR product formation is quantitatively monitored by determining the increase in fluorescence after binding a fluorescent DNA stain (SYBR green) to the amplicon (Higuchi et al., 1991), whereas the TaqMan probe assay determines the increase of PCR products by the release of a fluorescent moiety from specific oligonucleotide probes bound to the amplicon. TaqMan probes consist of a short oligonucleotide, which is labeled with a fluorescent chromophore and a quencher at the 5′ and 3′ ends, respectively. During template elongation, the probe is cleaved by the 5′→3′ exonuclease activity of *Taq* DNA polymerase, which releases the 5′-linked dye from the 3′-linked quencher, resulting in an increase in fluorescence with product formation. Both SYBR green and TaqMan probe assays were developed to quantify the *Bacteroides–Prevotella* groups from different hosts (Okabe et al., 2007; Santoro and Boehm, 2007; Shanks et al., 2008). Various q-PCR protocols have been developed to detect and enumerate other bacteria including *Staphylococcus aureus* (Hein et al., 2001), *Listeria monocytogenes* (Novga et al., 2000), *E. coli* (Huijsdens et al., 2002), and Bifidobacterium group (Malinen et al., 2003; Matsuki et al., 2004). Thus, molecular detection methods have provided specific detection of selected organisms as well as genetic indices to track the source of microbial contaminants in various environmental samples. Improved access to faster sequencing capabilities and development of primers have promoted wider use of these genetic detection approaches, but definitive attribution of particular indicator bacteria in a field sample to particular sources still requires caution and supporting data.

SAMPLING ISSUES

Assessment of microbiological threats to water quality is complicated by several considerations in addition to those described earlier. Almost all monitoring protocols are necessarily designed to be executed quickly, with relatively large numbers of samples, and by personnel with moderate levels of training. For most issues having to do with drinking water supplies, these protocols are generally sufficient. However, experience and published research have shown that water-quality hazards can occur outside the bounds of conventional sampling schemes and protocols, particularly when microbial contaminants grow or at least persist in aquatic habitats other than the water column.

Biofilms in water supply systems and sediment-associated pathogen populations in surface waters pose special hazards as potential sources of contamination and challenges for sampling. Biofilms are aggregates of bacteria and associated microbes that attach to solid surfaces or at the water surface (Preston, 2003) using excreted polysaccharides that form a complex matrix. Biofilms form when dissolved organic materials adsorb to otherwise clean surfaces, providing a substrate to attract and adsorb microbial cells, whose proliferation subsequently develops a mature biofilm. Flow past surfaces provide dissolved substrates that support microbial metabolism. Biofilms provide protection from many sources of mortality, including chlorination (Hallam et al., 2001), as well as mechanical resistance to disruption and suspension. However, microbial population growth can sustain steady release of microbes from biofilms, thus constituting a contamination source separate from and in addition to the usual suspected sources.

Biofilms have been implicated in the occurrence and persistence of many CCL-listed pathogens in both natural and man-made aquatic systems. Biofilm formation has been thought to be important in the survival and dispersal of many pathogenic bacteria in aquatic systems (Hallam et al., 2001; Hall-Stoodley and Stoodley, 2005), including *V. cholerae* (Reidl and Klose, 2002), *H. pylori* (Park et al., 2001), *L. pneumophila* and *S. typhimurium* (Armon et al., 1997). Fecal coliform bacteria have been found living in biofilms in corals offshore Florida, several kilometers from likely sources (Lipp et al., 2002). *Legionella* has been found in high percentages of surface films studied in aquatic ecosystems, and frequently cooccurs with important protozoan pathogens, such as *Naegleria* sp. and *Acanthamoeba* sp. (Declerck et al., 2007). These and other protozoans have also been found to be associated with bacterial biofilms formed in dental rinse units (Barbeau and Buhler, 2001). The ubiquitous nature, rapid formation, and microbe-friendly environments offered by biofilms therefore make them important, but poorly quantified reservoirs of human water-borne pathogens.

Sediments accumulate microbes in both viable and resting-stage forms (Van Donsel and Geldreich, 1971; Grimes, 1980; Sawyer, 1980; O'Malley et al., 1982; Izzo et al., 1983; Valiela et al., 1991; Doyle et al., 1992; Sawyer et al., 1998; Palmer, 2000; Lipp et al., 2001; Whitman and Nevers, 2003). Although bacterial cells and other very small microbes sink slowly at best, particle scavenging, floc formation, adsorption, deposition by turbulent flows, and filtration through sediments by oscillating flows can all drive recruitment of microbes to sediments (Rusch and Huettel, 2000; Fries and Trowbridge, 2003). As with biofilms, accumulation of organic matter, substrate supply by continuous flow, and protection from mortality sources can all act to support dense populations of microbes. Studies of sediment-associated fecal coliform and enterococcus bacteria yield estimates of concentrations in sediments an order of magnitude or greater than in the overlying water column (Lewis et al., 1986; Buckley et al., 1998; Obiri-Danso and Jones, 1999). Disturbance of sediments by humans, animals, or physical processes can subsequently resuspend microbial contaminants in the water column, constituting an apparently fresh source of contamination (Pettibone et al., 1996; Crabill et al., 1999; Baudart et al., 2000; An et al., 2002). One *caveat* must be advanced, however, there are very few data relating concentrations of the indicator bacteria to actual pathogen concentrations in sediments, so interpretation of indicator bacteria concentration data must be cautious when risk assessments are performed. Consequently, there are no standards yet established for sediment-associated fecal indicators, a topic that requires further research.

Sampling of biofilms and sediments is not yet standardized to the degree that water column sampling protocols have achieved, although methods have been developed more or less *ad hoc* by researchers. Development of standard sampling protocols must consider replicable sampling tools and techniques, statistically sound evaluation of the relationships between indicators and microbes of direct interest, and applicability of more sophisticated molecular tools to these sampling protocols.

CONCLUSIONS

Water-borne microbial pathogens pose significant health threats to humans, and challenges for investigators and water- quality managers. A variety of tools have been developed for estimating and identifying microbial sources of water-quality impairment, but significant improvements in the utility and specificity of these tools remain to be developed. The ecology itself of water-borne pathogens poses additional basic sampling challenges.

REFERENCES

APHA, 1998. Standard Methods for the Examination of Water and Wastewater, twentieth ed. American Public Health Association, Washington, D.C.

APHA, 2005. Standard Methods for the Examination of Water and Wastewater, twenty second ed. American Public Health Association, Washington, D.C.

An, Y.-J., Kampbell, D.H., Breidenbach, G.P., 2002. *Escherichia coli* and total coliforms in water and sediments at lake marinas. Environ. Pollut. 120, 771–778.

Armon, R., Starosvetzky, J., Arbel, T., Green, M., 1997. Survival of *Legionella pneumophila and Salmonella typhimurium* in biofilm systems. Water Sci.Technol. 35, 293–300.

Bagamboula, C.F., Uyttendaele, M., Debevere, J., 2002. Growth and survival of *Shigella sonnei* and *S. flexneri* in minimal processed vegetables packed under equilibrium modified atmosphere and stored at 7°C and 12°C. Food Microbiol. 19, 529–536.

Barbeau, J., Buhler, T., 2001. Biofilms augment the number of free-living amoebae in dental unit waterlines. Res. Microbiol. 152, 753–760.

Baudart, J., Grabulos, J., Barusseau, J.-P., Lebaron, P., 2000. *Salmonella* spp. and fecal coliform loads in coastal waters from a point vs. nonpoint source of pollution. J. Environ. Qual. 29, 241–250.

Bej, A.K., Steffan, R.J., DiCesare, J., Haff, L., Atlas, R.M., 1990. Detection of coliform bacteria in water by polymerase chain reaction and gene probes. Appl. Environ. Microbiol. 56, 307–314.

Bej, A.K., McCarty, S.C., Atlas, R.M., 1991. Detection of coliform bacteria and *Escherichia coli* by multiple polymerase chain reaction: comparison with defined substrate and plating methods for water quality monitoring. Appl. Environ. Microbiol. 57, 2429–2432.

Bernhard, A.E., Field, K.G., 2000a. Identification of nonpoint sources of fecal pollution in coastal waters by using host specific 16S ribosomal DNA genetic markers from fecal anaerobes. Appl. Environ. Microbiol. 66, 1587–1594.

Bernhard, A.E., Field, K.G., 2000b. A PCR assay to discriminate human and ruminant feces on the basis of host differences in *Bacteroides Prevotella* genes encoding 16S rRNA. Appl. Environ. Microbiol. 66, 4571–4574.

Bernstein, B.B., Griffith, J.F., Weisberg, S.B., 2002. Microbiological source tracking workshop: workgroup findings and recommendations. U.S. EPA Workshop on Microbial Source Tracking, February 5, 2002, Irvine, CA.

Boehm, A.B., Fuhrman, J.A., Mrse, R.D., Grant, S.B., 2003. Tiered approach for identification of a human fecal pollution source at a recreational beach: case study at Avalon Bay, Catalina Island, California. Environ. Sci. Technol. 37, 673–680.

Brion, G.M., Lingireddy, S., 1999. A neural network approach to identifying non-point sources of microbial contamination. Water Res. 33, 3099–3106.

Buckley, R., Clough, E., Warnken, W., Wild, C., 1998. Coliform bacteria in streambed sediments in a subtropical rainforest conservation reserve. Water Res. 32, 1852–1856.

Cohen, J., 2008. Pipe dreams come true. Science 319, 745–746.

Crabill, C., Donald, R., Sneling, J., Foust, R., Southam, G., 1999. The impact of sediment fecal coliform reservoirs on seasonal water quality in Oak Creek, Arizona. Water Res. 33, 2163–2171.

D'Aoust, J.-Y., 1994. *Salmonella* and the international food trade. Int. J. Food Microbiol. 24, 11–31.

Daly, K., Stewart, C.S., Flint, H.J., Shirazi-Beechey, S.P., 2001. Bacterial diversity within the equine large intestine as revealed by molecular analysis of cloned 16S rRNA genes. FEMS Microbiol. Ecol. 38, 141–151.

Declerck, P., Behets, J., van Hoef, V., Ollevier, F., 2007. Detection of *Legionella* spp. and some of their amoeba hosts in floating biofilms from anthropogenic and natural aquatic environments. Water Res. 41, 3159–3167.

Desmarais, T.R., Solo-Gabriele, H.M., Palmer, C.J., 2002. Influence of soil on fecal indicator organisms in a tidally influenced subtropical environment. Appl. Environ. Microbiol. 68, 1165–1172.

Dick, L.K., Field, K.G., 2004. Rapid estimation of numbers of fecal *Bacteroidetes* by use of a quantitative PCR assay for 16S rRNA genes. Appl. Environ. Microbiol. 70, 5695–5697.

Dick, L.K., Bernhard, A.E., Brodeur, T.J., Santo Domingo, J.W., Simpson, J.M., Walters, S.P., 2005a. Host distributions of uncultivated fecal *Bacteroides* bacteria reveal genetic markers for fecal source identification. Appl. Environ. Microbiol. 71, 3184–3191.

Dick, L.K., Simonich, M.T., Field, K.G., 2005b. Microplate subtractive hybridization to enrich for *Bacteroidales* genetic markers for fecal source identification. Appl. Environ. Microbiol. 71, 3179–3183.

Doyle, J.D., Tunnicliff, B., Kramer, R., Kuehl, R., Brickler, S.K., 1992. Instability of fecal coliform populations in waters and bottom sediments at recreational beaches in Arizona. Water Res. 26, 979–988.

Ekdahl, K., Andersson, Y., 2005. The epidemiology of travel-associated shigellosis—regional risks, seasonality and serogroups. J. Infect. 51, 222–229.

EPA, 1999. Method 1622: *Cryptosporidium* in Water by Filtration/IMS/FA EPA-821-R-99-006. US Environmental Protection Agency, Office of Water, Washington, D.C.

EPA, 2003. Bacterial Water Quality Standards for Recreational Waters (Freshwaters and Marine Waters) Status Report. EPA-823-R-03-008. U.S. Environmental Protection Agency, Office of Water (4305T), Washington, D.C.

EPA, 2008a. Candidate contaminant list microbes: identifying the universe, United States Environmental Protection Agency, Office of Water (4607M) EPA 815-R-08-005, report_ccl3_microbes_universe.pdf

EPA, 2008b. Contaminant candidate list 3 microbes: screening to the PCCL, United States Environmental Protection Agency, Office of Water (4607M) EPA 815-R-08-007, report_ccl3_microbes_screening.pdf

EPA, 2008c. Contaminant candidate list 3 microbes: PCCL to CCL process, United States Environmental Protection Agency, Office of Water (4607M) EPA 815-R-08-007, report_ccl3_microbes_pccl-to-ccl_classification.pdf

Field, K.G., Bernhard, A.E., Brodeur, T.J., 2003. Molecular approaches to microbiological monitoring: fecal source detection. Environ. Monit. Assess. 81, 313–326.

Fontaine, M., Guillot, E., 2002. Development of a TaqMan quantitative PCR assay specific for *Cryptosporidium parvum*. FEMS Microbiol. Lett. 214, 13–17.

Fries, J.S., Trowbridge, J.H., 2003. Flume observations of enhanced fine-particle deposition to permeable sediments. Limnol. Oceanogr. 48, 802–812.

Franks, A.H., Harmsen, H.J.M., Raangs, G.C., Jansen, G.J., Schut, F., Wellsing, G.W., 1998. Variation of bacterial populations in human feces measured by fluorescent in situ hybridization with group-specific 16S rRNA-targeted oligonucelotide probes. Appl. Environ. Microbiol. 65, 3336–3345.

Fukushima, M., Kakinuma, K., Kawaguchi, R., 2002. Phylogenetic analysis of *Salmonella*, *Shigella*, and *E. coli* strains on the basis of the gyrB gene sequence. J. Clin. Microbiol. 40, 2779–2785.

Gopal Rao, V., Padma, M.C., 1971. Some observations on the pathogenicity of strains of *Entamoeba histolytica*. Trans. R. Soc. Trop. Med. Hyg. 65, 606–616.

Grimes, D.J., 1980. Bacteriological water quality effects of hydraulically dredged contaminated upper Mississippi River bottom sediment. Appl. Environ. Microbiol. 39, 782–789.

Guan, S., Xu, R., Chen, S., Odumeru, J., Gyles, C., 2002. Development of a procedure for discriminating among *E. coli* isolates form animal and human sources. Appl. Environ. Microbiol. 68, 2690–2698.

Hagedorn Jr., C., Robinson, S.L., Filtz, J.R., Grubbs, S.M., Angier, T.A., Rebeau, R.B., 1999. Determining sources of fecal pollution in rural Virginia watershed with antibiotic resistance patterns in fecal streptococci. Appl. Environ. Microbiol. 65, 5522.

Hallam, N.B., West, J.R., Forster, C.F., Simms, J., 2001. The potential for biofilm growth in water distribution systems. Water Res. 35, 4063–4071.

Hall-Stoodley, L., Stoodley, P., 2005. Biofilm formation and dispersal and the transmission of human pathogens. Trends Microbiol. 13, 7–10.

Heid, C.A., Stevens, J., Livak, K.J., Williams, P.M., 1996. Real time quantitative PCR. Genome Methods 6, 986–994.

Hein, I., Lehner, A., Rieck, P., Klein, K., Brandl, E., Wagner, M., 2001. Comparison of different approaches to quantify *Staphylococcus aureus* cells by real-time quantitative PCR and application of this technique for examination of cheese. Appl. Environ. Microbiol. 67, 3122–3126.

Higuchi, R., Dollinger, G., Walsh, P.S., Gelfan, D.H., 1991. Simultaneous amplification and detection of specific DNA sequences. Biotechnology 10, 413.

Hold, G.L., Pryde, S.E., Russell, V.J., Furrie, E., Flint, H.J., 2002. Assessment of microbial diversity in human colonic samples by 16S rDNA sequence analysis. FEMS Microbiol. Ecol. 39, 33–39.

Horman, A., Rimhanen-Finne, R., Maunula, L., von Bonsdorff, C.-H., Torvela, N., Heikinheimo, A., 2004. *Campylobacter* spp., *Giardia* spp., *Cryptosporidium* spp., Noroviruses, and indicator organisms in surface water in southwestern Finland, 2000–2001. Appl. Environ. Microbiol. 70, 87–95.

Huijsdens, X.W., Linskens, R.K., Mak, M., Meuwissen, S.G.M., Vandenbroucke-Grauls, C.M.J.E., Savelkoul, P.H.M., 2002. Quantification of bacterial adherent to gastrointestinal mucosa by real-time PCR. J. Clin. Microbiol. 40, 4423–4427.

Izzo, G., Tosti, E., Volterra, L., 1983. Fecal contamination of marine sediments in a stretch of the Gulf of Naples. Water Air Soil Pollut. 20, 191–198.

Kandulski, A., Selgrad, M., Malfertheiner, P., 2008. *Helicobacter pylori* infection: a clinical overview. Dig. Liver Dis. 40, 619–626.

Lang, L., 2003. Acute gastroenteritis outbreaks on cruise ships linked to Norwalk-like viruses. Gastroenterology 124, 284–285.

Lappalainen, M., Chen, R.W., Maunula, L., von Bonsdorff, C.-H., Plyusnin, A., Vaheri, A., 2001. Molecular epidemiology of viral pathogens and tracing of transmission routes: hepatitis-, calici- and hantaviruses. J. Clin. Virol. 21, 177–185.

Leser, T.D., Amenuvor, J.Z., Jensen, T.K., Lindecrona, R.H., Boye, M., Moller, K., 2002. Culture-independent analysis of gut bacteria: the pig gastrointestinal tract microbiota revisited. Appl. Environ. Microbiol. 68, 673–690.

Lewis, G.D., Austin, F.J., Loutit, M.W., 1986. Enteroviruses of human origin and faecal coliforms in river water and sediments down stream from a sewage outfall in the Taieri River, Otago. N. Z. J. Mar. Freshwater Res. 20, 101–105.

Leung, K.T., Mackereth, R., Tien, Y.-C., Topp, E., 2004. A comparison of AFLP and ERIC-PCR analyses for discriminating *Escherichia coli* from cattle, pig and human sources. FEMS Microbiol. Ecol. 47, 111–119.

Lipp, E.K., Kurz, R., Vincent, R., Rodriguez-Palacios, C., Farrah, S.R., Rose, J.B., 2001. The effects of seasonal variability and weather on microbial fecal pollution and enteric pathogens in a subtropical estuary. Estuaries 24, 266–276.

Lipp, E.K., Jarrell, J.L., Griffin, D.W., Lukasik, J., Jacukiewicz, J., Rose, J.B., 2002. Preliminary evidence for human fecal contamination in corals of the Florida Keys, USA. Mar. Pollut. Bull. 44, 666–670.

Lofstrom, C., Knutsson, R., Axelsson, C.E., Radstrom, P., 2004. Rapid detection of *Salmonella* spp. in animal feed samples by PCR after culture enrichment. Appl. Environ. Microbiol. 70, 69–75.

Lubeck, P.S., Wolffs, P., On, S.L.W., Ahrens, P., Radstrom, P., Hoorfar, J., 2003a. Toward an international standards for PCR-based detection of food-borne thermotolerant Campylobacters: assay development and analytical validation. Appl. Environ. Microbiol. 69, 5664–5669.

Lubeck, P.S., Cook, N., Wagner, M., Fach, P., Hoorfar, J., 2003b. Toward an international standard for PCR-based detection of food-borne thermotolerant *Campylobacters*: validation in a multi-center collaborative trial. Appl. Environ. Microbiol. 69, 5670–5672.

MacKenzie, W.R., Hoxie, N.J., Proctor, M.E., Gradus, M.S., Blair, K., Peterson, D.E., et al., 1994. A massive outbreak in Milwaukee of *Cryptosporidium* infection transmitted through the public water supply. New England J. Med. 331, 161–167.

Malorny, B., Hoorfar, J., Bunge, C., Helmuth, R., 2003. Multicenter validation of the analytical accuracy of *Salmonella* PCR: towards an international standard. Appl. Environ. Microbiol. 69, 290–296.

Malinen, E., Kassinen, A., Rinttila, T., Palva, A., 2003. Comparison of real-time PCR with SYBR Green 1 or 50-nuclease assays and dot-blot hybridization with rDNA-targeted oligonucleotide probes in quantification of selected fecal bacteria. Microbiology 149, 269–277.

Mandaville, S.M., 2002. Bacterial source tracking (BST)—a review. Project H-2, Soil & Water Conservation Society of Metro Halifax. 46 p. Appendices A to T. http://lakes.chebucto.ort/H-2/bst.html

Matsuki, T., Watanabe, K., Fujimoto, J., Miyamoto, Y., Takada, T., Matsumoto, K., et al., 2002. Development of 16S rRNA-gene-targeted group-specific primers for the detection and identification of predominant bacteria in human feces. Appl. Environ. Microbiol. 68, 5445–5451.

Matsuki, T., Watanabe, K., Fujimoto, J., Kado, Y., Takada, T., Matsumoto, K., et al., 2004. Quantitative PCR with 16S rRNA-gene-targeted species-specific primers for analysis of human intestinal bifidobacteria. Appl. Environ. Microbiol. 70, 167–173.

McCorquodale, D., Burney, C., Row, M., 1996. Indicators for determining the sources and extent of fecal contamination in coastal waters: an annotated bibliography. Technical Report 96-06. Broward County, Department of Natural Resources Protection, Fort Lauderdale, FL.

Mead, P.S., Griffin, P.M., 1998. *Escherichia coli* O157:H7. Lancet 352, 1207–1212.

Meays, C.L., Broersma, K.L., Nordin, R., Mazumder, A., 2004. Source tracking fecal bacteria in water: a critical review of current methods. J. Environ. Manage. 73, 71–79.

MMWR, 1996. Lake-associated outbreak of *Escherichia coli* O157:H7I—Ilinois, 1995. Mortality and morbidity weekly report, May 31, 1996, 45(21), 437–439. Centers for Disease Control and Prevention, Atlanta, GA.

Novga, H.K., Rudi, K., Naterstad, K., Holck, A., Lillehaug, D., 2000. Application of 50-nuclease PCR for quantitative detection of *Listeria monocytogenes* in pure cultures, water, skim milk, and unpasteurized whole milk. Appl. Environ. Microbiol. 66, 4266–4271.

Obiri-Danso, K., Jones, K., 1999. Distribution and seasonality of microbial indicators and thermophilic campylobacters in two freshwater bathing sites on the River Lune in northwest England. J. Appl. Microbiol. 87, 822–832.

Okabe, S., Okayama, N., Savichtcheva, O., Ito, T., 2007. Quantification of host-specific *Bacteroides–Prevotella* 16S rRNA genetic markers for assessment of fecal pollution in freshwater. Appl. Microbiol. Biotechnol. 74, 890–901.

O'Malley, M.L., Lear, D.W., Adams, W.N., Gaines, J., Sawyer, T.K., Lewis, E.J., 1982. Microbial contamination of continental shelf sediments by wastewater. J. Water Pollut. Control Fed. 54, 1311–1317.

Palmer, M.D., 2000. Analyses of sediment bacteria monitoring data from two deep ocean raw wastewater outfalls, Victoria, BC. Can. Water Resour. J. 25, 1–18.

Panicker, G., Myers, M.L., Bej, A.K., 2004. Rapid detection of *Vibrio vulnificus* in shellfish and gulf of Mexico water by real-time PCR. Appl. Environ. Microbiol. 70, 498–507.

Park, S.R., Mackay, W.G., Reid, D.C., 2001. *Helicobacter* sp. recovered from drinking water biofilm sampled from a water distribution system. Water Res. 35, 1624–1626.

Parveen, S., Portier, K.M., Robinson, K., Edmiston, L., Tamplin, M.L., 1999. Discriminant analysis of ribotype profiles of *Escherichia coli* for differentiating human and nonhuman sources of fecal pollution. Appl. Environ. Microbiol. 65, 3142–3147.

Pettibone, G.W., Irvine, K.N., Monahan, K.M., 1996. Impact of a ship passage on bacteria levels and suspended sediment characteristics in the Buffalo River, New York. Water Res. 30, 2517–2521.

Peyrat, M.B., Soumet, C., Maris, P., Sanders, P., 2008. Recovery of *Campylobacter jejuni* from surfaces of poultry slaughterhouses after cleaning and disinfection procedures: analysis of a potential source of carcass contamination. Int. J. Food Microbiol. 124, 188–194.

Preston, T.M., 2003. The water–air interface: a microhabitat for amoebae. Eur. J. Protistol. 39, 385–389.

Rabsch, W., Tschäpe, H., Bäumler, A.J., 2001. Non-typhoidal salmonellosis: emerging problems. Microbes Infect. 3, 237–247.

Reidl, J., Klose, K.E., 2002. *Vibrio cholerae* and cholera: out of the water and into the host. FEMS Microbiol. Rev. 26, 125–139.

Rusch, A., Huettel, M., 2000. Advective particle transport into permeable sediments—evidence from experiments in an intertidal sandflat. Limnol. Oceanogr. 45, 525–533.

Sack, D.A., Sack, R.B., Nair, G.B., Siddique, A.K., 2004. Cholera. Lancet 363, 223–233.

Sankararamakrishnan, N., Guo, O., 2005. Chemical tracers as indicator of human fecal coliforms at storm water outfalls. Environ. Int. 31, 1133–1140.

Santoro, A.E., Boehm, A.B., 2007. Frequent occurrence of the human-specific *Bacteroides* fecal marker at an open coast marine beach: relationship to waves, tides and traditional indicators. Environ. Microbiol. 9, 2038–2049.

Sargeant, D., 1999. Fecal contamination source identification methods in surface water. Department of Ecology Report #99–345. Washington State Department of Ecology.

Sawyer, T.K., 1980. Marine amoebae from clean and stressed bottom sediments of the Atlantic Ocean and Gulf of Mexico. J. Protozool. 27, 13.

Sawyer, T.K., Nerad, T.A., Cahoon, L.B., Nearhoof, J.E., 1998. *Learamoeba waccamawensis*, n.g., n.sp. (Heterolobosea: Vahlkampfiidae), a new temperature tolerant cyst-forming soil amoeba. J. Eukaryot. Microbiol. 45, 260–264.

Schuster, F.L., Visvesvara, G.S., 2004. Amebae and ciliated protozoa as causal agents of water-borne zoonotic disease. Vet. Parasitol. 126, 91–120.

Scott, T.M., Rose, J.B., Jenkins, T.M., Farrah, S.R., Lukasik, J., 2002. Microbial source tracking: current methodology and future directions. Appl. Environ. Microbiol. 68, 5796–5803.

Seurinck, S., Deforidt, T., Verstraete, W., Siciliano, S.D., 2004. Detection and quantification of the human-specific HF183 *Bacteroides* 16S rRNA genetic marker with real-time PCR for assessment of human faecal pollution in freshwater. Environ. Microbiol. 7, 249–259.

Shanks, O.C., Atikovic, E., Blackwood, D., Lu, J., Noble, R.T., Santo Domingo, J., et al., 2008. Quantitative PCR for detection and enumeration of genetic markers of bovine fecal pollution. Appl. Environ. Microbiol. 74, 745–752.

Simmons, G., Jury, S., Thornley, C., Harte, D., Mohiuddin, J., Taylor, M., 2008. A Legionnaires' disease outbreak: a water blaster and roof-collected rainwater systems. Water Res. 42, 1449–1458.

Simpson, J.M., Santo Domingo, J.W., Reasoner, D.J., 2002. Microbial source tracking: state of the science. Environ. Sci. Technol. 24, 279–5288.

Solo-Gabriele, H.M., Wolfert, M.A., Desmarais, T.R., Palmer, C.J., 2000. Sources of Escherichia coli in a coastal subtropical environment. Appl. Environ. Microbiol. 66, 230–237.

Suarez, D.L., 2000. Evolution of avian influenza viruses. Vet. Microbiol. 74, 15–27.

Taylor, L.H., Latham, S.M., Woolhouse, M.E., 2001. Risk factors for human disease emergence. Phil. Trans. R. Soc. Lond. B. Biol. Sci. 356, 983–989.

Valiela, I., Alber, M., LaMontagne, M., 1991. Fecal coliform loadings and stocks in Buttermilk Bay, Massachusetts, USA, and management implications. Environ. Manage. 15, 659–674.

Van Donsel, D.J., Geldreich, E.E., 1971. Relationships of Salmonellae to fecal coliforms in bottom sediments. Water Res. 5, 1079–1087.

Wellinghausen, N., Frost, C., Marre, R., 2001. Detection of Legionellae in hospital water samples by quantitative real-time LightCycler PCR. Appl. Environ. Microbiol. 67, 3985–3993.

Whitman, R.L., Nevers, M.B., 2003. Foreshore sand as a source of *Escherichia coli* in nearshore water of a Lake Michigan beach. Appl. Environ. Microbiol. 69, 5555–5562.

Wiggins, B.A., Andrews, R.W., Conway, R.A., Corr, C.L., Dobratz, E.J., Dougherty, D.P., et al., 1999. Identification of sources of fecal pollution using discriminant analysis: Supporting evidence from large datasets. Appl. Environ. Microbiol. 65, 3483–3486.

Monitoring Inorganic Compounds

Pamela Heckel
College of Medicine, Department of Environmental Health, University of Cincinnati, Cincinnati, OH 45267-0056, USA

Tracy Dombek
NADP, Illinois State Water Survey, 2204 Griffith Drive, Champaign, IL 61820-7495, USA

Handbook of Water Purity and Quality
Copyright © 2009 by Elsevier Inc. All rights of reproduction in any form reserved.

OVERVIEW

The hydrological cycle describes the path of a water droplet from the time it falls to the ground until it evaporates and returns to our atmosphere (Purdue University, 2008). The difference in density between moist air and dry air allows moist air to rise through the troposphere until it reaches buoyant equilibrium. Microscopic particles of water suspended in our gaseous atmosphere bind to other particles called cloud condensation nuclei, attract water molecules to form clusters, and eventually form precipitation. Water precipitation includes rain, snow, sleet, and hail. When the cluster falls during a precipitation event, it collides with other atmospheric aerosols and removes them from the air. This process, called scavenging, is one way that inorganic compounds enter the water supply.

When rain hits Earth, some soaks into the ground and becomes available for plants. Some percolates through the soil to the groundwater table. Rainwater also flows overland as runoff into streams, rivers, lakes, and even the ocean. Fresh surface water includes flowing water such as streams and rivers and still water such as ponds. Water in the ocean contains ionic species; therefore, it is called salt water. Groundwater refers to all the water hidden in the ground. It may contribute to soil moisture or may be flowing through an aquifer. Artesian wells tap into groundwater trapped between two impermeable layers. Unconfined aquifers flow through deposits of rock, pebbles, sand, and other types of porous media. Humans and other animals consume both surface water and ground water.

The focus of this chapter is inorganic substances in surface water that must be monitored to ensure that it is suitable for drinking, i.e., it is potable (see later). In the United States, potable water, regardless of its source, must be cleaner than the maximum contaminant levels (MCLs) mandated by local, state (USEPA, 2008 "Local"), and federal guidelines (USEPA, 2008 "National") to protect human health. The United States Environmental Protection Agency (USEPA) not only enforces the guidelines but also is required to help communities establish wastewater treatment facilities to ensure compliance (USEPA, 2008 "Municipal"). These regulations specify the allowable concentration of microorganisms, disinfectants, and disinfection by-products (see Chapters 8 and 12), inorganic chemicals (discussed in this chapter), organic chemicals (see throughout), and radionuclides (see Chapter 10). Secondary contaminants (USEPA, 2008 "Secondary") such as iron and sulfur affect the smell, taste, or color of the water but do not cause illness. Some chemicals, such as the gasoline oxygenate methyl-*t*-butyl ether (MTBE), that are suspected to cause harm have not been included in the regulations (USEPA, 2008 "Unregulated"). Organic compounds in wastewater originate from sewerage, industrial processes, and the decomposition of living things. Inorganic compounds in wastewater originate from natural and anthropogenic sources.

TECHNIQUES USED TO IDENTIFY AND QUANTIFY INORGANIC CONSTITUENTS IN WATER

Surface water and groundwater contain many different contaminants. This chapter considers inorganic species that must be removed during the purification process. Some inorganic constituents, such as sulfur dioxide and mercury chloride, enter the air, and eventually the water cycle, during the fossil fuel combustion process. Nutrients, gases, and inorganic nonmetallic and metallic constituents in surface water originate in industrial processes, fertilizers, and natural mineral deposits. Chlorides leach from rocks and soils. The hydrogenion (pH) concentration affects the dissociation of minerals and the biological availability of these and other inorganic chemicals.

Table 1 shows the average freshwater concentrations of various inorganic compounds, the MCL in drinking water and the instruments which can detect them (Smith et al., 1987). Both the speciation and the concentration of an element determine whether that chemical is beneficial or toxic. For example, free metal ions, which are toxic to aquatic species, form nontoxic complexes with water. Some elements such as chromium (Cr^{3+}) are necessary nutrients for all

TABLE 1 Chemical Species in River Water

Base element	Species	Freshwater concentration	Biological nutrient	Priority pollutant	MCL (mg/L)	Instrument to detect
Hydrogen	H^+	7.8				pH meter
Lithium	Li^+	Trace				AAS
Beryllium	$BeOH^+$	Trace			2.000	AAS
Carbon	HCO_3^-	2.7				TOC
Nitrogen	NO_3^- NH_4^+	4.5			10.000	IC, UV-Vis
Oxygen	O_2	3.5				
Fluorine	F^-	5.3			4.000	IC
Sodium	Na^+	3.1	Yes			AAS
Manganese	Mg^{2+}	3.3	Yes			AAS
Aluminum	$Al(OH)_3$	6.0				AAS
Silicon	H_4SiO_4	4.5				AAS
Phosphorus	HPO_4^{2-}	5.4				IC, UV-Vis
Sulfur	SO_4^{2-}	3.4				IC
Chlorine	Cl^-	3.4				IC, UV-Vis
Potassium	K^+	4.1	Yes			AAS, ICP-OES
Calcium	Ca^{2+}	3.0	Yes			AAS
Chromium	$Cr^{6+,3+}$	6.7	Yes	Yes	0.100	AAS

(Continued)

TABLE 1 (Continued)

Base element	Species	Freshwater concentration	Biological nutrient	Priority pollutant	MCL (mg/L)	Instrument to detect
Manganese	$Mn^{4+,2+}$	6.4	Yes			AAS
Iron	$Fe^{3+,2+}$	6.0	Yes			AAS, UV-Vis
Cobalt	Co^{2+}	Trace	Yes			AAS
Nickel	Ni^{2+}	7.3	Yes			AAS
Copper	$Cu^{2+,+}$	7.0	Yes			AAS
Zinc	Zn^{2+}	6.6	Yes			AAS
Arsenic	$HAsO_4^{2-}$	7.9		Yes	0.010	AAS, ICP-OES
Selenium	SeO_3^{2-}	8.6	Yes	Yes	0.050	AAS
Bromine	Br^-	5.9				IC
Strontium	Sr^{2+}	Trace				AAS
Cadmium	Cd^{2+}	8.1			0.005	AAS, ICP-OES
Tin	Sn^{2+}	Trace				AAS
Iodine	I^-, IO_3^-	Trace				IC
Cesium	Cs^+	Trace				AAS
Barium	Ba^{2+}	6.0		Yes	2.000	AAS, ICP-OES
Mercury	$Hg(OH)_2^+$	8.0		Yes	0.002	AAS
Lead	Pb^{2+}, Pb^+	7.7		Yes	0.015	AAS, ICP-OES

biological organisms; yet hexavalent chromium is a carcinogenic priority pollutant (Crites and Tchobanoglous, 1998). The MCL for nitrate is 10.0 mg/L, whereas that for nitrite is only 1.0 mg/L. The MCL for arsenic has been lowered to 0.01 mg/L, according to the *Federal Register* 68(57), 14501-145, March 25, 2003. Some communities add fluoride to their drinking water to protect teeth from decay; however, the MCL for fluoride is 4.0 mg/L. The United States Department of Agriculture recommends 70 mg Selenium (Se) per day. A dose of 800 mg Se/day has toxic effects. Thus, Se is both a biological nutrient and a priority pollutant.

Clearly, the regulation and monitoring of chemical species in drinking water is essential to protect public health. Municipal drinking water treatment facilities monitor the contaminants listed in Table 1, where MCL is the level permitted by the USEPA. The agency has also set maximum contaminant level goals (MCLG) for many substances, reflecting the concentration of a substance for which there is no known adverse health effect.

Under the provisions of the Clean Water Act, the USEPA has developed water quality standards and provides training for state and tribal entities. The *Water Quality Standards Handbook: Second Edition* (EPA-823-B-94-005) issued in August 1994 and updated in part in June 2007 is available in its entirety on the EPA Web site (http://www.epa.gov/waterscience/standards/handbook/index.html). A complete list of USEPA analytical methods approved for drinking water compliance monitoring is available at http://www.epa.gov/safewater/methods/pdfs/methods/methods_inorganic.pdf. The USEPA has staff available to assist local agencies with wastewater treatment facilities and procedures. For more information, visit http://www.epa.gov/safewater/smallsystems/index.html.

The following sections describe instruments used to identify various chemical species in water. The manufacturer may specify a QC/QA protocol for the instrument. These procedures ensure that the instrument functions as intended. All instruments are to be calibrated prior to use. Analyzer drift is to be checked during the analysis using industry standard samples and blanks. Most procedures also specify the minumum number of specimens, typically three or more, which are required to establish the representative levels of a substance in a sample. The use of these standards, procedures and protocols is essential.

Atomic Absorption Spectrophotometry

Atomic absorption spectrophotometry (AAS) is used to quantify elements based on the amount of light that they absorb. The analysis requires destruction of the sample, usually through acid digestion. For flame atomization, the resulting solution is nebulized to form fine droplets that are sprayed into the flame. For graphite furnace atomization, a small volume ($50\,\mu L$) is placed in a small carbon tube that is heated by the passage of an electric current. In both cases, a complex series of physical and chemical processes occur to produce free gaseous atoms in the light path of the spectrometer. The amount of light absorbed is proportional to the concentration of the element in the solution. Flame AAS has detection limits at the parts-per-million level. Elements identified in Figure 1 can be determined by AAS (Sorial, 2005).

The Basics of AAS

Every element absorbs and emits a unique set of wavelengths of light. The spectrometer contains a hollow cathode lamp that emits light at the wavelength of interest. The light is directed through the gaseous vapor of the element where the atoms absorb energy from the incident light beam. A monochromator filters the incident light so that a narrow range of wavelengths pass through to the detector that measures the unabsorbed light and light emitted by the hot atoms. The light

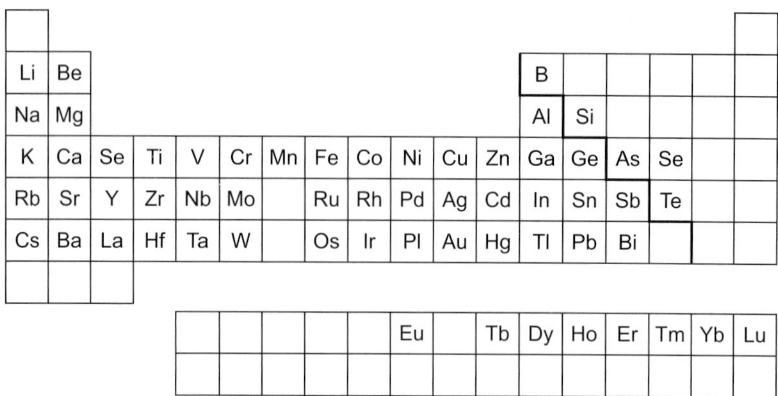

FIGURE 1   Elements that can be determined by AAS.

source emission is modulated and the detector electronics are locked in to this frequency so that only changes in the light intensity of the source are monitored. This provides a measure of the concentration of the gaseous metal vapor.

Absorption and Emission of Light

By definition, an atom at its ground state has its electrons in their lowest energy orbital configuration. When an atom is excited, an electron moves into a higher energy orbital. Light can excite atoms. Electrons may release light at a longer wavelength if they return to the ground state in more than one step. The difference between the ground and excited states is called the energy gap (E). The frequency (ν) equals the speed of light (c) divided by the wavelength (λ) Plank's constant (h) is used to relate energy to frequency. $E = h\gamma = hc/\lambda$

The Lambert–Beer law describes transmittance (T) as the relationship between light that travels parallel to an aerosol without being absorbed (I) to the incident light (I_0), which is the light emitted by the source (Hinds, 1999). Thus, transmittance is defined as the ratio of the transmitted intensity (I) to the incident intensity (I_0), $T = I/I_0$.

Relating Absorbed Light to Concentration: Beer's Law

Absorbance (A) of a sample is related to concentration as follows:

$A = -\log T$, where T is the transmittance. ($T = I/I_0$ as previously explained.)

$A = \varepsilon bc$, where ε is an experimentally determined constant for each species and instrument.

The length the light travels through the sample (about 10 cm for flame atomization) is given by b, and c is the concentration of the substance in the original solution. Each time the instrument is used, it must be calibrated against known standards. The slope obtained from plotting absorbance vs. concentration is εb.

Cold Vapor Atomic Absorption Spectrometry

Mercury is one of the few elements that has a monatomic vapor pressure at room temperature that is sufficiently high to allow determination by AAS at this "cold" temperature. Chemical reduction of mercury ions in solution with stannous chloride or sodium borohydride converts the species of interest to a mercury vapor. An aerator separates the gaseous atoms into the gas stream that passes through the optical cell. There are hundreds of applications for this technology in many different fields. For example, the USEPA has promulgated several cold vapor AAS (CVAAS) procedures to detect mercury in fish.

Flame Atomic Absorption Spectrometry

Flame atomic absorption spectrometry (FAAS) is used to analyze solutions containing dissolved metallic salts. A nebulizer generates an aerosol that mixes with an oxidant gas such as air and fuel such as acetylene.

Consider a solution of NaCl. The nebulization step creates an aerosol containing ionic species Na^+ and Cl^-. The heat from the flame evaporates the water, creating a salt NaCl. Additional heat melts the salt. Further heating vaporizes the NaCl liquid, and the resulting gaseous NaCl molecules decompose creating Na^0 and Cl^0. Light from a sodium hollow cathode lamp excites the sodium atoms and is absorbed. Flame AAS can measure many elements at the single-digit ppm (mg/L) level.

Gas Chromatography

Gas chromatography (GC) separates volatile sample components based on their relative affinities for two phases. One of these is the mobile phase, a gas, and the other, the stationary phase, is the coating on the column through which the mobile phase is passing. GC can be used to identify halogenated pesticides and organomercuric compounds in drinking water, but it is not suitable for ionic species or salts. Typically, for the analysis of waters, the analytes are extracted into a suitable solvent before injection into the chromatograph. One microliter of a solution containing the chemical compounds of interest is heated to 300°C in the injection port, causing the solution to become a gas as it enters the column. The sample travels through the column in an inert gas. Helium, which is both inert and highly diffusive, is often used. The column coating adsorbs the sample compounds. The affinity of each compound in the sample for the column coating and chemical volatility determine its residence time inside the column and the choice of column. More information on chromatography is available at http://www.chromatography-online.org/resources.html.

Gas Chromatography-Mass Spectrometry

There are several detector devices available for GC, and the mass spectrometer is probably the most widely used. The typical configuration is electron

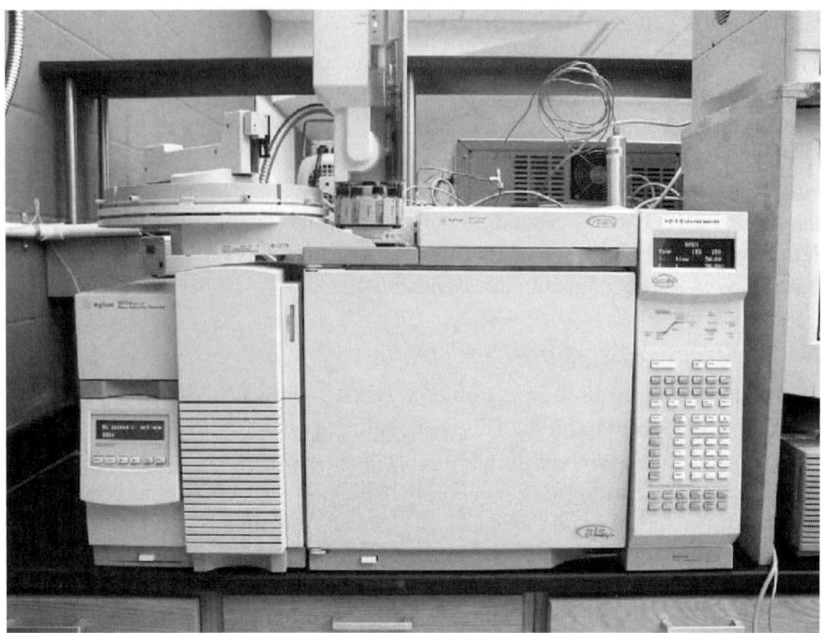

FIGURE 2   GC-MS equipment.

impact ionization with a quadrupole mass analyzer, though other configurations have recently been introduced. As a mass spectrum of a compound is a unique fingerprint, qualitative analysis is possible by matching the spectrum of an unknown compound with those of authentic compounds stored in a library in the computer's memory. An example of the typical GC-MS combination is the Agilent 6840 series gas chromatograph, and the Agilent 5973 network mass selective detector shown in Figure 2.

High Performance Liquid Chromatography

High performance liquid chromatography (HPLC) is used to separate the components in a sample solution based on their relative affinities for the liquid mobile phase and the solid stationary phase. Separated components are quantified by measurement in a flow-through detector mounted as close to the exit of the column as possible. Mass spectrometry, although more difficult to interface with LC than with GC, can be used for qualitative analysis. Each substance produces a unique pattern of peaks.

Basic Hardware

A typical HPLC system contains a solvent reservoir, a pump, an injector, a column, a detector, and a waste reservoir. A high-pressure pump pushes the solvent into the system at pressures of 2,000–6,000 psi. The injector delivers a small volume (20 μL) of the sample solution as a discrete plug, which then mixes with the solvent and travels through the column. Compounds in the sample are partitioned between the column and the solvent and are eventually washed into the detector flow cell where concentration is measured.

Ion Chromatography

A special type of HPLC in which the separation is predominantly by ion-exchange is known as ion chromatography (IC). IC technology involves introducing a sample into an eluent stream via a sample loop or syringe and pumping that stream through a guard column and separator column. Analytes are detected, using a conductivity detector, and a suppressor is utilized to lower eluent background conductivity. The order in which ions are retained and elute from the column are dependent upon their respective charges and sizes. This instrument can detect cations such as lithium, sodium, ammonium, potassium, calcium, and magnesium at the parts-per-billion (ppb) level. Anions in drinking water including bromide, chloride, fluoride, nitrate, nitrite, ortho-phosphate-P, sulfate, disinfection by-products including bromate, bromide, chlorite, and chlorate can also be detected. Various detectors may be used with IC, though those based on conductivity for the quantification of anions are probably the most widely used. Dionex Corporation recommends IC to identify "Oxyhalides and Bromate, Perchlorate, Haloacetic Acids, Selenium and Arsenic, Hexavalent Chromium, Cyanide and Metal Cyanides, Pesticides and Herbicides, Phenols, Ammonia and Amines, Metals, and Organic Acids" (http://www1.dionex.com/en-us/Environmental/Water_Analysis/lp40893.html) Lachat Instruments recommends IC for the analysis of brackish waters. (http://www.lachatinstruments.com/applications/AppsSearch.asp).

Inductively Coupled Plasma-Optical Emission Spectrometry

Inductively coupled plasma-optical emission spectrometry (ICP-OES) is a spectrophotometric technique that can be used to quantify contaminants in drinking water. ICP-OES is capable of analyzing many elements simultaneously using multiple wavelengths and calibration ranges. An ICP-OES system consists of a sample introduction system, an excitation source, a spectrometer, and a detection system. The ICP energy source is a pulsed alternating radio-frequency magnetic field that interacts with electrons seeded into a carrier gas such as argon, causing them to move in circular paths with sufficient kinetic energy to cause collsional ionization. The sample is introduced into the instrument using an autosampler and peristaltic pump. The peristaltic pump has three channels that allow for the mixing of the sample stream with standards or ionization suppressants. The combined sample moves into a nebulizer where it is mixed with gaseous argon to form an aerosol. The aerosol mixture moves into the spray chamber and ultimately through a plasma stream at approximately 6,000 K. As the aerosol evaporates and vaporizes, molecules dissociate into atoms. Electrons associated with the atoms are excited to higher energy levels and their subsequent return to the ground state produces photons, which are measured by the spectrometer using a charge-coupled device.

The instrument manufacturer can provide a general analytical method for the instrument when it is installed for aqueous analyses. ICP-OES has been used to quantify antimony, arsenic, barium, cadmium, chromium, lead, and selenium with detection limits in the range of 0.2–100 ppb.

UV–Vis Absorption Spectroscopy

UV–Vis absorption spectroscopy measures the attenuation of a beam of light as it passes through a sample. Each substance absorbs radiation at different wavelengths. The concentration of an analyte in a solution can be determined by measuring the absorbance at some particular wavelength and applying Beer's law.

$$A = \varepsilon bc$$

where A: measured absorbance

b: path length

ε: wavelength-dependent molar absorptivity

c: analyte concentration

The major instrumental components of the UV–Vis spectrophotometer are as follows:

1. Source of light (UV or visible range)
2. Monochromator i.e., wavelength selector
3. Sample container
4. Detector and signal readout

Flow Injection Analysis

Flow injection analysis (FIA) is a technique used for the addition of reagents to sample solution. A small volume of sample (100 μL) is injected into a carrier stream flowing in a narrow bore (0.8 mm i.d.) nonwettable tubing and merged with reagent solution at a confluence point. After flowing through a reaction coil (100 cm), the product is measured in a flow-through cell. Automated, serial sampling reduces the amount of time required in the laboratory to perform various spectrophotometric determinations. Flow injection instruments with either a UV–Vis or a Vis–NIR spectrophotometer detector can be used to detect ammonia, chloride, iron, nitrate, nitrite, and phosphate according to USEPA promulgated methods.

METHODS TO IDENTIFY TYPICAL INORGANIC COMPOUNDS IN WATER

USEPA maintains an extensive listing of promulgated methods and applications. This section is arranged by chemical species of interest. More methods are available at http://www.epa.gov/safewater/mdbp/implement.html

Arsenic

Technologies and costs for the removal of arsenic from drinking water 10-17-2000 arsenic in drinking water (http://www.epa.gov/safewater/arsenic/pdfs/techcosts.pdf).

Arsenic mitigation strategies (http://www.epa.gov/safewater/arsenic/pdfs/arsenic_training_2002/train5-mitigation.pdf).

Copper

Revised guidance manual for selecting lead and copper control strategies: provides guidance for selecting lead and copper control strategies (http://www.epa.gov/safewater/lcrmr/pdfs/guidance_lcmr_control_stratageis_revised.pdf).

Disinfection By-products

The disinfection by-products are TTHM/HAAS, chlorite, chlorine dioxide, and bromate.

Stage 1 disinfectants and disinfection by-products rule: Laboratory quick reference guide (EPA 816-F-02-021) (http://www.epa.gov/safewater/publicoutreach/quickreferenceguides.html).

Method 317.0: Determination of inorganic oxyhalide disinfection by-products in drinking water uses ion chromatography with the addition of a postcolumn reagent for trace bromate analysis: This method is used to quantify inorganic oxyhalide disinfection by-products (http://www.epa.gov/safewater/methods/pdfs/met317rev2.pdf).

Method 326.0: Revision 1.0 June 2002-EPA document # EPA 815-R-03-00702-06-2003: The concentration of bromate, bromide, chlorate, and chlorides can by determined by IC (http://www.epa.gov/safewater/methods/pdfs/met326_0.pdf).

Fluoride

Determination of inorganic anions in drinking water by IC: the presence of fluoride can be determined by IC (http://www.epa.gov/safewater/methods/pdfs/met300.pdf).

Iron

Arsenic removal from drinking water by iron removal plants EPA 600/R-00/086/Risk Management Research/Risk Management Research/USEPA: Iron can be used to remove arsenic from drinking water (http://www.epa.gov/nrmrl/pubs/600r00086/600r00086.htm).

Lead

Revised guidance manual for selecting lead and copper control strategies: the USEPA procedure for lead and copper control strategies (http://www.epa.gov/safewater/lcrmr/pdfs/guidance_lcmr_control_stratageis_revised.pdf).

Manganese

Drinking water health advisory for manganese communicates health risks and avoidance strategies for communities with manganese in their drinking water (http://www.epa.gov/safewater/ccl/pdfs/reg_determine1/support_cc1_magnese_dwreport.pdf).

Mercury

USEPA removal of chemical contaminants in drinking water EcoWater Systems Incorporated ERO-R450E drinking water treatment system—environmental technology verification report: this report describes the use of the EcoWater Systems Incorporated ERO-R450E drinking water treatment system (http://www.epa.gov/etv/pubs/600r05122.pdf).

USEPA removal of chemical contaminants in drinking water Kinetico Inc. Pall/Kinetico Purefecta drinking water treatment system—environmental technology verification report: Pall/Kinetico Purefecta POU is a drinking water treatment system (http://www.epa.gov/etv/pubs/600r05108.pdf).

USEPA environmental technology verification report removal of chemical contaminants in drinking water Watts Premier Inc., WP-4 V drinking water treatment system: the Watts Premier WP-4 V POU drinking water treatment system removes mercury, cesium, cadmium, and other elements (http://www.epa.gov/etv/pubs/600r06005.pdf).

USEPA ETV removal of chemical and microbial contaminants in drinking water Watts Premier, Inc. M-2400 point-of-entry reverse osmosis drinking water treatment system: the Water Watts reverse osmosis system removes mercury (http://www.epa.gov/etv/pubs/600r06101.pdf).

Nitrate

G:MSEMwtp73.PDF02-08-1999 final report—photo-assisted electron transfer reactions of application to mine wastewater cleanup: nitrate and cyanide mine waste technology program activity iv, project 3: this procedure can be used to clean wastewaters from mining operations (http://www.epa.gov/hardrockmining/a4/a4p3.pdf[PDF]).

Sodium

Contaminant candidate list regulatory determination support document for sodium: sodium is not regulated as a contaminant (http://www.epa.gov/safewater/ccl/pdfs/reg_determine1/support_cc1_sodium_cc1regdet.pdf).

Sulfate

Data quality control methods manual: it gives consumer advice concerning the tolerable levels of sulfate in drinking water (http://www.epa.gov/safewater/ccl/pdfs/reg_determine1/support_cc1_sulfate_dwreport.pdf).

LONG-TERM MONITORING NETWORK FOR ATMOSPHERIC DEPOSITION

The National Atmospheric Deposition Program/National Trends Network (NADP/NTN) is a nationwide network of sites that has collect wet deposition samples since 1978 to provide a long-term monitoring system for atmospheric deposition. The network includes over 250 sites located throughout the United States. Weekly wet deposition samples sent to the Central Analytical Laboratory (CAL) located at the Illinois State Water Survey are analyzed for pH, conductance, sodium, potassium, calcium, magnesium, ammonium, orthophosphate, sulfate, nitrate, and chloride. Samples are filtered through 0.45 μm polyethersulfone filters prior to analysis. Once samples are filtered, they are analyzed by IC for sulfate, nitrate, and chloride; by ICP-OES for sodium, potassium, calcium, and magnesium; and by FIA for ammonium and orthophosphate.

The quality of the data is extremely important. A quality assurance (QA) plan contains basic guidelines (Rothert et al., 2002) for data quality. Standard operating procedures (SOPs) are followed for each instrument and each analysis. Bias and interference from reagent solutions is critical; thus only deionized (DI) water with a specific resistance of 18.0 megohms cm or better can be used to prepare reagents and standards and check solutions. DI water is also used for all supply cleaning and rinsing procedures. New calibration curves are prepared daily. The calibration is verified with two in-house prepared quality check (QC) solutions. All calibration curves must produce $r^2 \geq 0.999$. QC solutions are prepared annually by a QA chemist. These solutions, along with a low standard and a high standard, are used throughout the analytical run to verify data. All data must be bracketed by two QC samples falling within acceptable limits, which are analyzed before and after every 12 samples. If QC checks fail, the samples are reanalyzed. Blank samples are analyzed at the beginning and end of analyses to detect background contamination and instrument drift. In addition to standard checks, the QA chemist submits samples that are "blind" to the analyst (i.e., the concentration and the sample type are unknown to the analyst).

Chloride, *nitrate*, and *sulfate* are measured on an IC following the methods adapted from those developed by the instrument manufacturer, Dionex™. The instruments in use by the NADP utilize a potassium hydroxide (KOH) eluent generator, which generates a 35 mM KOH eluent solution. Samples are introduced into the system via an autosampler into a 25 μL sample loop and then onto the AG-18/AS18 guard and separator columns. Both systems utilize the ASRS (ULTRA II—4 mm) suppressors. The instruments are calibrated from 0.025 to 1.5 ppm for chloride, and from 0.050 to 6.000 ppm for nitrate and sulfate.

Background conductivity and column back pressure readings are recorded and monitored daily. As the columns and eluent generators age, both readings will increase. These readings help the analyst determine when changes are required to achieve the optimal conditions for analyses. During the data review, the analyst checks each chromatogram to verify that all analytes present have been marked for detection and quantified.

Interferences result from impurities in reagents or contamination of DI water. Impurities produce unidentified peaks that may overlap with the peaks of interest. Retention times are verified by standards either externally or internally. As the columns age, shifts occur in retention time or more impurities may be eluted from the column.

Sodium, *potassium*, *calcium*, and *magnesium* are determined on a Varian Vista Pro ICP-OES. The sample is introduced into the instrument, using an autosampler and peristaltic pump. The peristaltic pump has three channels that allow for the mixing of the sample stream with standards or ionization suppressants. The instrument is calibrated from 0 to 10.0 ppm for calcium and sodium and from 0 to 2.0 ppm for potassium and magnesium. Yttrium is used as an internal standard and cesium is used for ionization suppression. The instrument is calibrated at the beginning of each day and analysis of samples starts when QC solutions are verified to be good. In addition to QCs, the analyst can monitor percent relative standard deviations (% RSD) produced from triplicate readings for each sample. Increasing % RSDs indicate that there is a problem in the sample introduction system, which is comprised of the pump, tubing, nebulizer, and spray chamber. The internal standard is monitored throughout the analytical run and helps the analyst determine if instrument drift is occurring. Should the instrument drift, the analyst will verify that quality control solutions are good, then recalibrate the instrument and check quality control solutions again for verification. Routine maintenance is required to keep optimum operating conditions. Stretched or smashed pump tubing must be replaced. The torch and spray chamber are cleaned weekly. Water levels in the chiller are checked daily. The water in the chiller is changed periodically and the filters are cleaned.

Interferences that occur can be of the chemical, physical, or spectral nature. The high temperature of the plasma eliminates chemical interferences. Physical interferences are due to matrix differences and ionization effects. There are also spectral interferences resulting from overlaps of analytical lines. These can be avoided by selecting an alternate wavelength.

Ammonium and *orthophosphate* are analyzed using automated colorimetric methods. These automated methods involve introducing a specific volume of sample into a carrier stream onto a manifold set up with mixing coils. Reagents are added to the mixing coils, and they combine with the sample and react to produce a compound that can be detected colorimetrically. Methods for a vast array of analytes are provided by the instrument manufacturer. The instrument used for the wet deposition analysis is a Lachat FIA instrument, and the methods are 10-107-06-1B for ammonia and 10-115-01-1B for orthophosphate. The methods provide information necessary to analyze samples by FIA. These methods include parameters to set up the instrument to achieve the published detection limits for the specified analytical ranges. The instrument manufacturer provides reagent recipes and their recommended shelf life.

Prepared reagents are degassed for a minimum of 30 min, then transferred to polyethylene bottles, placed inline, and allowed to flow through the instrument manifold. The analyst monitors the baseline for a minimum of 20 min. The two methods (ammonia and orthophosphate) are run simultaneously using different manifolds for each method. The instrument is equipped with sample loops on each manifold. The timing is adjusted so that each sample loop is filled and rinsed before the sample is injected into the manifold. The analyst monitors the analytical run and checks all spectra to ensure that there are no air spikes. He/She marks samples with computed results that exceed the highest calibration standard. The worksheet can be modified before the run has been completed. This gives the analyst flexibility and allows for dilutions to be completed on the same day as the original sample is analyzed. Interferences are specific and documented in each method provided by Lachat.

CONCLUSIONS

Inorganic compounds in wastewater originate from natural and anthropogenic sources. They can be monitored satisfactorily by various instrumentation techniques described in this chapter.

REFERENCES

Ahuja, S., 1986. Ultratrace Analysis of Pharmaceuticals and Other Compounds of Interest. Wiley & Sons, New York.

Crites, R., Tchobanoglous, G., 1998. Small and Decentralized Wastewater Management Systems. McGraw Hill, New York.

Hinds, W.C., 1999. Aerosol Technology: Properties, Behavior and Measurement of Airborne Particles, second ed. John Wiley & Sons, Inc., New York ISBN 0-471-19410-7, p. 352.

Lachat Method 10-107-06-1-B Revised by P. Smith 27 August 2001. Written and copyrighted by William R. Prokopy 24 December 1993 by Lachat Instruments, 5600 Lindburgh Drive, Loveland, CO 80539, USA.

Lachat Method 10-115-01-1-B [8] Revised by L. Egan 29 November 2007. Written and copyrighted by W. Prokopy 4 August 1994 by Lachat Instruments, 5600 Lindburgh Drive, Loveland, CO 80539, USA.

Metcalf & Eddy, 2003. Wastewater Engineering, Treatment and Reuse. McGraw Hill, New York.

Purdue University. Groundwater & Surface Water. http://www2.ctic.purdue.edu/KYW/Brochures/GroundSurface.html (accessed 17.03.08).

Rothert, J., Harlin, K., Douglas, K., 2002. Quality Assurance Plan, Central Analytical Laboratory. NADP Program Office, Illinois State Water Survey Champaign, IL, August 2002.

Schnoor, J.L., 1996. Environmental Modeling, Fate and Transport of Pollutants in Water, Air and Soil. Wiley-International, New York.

Smith, R.A., Alexander, R.B., Wolman, M.G., 1987. Water-quality trends in the nation's rivers. Science 235 (4796), 1607–1615.

Sorial, G., 2005. Environmental Instrumentation. University of Cincinnati, Cincinnati, OH.

USEPA local drinking water reports. http://www.epa.gov/safewater/ccr/whereyoulive.html? Open View (accessed 17.03.08).

USEPA municipal assistance. http://www.epa.gov/owmitnet/mtb/index.htm (accessed 17.03.08).

USEPA national primary drinking water regulations. http://www.epa.gov/safewater/contaminants/index.html#listmcl (accessed 17.03.08).

USEPA national secondary drinking water regulations. http://www.epa.gov/safewater/contaminants/index.html#listsec (accessed 17.03.08).

USEPA unregulated contaminants. http://www.epa.gov/safewater/contaminants/unregulated/mtbe.html (accessed 17.03.08).

Chapter 10

Radionuclides in Surface Water and Groundwater

Kate M. Campbell

U.S. Geological Survey, 345 Middlefield Rd, MS 495, Menlo Park, CA 94025, USA

RADIOACTIVE COMPOUNDS AND RADIOACTIVITY

Unique among all the contaminants that adversely affect surface and water quality, radioactive compounds pose a double threat from both toxicity and damaging radiation. The extreme energy potential of many of these materials makes them both useful and toxic. The unique properties of radioactive materials make them invaluable for medical, weapons, and energy applications. However, mining, production, use, and disposal of these compounds provide potential pathways for their release into the environment, posing a risk to both humans and wildlife. This chapter discusses the sources, uses, and regulation of radioactive compounds in the United States, biogeochemical processes that

Handbook of Water Purity and Quality

Copyright © 2009 by Elsevier Inc. All rights of reproduction in any form reserved.

control mobility in the environment, examples of radionuclide contamination, and current work related to contaminated site remediation.

Radioactive Decay

A radionuclide is an unstable element that emits high energy radiation from the atomic nucleus. Energy is released from the nucleus as it relaxes into a more stable state in the form of ionizing radiation, capable of stripping at least one electron from another atom or molecule. This emission is coupled to a change in the atomic number, due to the nature of the radiation emitted. The original element is referred to as the parent nuclide, and the decay product is called a daughter nuclide. Daughter nuclides can also be radioactive and continue to decay according to a well-established sequence. An example of an important radioactive decay sequence is presented in Figure 1, showing the daughter products of uranium-238 decay.

There are three types of ionizing radiation resulting from decay of radioactive elements. Alpha (α) particles consist of two protons and two neutrons; this is equivalent to a nucleus of a helium atom (He^{2+}). Beta (β) particles are electrons or positrons emitted from the nucleus during the decay of a neutron.

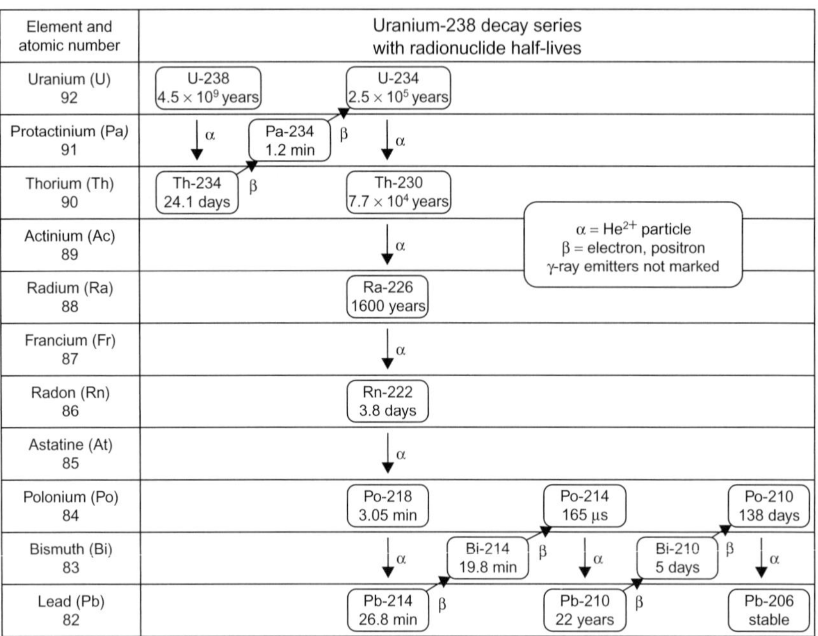

FIGURE 1 Diagram of uranium-238 decay series. The atomic number is determined by the number of protons in the nucleus, and the isotope is named by the atomic weight. Unstable nuclei release the instability by emitting alpha, beta, and/or gamma radiation from the nucleus of the atom. Because of the uranium-238 decay sequence, many of the elements shown here co-occur in contaminated areas with uranium as daughter nuclides of the dominant uranium isotope (based on Abdelouas, 2006; Focazio et al., 1998, and references therein).

In this type of decay, the atomic mass does not change, but the atomic number increases by one. Gamma (γ) radiation consists of photons with extremely high energy and the shortest wavelength in the electromagnetic spectrum. This type of radiation comes from transitions that occur within atomic nuclei and is often coupled to the emission of alpha or beta particles.

The SI unit of radioactive decay is the becquerel (Bq), which is defined as one decay per second. Because the Bq is a very small unit of radioactivity, another common unit is the curie (Ci), equal to 3.7×10^{10} Bq. Natural water samples are often reported in picocuries, which is equivalent to 2.22 radioactive disintegrations per minute (Focazio et al., 1998). The frequency of radioactive decay can vary from fractions of a second to millions of years. The accepted measure of the rate of radioactive decay is an isotopic half life, defined as the amount of time taken for a given quantity of radioactive material to decay to half of its original amount. Although it is not possible to determine exactly when a particular atom will decay, half life is a probabilistic and reliable measure of decay for a given quantity of radioactive material.

For the purpose of this chapter, the above discussion of radioactive decay is sufficient. However, radionuclide physics and decay is a complex and well-studied subject with many excellent texts and articles available for those wishing to delve into more detail.

Exposure to Radiation

Alpha, beta, and gamma radiation have very different routes of exposure and effects on tissue. All ionizing radiations can be mutagenic and exposure increases the risk of cancer (WHO, 2005; EPA, 2003). Alpha particles are highly ionizing but have very low tissue penetration; an alpha particle cannot penetrate the upper layers of the skin. However, if an alpha-emitting source is ingested, the health effects can be severe as alpha particles are the most damaging form of radiation. Beta particles are less damaging but have greater penetration than alpha particles, and can cause DNA mutation and cell damage. The effects of this type of radiation have been harnessed for medical radiation therapy to kill cancerous cells. Gamma radiation is very penetrating and can cause severe cell damage and mutagenesis even when the source is not taken internally. As a result, gamma emitters should be well-shielded to prevent direct tissue exposure. Sealed gamma sources are often used for the sterilization of medical and scientific equipment as well as food products. Several examples of environmentally relevant alpha, beta, and gamma emitters are given in Table 1.

Exposure to radiation is often measured as a dose, or the amount of radiation exposure to a human body over a given amount of time. Although there are several units of radiation dose, the most commonly used unit is a rem, which takes into account the amount of radiation absorbed by the body and the biological effect. This directly incorporates the effects of different types of particles emitted by radionuclide decay. The Environmental Protection Agency

TABLE 1 Properties and Sources of Environmentally Relevant Radionuclides (ANL, 2007; EPA, 1996, 2007; NRC, 2004, 2006, 2007a,b; Tso et al., 1964)

Element	Number of Isotopes	Environmentally Important Radioisotopes	Naturally Occurring	Half-life	Source	Applications	Radiation Type
Americium (Am)	16, all radioactive	^{241}Am	No	432.7 years	Neutron capture by plutonium	Smoke detectors, industrial gauges	α, γ
Carbon (C)	3, ^{12}C, ^{13}C are stable	^{14}C	Yes	5,700 years	Cosmic ray reaction with N_2, nuclear weapons testing	Radiocarbon dating, medical tracer	β
Cesium (Cs)	39, only ^{133}Cs is stable	^{137}Cs	Yes	30.17 years	Fission product	Industrial gauges, well logging, therapeutic nuclear medicine	β, γ
		^{134}Cs	No	2 years	Neutron activation of ^{133}Cs	–	β, γ
		^{135}Cs	No	2.3 million years	Fission product	–	β, γ
Cobalt (Co)	59, 22 are radioactive	^{60}Co	No	5.27 years	Fission product	Radiotherapy, sterilization, radiography, γ-ray source	β, γ
Hydrogen (H)	3, one is radioactive	^{3}H (tritium)	Yes	12.3 years	Cosmic ray collision in upper atmosphere, fission product	Soluble in water as tritiated water (HTO), tracer, groundwater dating	Weak β

Element	Number of isotopes	Isotope	Naturally occurring	Half-life	Origin	Uses	Emission
Iodine (I)	37, only ^{127}I is stable	^{129}I	No	15.7 million years	Fission product	Radiometric dating	β, weak γ
		^{131}I	No	8.02 days	Fission product	Thyroid treatment, drug metabolisms	β, weak γ
		^{123}I	No	13.2 hours	Fission product	Medical imaging	β, weak γ
		^{124}I	No	4.18 days	Fission product	Immunotherapy	β, weak γ
Neptunium (Np)	17, all radioactive	^{237}Np	No	2.14 million years	Decay product of ^{241}Am, fission product	–	α
Plutonium (Pu)	15, all radioactive	^{238}Pu	No	87.7 years	U reaction with neutrons	Heat (energy)	α
		^{239}Pu	No	24,100 years	U reaction with neutrons	Nuclear weapons	α
		^{240}Pu	No	6,560 years	U reaction with neutrons	–	α
Polonium (Po)	33, all radioactive	^{210}Po	Yes	138.38 days	^{238}U decay, fission product	Power source for satellites, found in tobacco from fallout	α
Radium (Ra)	25, all radioactive	^{224}Ra	Yes	3.7 days	Th decay	–	α
		^{226}Ra	Yes	5.76 years	U decay	Luminescent dials, radiography, found in tobacco from fallout	α, γ
		^{228}Ra	Yes	3.66 years	Th decay	–	α

(Continued)

TABLE 1 (Continued)

Element	Number of Isotopes	Environmentally Important Radioisotopes	Naturally Occurring	Half-life	Source	Applications	Radiation Type
Radon (Rn)	34, all radioactive	^{222}Rn	Yes	3.8 days	Decay of ^{226}Ra	Nuclear medicine	α
Thorium (Th)	27, all radioactive	^{232}Th	Yes	14 billion years	U decay series, nuclear reactors	Lantern mantles, chemical catalyst, ceramics, welding rods	α, γ
Strontium (Sr)	16, ^{84}Sr, ^{86}Sr, ^{87}Sr, ^{88}Sr are stable	^{90}Sr	No	29.1 years	Fission product, fallout	Radiological tracer for medical use, power source (heat)	β
Technetium (Tc)	22, all radioactive	^{99}Tc	No	212,000 years	Fission product	Medical applications (metastable form, Tc – 99 m, half-life = 6 h)	β, γ
Uranium (U)	6, all radioactive	^{238}U	Yes	4.47 billion years	Natural mineral ores	Shielding, bullets, missiles, weights, yellow color in ceramic glaze	α, γ
		^{235}U	Yes	700 million years	Natural mineral ores	Nuclear reactors, weapons	α, γ
		^{234}U	Yes	246,000 years	Natural mineral ores	–	α, γ

(EPA) sets an annual dose limit for radiation exposure through drinking water sources and calculates the maximum contaminant concentration allowable for a particular radionuclide (see Section Radioactive Compounds in Water). The average background radiation dose is approximately 300 mrem per person, primarily from naturally occurring cosmic and terrestrial sources of radiation and routine medical procedures.

Radionuclide Toxicity and Metabolism

Many radioactive elements are primarily dangerous because of the release of ionizing radiation during the decay process. However, a number of radioactive metals exhibit toxicity similar to nonradioactive metals. For example, exposure to elevated concentrations of uranium through drinking water causes increased risk of kidney failure and reproductive organ damage (WHO, 2005, and references therein). The toxicity and radioactive damage to tissues due to exposure to a particular element is dependent on the radiological and chemical activity of the element, the biodistribution, and the metabolic removal of the element by the body. Many radioactive elements are more dangerous when inhaled than when ingested through drinking contaminated water because the body can eliminate the compound more readily through the digestive tract, liver, and kidneys. However, some elements target particular organs when ingested because of their chemical behavior. For example, strontium-90 accumulates in bone because of its chemical similarity to calcium, whereas the radioisotopes of iodine are concentrated in the thyroid gland along with stable isotopes of iodine (ANL, 2007).

PRIMARY RADIOACTIVE CONTAMINANTS: SOURCES, USES, AND DISPOSAL

While there are thousands of naturally occurring and synthetic (anthropogenic) radionuclides, this section discusses some of the most commonly encountered radionuclides in the environment. Many radionuclides are naturally occurring in minerals and rocks, but the fallout from nuclear weapons testing and waste generated from weapons and energy production is a significant source of contamination to the environment. Mill tailings—the waste products generated during the extraction of naturally occurring radioactive elements from mineral ores—generally have relatively low radionuclide concentrations. However, mill tailings can easily serve as a source of contamination to surface water and groundwater because they are often extensive and can be in direct hydrologic contact with rivers, lakes, and groundwater. Radioactive waste generated by human activity is categorized as either high- or low-level waste, depending on the origin and use of the material. High-level waste is primarily spent uranium fuel from nuclear power reactors. Low-level waste is essentially waste from any source other than spent nuclear fuel (SNF).

Low-level waste can have radiation levels that range from only slightly above natural abundance to very radioactive, depending on the source material (NRC 2007).

Nuclear Reactors

Nuclear power plants and other reactors utilize fuel rods containing uranium-235, which is only slightly radioactive. The three uranium isotopes must be separated in a process called enrichment before uranium-235 can be used as fuel, since the natural abundance of uranium-235 is only 0.72% (Table 1). The leftover uranium is primarily uranium-238, referred to as depleted uranium. When uranium-235 is irradiated with neutrons, the uranium atoms split through a process called fission, releasing more neutrons, atoms with smaller atomic numbers, and large amounts of heat. The heat is harnessed to create electricity, similar to conventional fossil fuel power plants. During fission, some uranium-235 atoms capture neutrons to form heavier radioactive elements (transuranic elements) such as plutonium. The by-products of fission are highly radioactive and often long lived (e.g., cesium-137, strontium-90, and plutonium; Table 1) and are the primary reason that spent fuel is significantly more hazardous than the original unreacted fuel rod (NRC, 2007).

Long-term disposal of SNF is the subject of intense research. There are many ways to sequester the radionuclides in SNF, including nuclear waste glasses (vitrification), ceramics, and advanced storage cask technology. Discussion of these technologies and the related research of SNF sequestration are beyond the scope of this chapter. However, the stability of these materials has significant consequences for the success of long-term storage of SNF at a permanent disposal facility by keeping long-lived radionuclides out of accessible water resources for thousands of years.

Medical Applications

Radionuclides are used in medical applications for diagnostic procedures (imaging) and therapeutic treatments of diseased tissue such as cancer. For diagnostic tests, a small amount of radioactive material, such as technetium-99m (metastable form, see Table 1) or iodine, is taken internally to image a particular organ. Therapeutic uses of radioisotopes include external exposure by utilizing a focused external cobalt-60 gamma ray source, the placement of removable radiation sources in or near a tumor, and ingestion of high dosages of radioactive materials to target a particular organ (e.g., iodine for thyroid disease). Medical facilities also utilize sealed radioactive sources for imaging and sterilization devices (NRC, 2004).

Mill Tailings

The extraction of useful quantities of radionuclides requires the processing of enormous amounts of mineral ores, resulting in large quantities of mill tailings.

Mill tailings do not have high enough concentrations to be economically valuable, but contain enough residual material to contaminate surrounding surface water and groundwater resources. The most commonly mined radioactive element is uranium, and the milling of uranium ore is a significant source of environmental contamination. There are three types of processes used for uranium mining: open pit mining, underground mining, and *in situ* leaching. Tailings are produced during open pit and underground mining, and can contain low but measurable amounts of uranium and other radioactive elements that co-occur in the mineral ore phase such as radium and radon. *In situ* leaching involves injecting either sulfuric acid or a strong alkali solution, depending on the geology, into a subsurface ore body. Uranium is directly extracted into the injectate and pumped out of the ground for processing. Although *in situ* leach mining does not produce mine tailings, there is still potential for groundwater contamination.

Historically, uranium mining was most commonly performed with open pit or underground techniques, resulting in large quantities of mine tailings that were often in contact with surface water and groundwater. The United States government has invested billions of dollars in remediating uranium mill tailings-impacted sites by removing the tailings to lined and capped disposal locations. Nevertheless, there is still a legacy of uranium and other metal contamination in groundwater because of uranium mill tailings. As a result, there continues to be an extensive research into groundwater remediation technologies for sites contaminated by uranium mill tailings (Section Remediation of Uranium Contamination). Because of the variable geology and ore composition, the groundwater contamination from mill tailings varies appreciably between sites. The processes used to extract uranium from the ore also varied from site to site and can impact the current aquifer conditions. Radioactive compounds of concern at mill tailings sites are primarily uranium, radium, and radon, although other radioactive elements can be present as well. Nonradioactive elements that can co-occur in uranium mill tailings are arsenic, selenium, cadmium, chromium, mercury, molybdenum, lead, and vanadium. In addition, sulfide minerals are commonly associated with uranium ores and can lead to acid drainage when in contact with water (Abdelouas, 2006; NRC, 2006b).

Radon Emissions from Soil Minerals and Groundwater

Radon is a naturally occurring decay product of radium (Figure 1). Because of its electronic structure, it is a heavy noble gas and is relatively nonreactive. However, it poses a large threat to human health through inhalation because it is very stable in the gaseous form and tends to accumulate indoors. The primary route of exposure is radon emissions from soil, but it can also dissolve in groundwater. When the groundwater is brought to the surface for human consumption, the radon will degas into the atmosphere. This is mainly a concern for privately-owned well users who use the groundwater with minimal treatment and may be exposed to radon as it degasses from the water while bathing.

RADIOACTIVE COMPOUNDS IN WATER

Regulation of Radioactive Wastes and Water Standards

Radionuclides are ubiquitous in naturally occurring rocks and soils. In addition, traces of radionuclides have been dispersed worldwide as fallout from the detonation of nuclear weapons. Exposure to a background level of radiation from natural sources and fallout via drinking water and soil contact is normal and has been incorporated into the Nuclear Regulatory Commission (NRC) and EPA calculations of radiation exposure risk. Contaminated water has radiation levels above the normal acceptable background amount of radiation.

The responsibility for proper handling, use, and storage of radioactive materials bridges a large number of federal, state, and local agencies. The NRC is responsible for regulation of storage and disposal of all commercially generated waste as well as long-term storage of high-level waste from the Department of Energy (DOE). The NRC also sets dose limits for different radioactive compounds for the general public, radiation workers, and medical personnel (NRC, 2007). The DOE is responsible for developing technologies and storage facilities for waste generated by DOE activities (primarily high-level waste), and must be licensed by the NRC. The EPA is responsible for water standards outside the boundaries of NRC and DOE sites, which includes drinking water standards. The DOE is responsible for safe handling of radioactive materials at production facilities supplying military weapons or energy applications as well as disposal of transuranic wastes from military activity, but the Department of Defense (DOD) internally regulates the use of radioactive materials during active military application. The Department of Transportation (DOT) regulates the transportation of all radioactive wastes with standards set by the NRC. The Department of Interior (DOI) provides scientific support for DOE disposal programs through the U.S. Geological Survey (USGS). The Bureau of Land Management (BLM) manages certain sites for the DOE. Food and Drug Administration (FDA) oversees approval for safe use of radiopharmaceuticals. Licensing and medical use of radioactive compounds is overseen by the NRC (NRC 2004, 2006a,b, 2007).

High-Level Waste

Because the decay of high-level waste is extremely slow and the waste will pose a human health threat for hundreds of thousands of years, long-term storage is necessary. Currently, there are no permanent locations approved for high-level waste disposal in the United States, and high-level waste is being stored in numerous temporary storage facilities around the country. However, Yucca Mountain, Nevada, is being explored as permanent disposal site for commercial (nonmilitary) high-level nuclear waste.

Low-level Waste

For elements with short half-lives and nonradioactive daughter products, waste can be "treated" by allowing it to decay and then disposing of the daughter products as regular hazardous waste. Otherwise, it must be taken to low-level waste disposal sites regulated by the NRC.

Mill Tailings

Congress enacted the Uranium Mill Tailings Radiation Control Act (UMTRCA) in 1978 to remediate the sites of uranium mill tailings. The DOE is responsible for the remediation of sites where uranium was used for the weapons program, regardless of the current ownership of the land. Remediation must be evaluated by the NRC and meet EPA standards. The first phase is to physically move the tailings into a lined repository and clear the site so that no radiation hazard is present at the surface. Phase 2 requires determination and implementation of remedial action for groundwater clean up. The EPA has set groundwater concentration limits for elements and other contaminants commonly found in mill tailings (Table 2). For some elements, these limits are higher than drinking water standards.

Drinking water standards are summarized in Table 3. Maximum contaminant levels (MCLs) take into consideration the available treatment technologies as well as the health effects of exposure to a particular contaminant. MCLs are enforceable standards, and all drinking water supplies are required to comply with these standards. Only uranium has an MCL that includes chemical toxicity in addition to radiation dose. Gamma emitters are not regulated separately because gamma radiation co-occurs with either alpha or beta particles. Public health goals are nonenforceable concentration thresholds below which ingestion holds no health risk (EPA, 2003). For all radionuclides, the public health goals are zero.

Standard iron or aluminum coagulation, anion exchange, and reverse osmosis are all effective technologies for removing radionuclides, especially uranium, from water during conventional municipal water treatment (WHO, 2005, and references therein).

Analysis of Radioactive Compounds in Water

Screening of water samples for radioactive elements can be achieved with a gross alpha and beta measurement. This method counts the total amount of alpha or beta particles emitted from a sample over a specific period of time. Although these measurements are not useful for determining the composition of the sample and radiation dose, and can be impaired by various interferences, a gross alpha and beta screen can semiquantitatively indicate whether further testing is necessary for a particular water sample. This measurement is sometimes performed with a dried sample, which can yield artificially low results if the dominant radionuclides are volatile (e.g., iodine, radon, tritium, and carbon-14).

Alpha spectrometers, gamma spectrometers, and liquid scintillation counters for beta emitters are commonly used for quantifying concentrations of different

TABLE 2 NRC and EPA Limits for Uranium Mill Tailings—Impacted Groundwater (EPA, 2003; NRC, 2008)

Contaminant	Groundwater Limit	Drinking Water Standard
Ra (226 + 228)	0.185 Bq/L (5 pCi/L)	5 pCi/L
U (234 + 238)	1.11 Bq/L (equivalent to 0.044 mg U/L)	0.03 mg/L
β particles/photons	–	4 mrem/year[*]
Gross α activity	0.555 Bq/L (15 pCi/L)	15 pCi/L
Ag	0.05 mg/L	–
As	0.05 mg/L	0.01 mg/L
Ba	1.0 mg/L	2.0 mg/L
Cd	0.01 mg/L	0.005 mg/L
Cr	0.05 mg/L	0.1 mg/L
Hg	0.002 mg/L	0.002 mg/L
Mo	0.1 mg/L	–
Pb	0.05 mg/L	0.015 mg/L
Se	0.01 mg/L	0.05 mg/L
NO_3^-	10 mg/L as N	10 mg/L as N
Endrin	0.0002 mg/L	0.002 mg/L
Lindane	0.004 mg/L	0.0002 mg/L
Methoxychlor	0.1 mg/L	0.04 mg/L
Toxaphene	0.005 mg/L	0.003 mg/L
2,4-D (2,4-dicholorphenoxyacetic acid)	0.1 mg/L	0.07 mg/L

[*]*No further analysis needed if gross beta emission is less than 50 pCi/L and if the concentrations of tritium and strontium-90 are less than 20,000 and 8 pCi/L, respectively. If gross beta activity exceeds 50 pCi/L, then all major radioactive contaminants must be quantified and the overall dose should not exceed 4 mrem/year.*

radionuclides depending on the type of particle and energy emitted. The decay of a radionuclide results in a particle emission with a specific energy signature that can be used as a "fingerprint" for the parent isotope. By counting the particles emitted from a sample at specified energy levels, the concentration of

TABLE 3 EPA Limits for Radionuclides in Drinking Water (EPA, 2003; NRC, 2006a).

Contaminant Type	Maximum Contaminant Level	Public Health Goal
Alpha particle emitters	15 pCi/L	Zero
Beta particle emitters	4 mrem/year	Zero
Radium-226 and Radium-228 (combined)	5 pCi/L	Zero
Uranium (sum of all three isotopes)	30 μg/L	Zero
Tritium	20,000 pCi/L (equivalent to 4 mrem/year)	Zero

various radionuclides can be calculated. Detectors for alpha and gamma spectrometry operate on the general principle that incoming radiation ionizes the atoms in the detector, which can be measured as a voltage proportional to the energy of the original radiation particle. There are several types of detectors, and selection of the appropriate system depends on the concentration range, number of radionuclides, and matrix, and self-absorption effects. Multichannel instruments have the capability of measuring several energies simultaneously, allowing for the detection of several elements at the same time. Liquid scintillation counting is very useful for low-energy beta emitters, such as tritium and carbon-14, but can be used for other beta emitters as well. The water sample is mixed with an organic liquid scintillator, resulting in light emission, which can be measured with a photomultiplier tube. Hand-held Geiger–Mueller counters are generally not sensitive enough to be useful for environmental water samples.

When the concentrations of target radionuclides are too low for direct counting, several approaches can be taken. Some analytes can be measured with an alternate instrumental method with lower detection limits, such as inductively coupled plasma mass spectrometry (ICP-MS). Alternately, a chemical precipitation method can be used to purify and concentrate a known amount of sample into a solid pellet that can be counted as described earlier. Several elements can be precipitated in this manner, such as Cs (as Cs_2PtCl_6), I (as PdI_2), Ra (in a Ba-Ra sulfate), and Sr (as $SrCO_3$), although these precipitation methods can be susceptible to interferences.

Uranium can be measured using an ICP-MS or a Kinetic Phosphorescence Analyzer (KPA, Chemcheck Instruments, WA). Multiple analytes can be simultaneously measured on an ICP-MS, whereas the KPA is specific to uranium. Both instruments have very low detection limits (<1 ppb). Uranium can also be calculated by measuring the decay of thorium-234 (decay product of uranium-238, Figure 1) by gamma spectrometry or directly by alpha spectrometry.

A useful guide to analysis of radionuclides is included in *Standard Methods for the Examination of Water and Wastewater*, which provides standardized chemical and instrumental methods, as well as interference information (Eaton and Franson, 2005). It is important to note that all laboratories handling radioactive compounds must be licensed by the NRC and have appropriate safety precautions specific to radionuclides to avoid contamination of laboratory equipment and personnel.

Examples of Radionuclide Contamination in Groundwater

The extent of contamination by a particular radionuclide depends on the biogeochemical processes that affect mobility in aqueous solution at environmentally relevant conditions as well as the physical characteristics of the aquifer soil and fluid flow. As uranium is the most widespread radionuclide contaminant in groundwater, Sections Biogeochemical Processes Controlling Uranium Fate and Transport and Remediation of Uranium Contamination will focus on uranium biogeochemistry and remediation technology. However, other radionuclides also pose a threat to drinking water resources, and several examples will be briefly discussed later and in Section Biogeochemistry of Other Radionuclides.

There is a legacy of contamination associated with the production of weapons at four major DOE sites (Hanford, WA; Savannah River, GA; Oak Ridge National Laboratory, TN; and Idaho National Laboratory, ID) and other processing facilities such as Rocky Flats, Colorado. There has been a substantial amount of research at each of these sites because the extent of contamination is large and complex. For example, the vadose zone and aquifer sediment at the Hanford site have been contaminated with cesium-137, strontium-90, and uranium, along with other toxic metals below processing ponds and buried waste tanks (e.g., Liu et al., 2003, 2006; McKinley et al., 2006; Arai et al., 2007, and references therein). The Hanford site sits along the edge of the Columbia River, posing a threat to both groundwater and surface water quality. In addition to treating the radioactive waste repositories, significant progress has been made in remediating contaminated groundwater, but the complexity and extent of contamination is extensive. Ongoing research at Hanford is currently developing novel treatment mechanisms for immobilizing radionuclide contamination. Drinking water sources in the areas surrounding the Hanford site are monitored carefully for contamination.

There have been several reported incidents of accidental release of tritium into groundwater from commercial nuclear power plants. Tritium is a by-product of boron irradiation, which is used to control the fission reaction in nuclear power plants. Although the measured impact on drinking water supply was very low, these incidents point to the importance of controlled disposal of all by-products of fission reactors (NRC, 2006a).

Several studies have found naturally occurring deposits of uranium, radium, and radon affecting drinking water and public supply wells. In the Kirkwood–Cohansey aquifer in southern New Jersey, elevated concentrations of radium were found near agricultural areas. Use of lime and fertilizers in the fields altered the groundwater chemistry by lowering the pH to less than 5, increasing the mobility of radium into the groundwater (Szabo and DePaul, 1998; Hirsch et al., 2008). In the San Joaquin Valley, California, uranium concentrations exceeding the MCL have been measured in public supply wells, partly because of the influx of higher dissolved oxygen from increased well pumping (Burow et al., 2005). The EPA and the USGS did a reconnaissance survey of radium isotopes, polonium-210, and lead-210 in groundwater used for drinking water in 27 states in areas suspected of having elevated radium concentrations (Focazio et al., 1998). Out of 99 samples, 21% exceeded the MCL for radium. Although this report does not provide information on the population exposed, it does indicate that naturally occurring radioisotopes can be found in drinking water supplies under favorable geochemical conditions.

BIOGEOCHEMICAL PROCESSES CONTROLLING URANIUM FATE AND TRANSPORT

Because of the reliance on uranium as a reactant in the nuclear fuel cycle and weapons manufacture, uranium-contaminated water and soil is one of the most important and widespread challenges facing environmental radionuclide science. Uranium mobility is controlled by a complex network of chemical, biological, and physical processes in an aquifer system (Figure 2). The later discussion will focus on key chemical and biological processes, but it is important to note that the mobility of uranium is also affected by the physical characteristics of the soil, such as aquifer hydraulic conductivity and lithology.

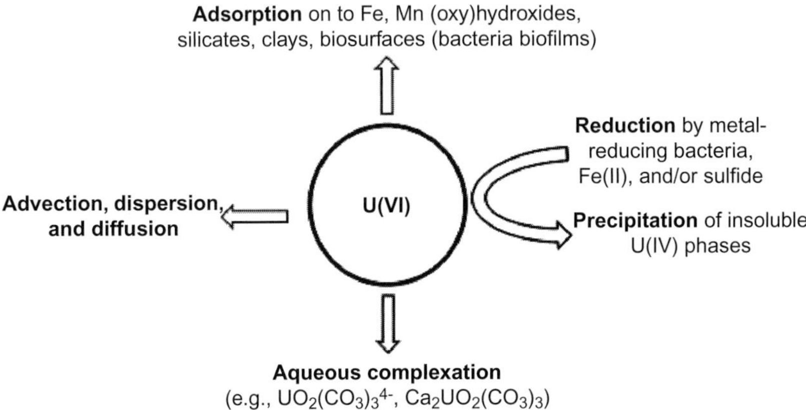

FIGURE 2 Important chemical, biological, and physical processes in uranium(VI) bioremediation.

Uranium (U) can exist in four oxidation states ($+$III, $+$IV, $+$V, and $+$VI), although only U(VI) and U(IV) occur widely in the environment. Under oxic conditions, U(VI) forms a large number of aqueous (oxo)anionic complexes and can be very mobile in groundwater. Common U(VI) species at circumneutral pH can inlcude the uranyl ion (UO_2^{2+}), hydroxide species (e.g., $UO_2(OH)_2^0$), carbonate species (e.g., $UO_2(CO_3)_3^{4-}$), ternary metal-carbonate species (e.g., $Ca_2UO_2(CO_3)_3^0$), and other minor species with nitrate, chloride, sulfate, and phosphate, depending on the aqueous geochemical conditions (pH, alkalinity, etc.) (Guillaumont et al., 2003; Dong and Brooks, 2006). Uranium(VI) also forms inner-sphere adsorption complexes with iron and manganese (oxy)hydroxides, silicates, and clays (e.g., Barnett et al., 2002; Waite et al., 1994; Payne et al., 1996; Krepelova et al., 2006). Adsorption depends on pH and the presence of aqueous species such as carbonate and calcium (Fox et al., 2006). Because groundwater is often rich in carbonate and calcium, the carbonate and ternary complexes are often dominant, lowering U(VI) adsorption to mineral surfaces and making U(VI) very mobile in oxic groundwater. Precipitation with carbonate and phosphate minerals (e.g., autunite, $Ca(UO_2)_2(PO_4)_2 \cdot nH_2O$) can sequester U(VI) in the solid phase under specific but environmentally relevant conditions (Abdelouas et al., 1998). Under the conditions of most oxic aquifers, however, U(VI) tends to be quite mobile because of the strong aqueous carbonate complexation. When U(VI) is reduced to U(IV), a highly insoluble precipitate forms, often as UO_2. Even in the presence of carbonate and calcium, U(IV) remains relatively immobile (Guillaumont et al., 2003). Thus, the mobility of uranium is affected strongly by the processes controlling U oxidation state and the redox conditions in the aquifer (e.g., van Hullebusch et al., 2005).

Uranium(VI) Reduction

There are both abiotic (chemical) and biotic (microbial) pathways for U(VI) reduction to insoluble U(IV). Aqueous ferrous iron (Fe^{2+}) is a weak reductant of U(VI), but adsorbed Fe(II) and ferrous iron-bearing minerals are effective at reducing adsorbed U(VI) (Liger et al., 1999; Jeon et al., 2005). Iron(II) adsorbed to the surface of Fe oxides, clays, and natural mixtures as well as mixed Fe(II)–Fe(III) phases such as green rust and magnetite ($<$pH 5) have been shown to reduce U(VI) to U(IV) (O'Loughlin et al., 2003; Missana et al., 2003; Scott et al., 2005; and Dodge et al., 2002). Sulfide mineral surfaces also partially reduce adsorbed U(VI) to form mixed U(IV)–U(VI) solid phases (Wersin et al., 1994).

Biotic reduction of U(VI) can be carried out by various commonly occurring dissimilatory metal-reducing and sulfate-reducing groundwater bacteria including *Geobacter, Shewanella*, and *Desulfovibrio* species (Lovley et al., 1991; Gorby and Lovley, 1992; Lovley and Phillips, 1992; Fredrickson et al., 2000; Marshall et al., 2006). The rate of microbial U(VI) reduction may be limited by complexation with Ca and carbonate (Brooks et al., 2003). Conversely, natural organic matter (NOM) may act as an electron donor and/or electron shuttle, enhancing the microbial reduction of U(VI) (Gu and Chen, 2003). Uranium(IV) precipitates formed from microbial enzymatic reduction tend to be aggregates of nanoparticles and may be more easily re-oxidized than bulk $UO_{2(s)}$ (Suzuki et al., 2003; Singer et al., 2007). Iron- and sulfate-reducing bacteria may also stimulate U(VI) reduction indirectly by producing reducing agents (sulfide and Fe(II)).

Uranium(IV) Oxidation

Oxidation of $UO_{2(s)}$ can occur via abiotic and biotic pathways, although little is known about direct microbial $UO_{2(s)}$ oxidation. Potential important oxidizing agents under environmentally relevant conditions include dissolved oxygen $(O_{2(aq)})$, manganese(III/IV) minerals, Fe(III)(oxy)hydroxides, nitrate, microbial denitrification intermediates (NO_2^-, NO, and N_2O), and some types of NOM (Ginder-Vogel et al., 2006; Peper et al., 2004; Istok et al., 2004; Gu et al., 2005). Oxidative dissolution of $UO_{2(s)}$ has been well-studied under different environmental conditions. As pH decreases or $O_{2(aq)}$ and bicarbonate concentrations increase, the rate of UO_2 oxidation increases (e.g., Peper et al., 2004). Since microbially precipitated U(IV) is smaller than the particles of bulk $UO_{2(s)}$ used in many studies by several orders of magnitude, the rates of oxidative dissolution for nanoparticulate U(IV) may be faster than bulk UO_2 (Suzuki et al., 2002, and Urlich et al., 2008). Further research is needed to test whether this observation is relevant in natural sediments and soils.

REMEDIATION OF URANIUM CONTAMINATION

Remediation Technologies

Once soil and groundwater have been polluted by radionuclides, the process of remediating the contaminated area can be complicated and costly. One approach is to "dig and dump" or physically remove the contaminated sediment to a lined and clay-capped disposal site. This has been extensively employed for mill tailings, although it is often impractical for treating the contaminated soil beneath the tailings. Another strategy is "pump and treat," or actively pumping the contaminated groundwater to an above ground treatment unit, and reinjecting the treated water into the aquifer. This technique is useful for very mobile elements, but can be expensive, requires harsh additives, or takes a prohibitive amount of time if the contaminant is strongly associated with the soil phase. Soil washing involves flushing the sediment system to remove the small, mobile particles that tend to contain elevated concentrations of contaminant. Similar to the pump and treat approach, soil washing can be expensive and slow, depending on the characteristics of the contaminant and site (Tabak et al., 2005). Permeable reactive barriers (PRBs) such as zero valent iron and bone charcoal concentrate the contaminant in the barrier material through adsorption, precipitation, and chemical transformation. The barrier can be removed for disposal once the reactive sites have been poisoned. Injectable phosphate amendments are being explored as a way to sequester dissolved uranium in uranium-phosphate precipitates such as autunite (Abdelouas et al., 1998). Although these strategies are effective under the right circumstances, they are often impractical because of the large quantity of sediment or water to be treated.

Alternatives to these conventional techniques are currently being explored. Bioremediation has become an attractive alternative because the treatment can be very economical and is relatively noninvasive. It is defined as a technology that uses microorganisms (bacteria, archaea, fungi, or algae) to reduce, contain, or transform the target compound to eliminate water contamination (Tabak et al., 2005). The technology has been enormously successful in treatment of organic contaminants and is currently under investigation for large-scale treatment of radionuclide-contaminated groundwater (Tabak et al., 2005; Hazen and Tabak, 2005). Bioremediation strategies for metals and radionuclides can take one of the several approaches: permanently sequestering the element in an immobile form, enhancing mobilization of the element from the solid phase to flush it from the system, or a combination of concentrating the element (e.g., as a solid phase) and remobilizing it in a pulse that can be easily treated *ex situ*. There has also been some research into the possibility of using phytoremediation (accumulating contaminants in plants), but these treatment options are limited to the root zone. Most treatment strategies are focused on bioremediation with bacteria because of the wealth of microbiological studies available and the versatility of using bacteria *in situ* in contaminated aquifers.

Bioremediation

Ex Situ Techniques

For sites with very high concentrations of radionuclide contaminants or where pump and treat systems are in operation, an *ex situ* bioremediation strategy may be an effective and economically viable option. *Ex situ* treatment processes include enzymatically catalyzed biotransformation (e.g., U(VI) reduction), biosorption/bioaccumulation, and biodegradation of organic complexing agents such as EDTA. Biosorption is the accumulation of radionuclides on biomass particles and is independent of bacterial metabolism. Bioremediation is also being applied to industrial and medical waste streams to consolidate radioactive waste, thus decreasing both the storage cost and the risk of future environmental contamination. However, the radiation toxicity of radionuclides often prevents *ex situ* bioremediation from being effective for highly concentrated radionuclide wastes (Tabak et al., 2005).

In Situ Techniques

In situ bioremediation is an attractive option for groundwater with lower contaminant concentrations because the treatment occurs directly in the subsurface aquifer. Ideally, *in situ* bioremediation only requires injection of an electron donor to stimulate activity of indigenous dissimilatory metal reducing or sulfate reducing bacterial communities. This is often achieved by injecting an organic carbon source such as acetate, ethanol, or molasses. The process of stimulating microbial growth in this manner is called *in situ* biostimulation. The success of biostimulation depends on targeting groups of naturally occurring bacteria whose physiology is capable of directly metabolizing the radionuclude (e.g., U(VI) reduction to insoluble U(IV)) and/or creating geochemical conditions favorable for the abiotic reactions to sequester the contaminant (e.g., dissimilatory Fe(III) reduction producing reactive Fe(II)).

Field-Scale Application of In Situ Bioremediation

The efficacy of *in situ* biostimulation for the removal of U(VI) from groundwater has been demonstrated at the field-scale at Old Rifle, Colorado, and Oak Ridge National Lab, Tennessee (Hazen and Tabak, 2005; Anderson et al., 2003; Stucki et al., 2007; Vrionis et al., 2005; Wu et al., 2006; Wu et al., 2007), and in the laboratory with synthetic mineral phases and natural sediments (Finneran et al., 2002; Abdelouas et al., 2000; Jeon et al., 2004; Suzuki et al., 2003). The following example demonstrates the potential of bioremediation as a treatment strategy for large, dilute plumes of uranium-contaminated groundwater.

The Old Rifle site in western Colorado is part of the DOE uranium mill tailings remedial action (UMTRA) program. The mill tailings have been removed and the site has been capped with a clay layer (DOE, 2006). Nevertheless, groundwater often exceeds the maximum allowable uranium concentration for a mill tailings site with average concentrations between 0.4 and 1.4 μM (Anderson et al., 2003). Preliminary laboratory studies showed that U(VI) could be removed from solution in live sediments with the addition of acetate (Finneran et al., 2002). Subsequent field-scale studies were conducted by injecting an acetate-amended groundwater with a conservative bromide tracer and monitoring the changes in groundwater geochemistry and microbiology before, during, and after the injection (Anderson et al., 2003; N'Guessan et al., 2008; Hazen and Tabak, 2005). The acetate injection stimulated microbial Fe(III) reduction concurrent with decreasing dissolved uranium concentrations. The groundwater microbial community became enriched in *Geobacter*-type organisms, based on 16S DNA analysis and specific phospholipid fatty acid analysis. *Geobacter* organisms are common soil bacteria capable of both Fe(III) and U(VI) reduction. As conditions in the study plot became more reducing, the community shifted from iron reducers to sulfate reducers, probably because sedimentary bioavailable Fe(III) pools were depleted. Near the injection wells, the acetate became limiting as sulfate reduction became the dominant microbial process, and the microbial community shifted to sulfate-reducing bacteria such as *Desulfovibrio*-type organisms. Uranium(VI) removal from groundwater substantially decreased during sulfate reduction, and dissolved uranium concentrations rebounded. This is consistent with the observation that acetate-oxidizing sulfate-reducing bacteria do not typically reduce uranium. The success of *in situ* biostimulation at this site depends on sustainable metal-reducing conditions in the aquifer. This case study highlights the complexity and importance of thoroughly understanding the key biogeochemical processes during *in situ* bioremediation.

BIOGEOCHEMISTRY OF OTHER RADIONUCLIDES

Naturally occurring radioisotopes and radioactive by-products of nuclear fission blanket virtually the entire periodic table, and the varied chemical behavior of radionuclides reflects this diversity. Location within the periodic table is a good indicator of the chemistry exhibited by a given radionuclide. For example, radon is relatively inert like the other noble gases, and it poses a threat to human health primarily through inhalation in its gaseous form. Tritium in solution essentially behaves like a proton in water (^1H-^3H-O). The radioisotopes of cesium and strontium are not redox active in the environment and have similar chemical behavior to other alkali and alkaline earth metals, respectively.

Multiple elements in the lanthanide and actinide series can exist in many oxidation states and can exhibit complex redox chemistry. Technetium(VII) and plutonium(VI/V) can both be reduced to insoluble precipitates by bacteria and/or abioic reducing agents (Tc(IV) and Pu(IV)) (Peretyazhko et al., 2008; Boukhalfa et al., 2007, and references therein). Interestingly, Pu(IV) can be further reduced by bacteria to Pu(III), which is relatively soluble.

The fate and transport of many radioisotopes becomes significantly more complex when interactions with microbes, mineral surfaces, and redox reactions are considered, as demonstrated for uranium in Sections Biogeochemical Processes Controlling Uranium Fate and Transport and Remediation of Uranium Contamination. The fate and transport of many radioactive elements in the environment is currently an area of intense research, particularly for elements in the lanthanide and actinide series. As more information becomes available about the biogeochemical reactions in the environment, bioremediation may become a treatment strategy for other radionucludes besides uranium, such as technetium, plutonium, and neptunium (Lloyd and Renshaw, 2005; Morris and Raiswell, 2002).

CONCLUSIONS

The legacy of nuclear weapons and the continued use of nuclear power and medicine have left the United States with a number of sites contaminated with various radionuclides. If the demand for nuclear energy increases, proper mining, milling, use, and storage of the radioactive products will continue to be a top priority in order to protect natural water resources. Although natural and anthropogenic sources of radionuclides tend to be relatively localized, clearly there are areas where radionuclide contamination threatens surface and groundwater sources. The complexity of soil–microbe–radioisotope interactions makes remediation strategies both scientifically intriguing and logistically difficult. There are many areas of research that would benefit not only the scientific community, but also organizations working to restore impacted areas to safe levels of radiation. One important research area is the role of microbial communities, geochemistry, biosurfaces, and biofilms during *in situ* remediation. A detailed understanding of biogeochemical processes controlling radionuclide fate in groundwater is of vital importance to maximizing and stabilizing field-scale treatment.

ACKNOWLEDGMENTS

The author is grateful to Jim Davis, Michael Hay, and Patricia Fox at the USGS for their insightful comments, and to the editor, Sut Ahuja, for the invitation to contribute to this work.

REFERENCES

Abdelouas, A., 2006. Uranium mill tailings: geochemistry, mineralogy, and environmental impact. Elements 2, 335–341.

Abdelouas, A., Lutze, W., Nuttall, E., 1998. Chemical reactions of uranium in ground water at a mill tailings site. J. Contam. Hydrol. 34, 343–361.

Abdelouas, A., Lutze, W., Gong, W., Nuttall, E.H., Strietelmeier, B.A., Travis, B.J., 2000. Biological reduction of uranium in groundwater and subsurface soil. Sci. Total Environ. 250, 21–35.

Anderson, R.T., Vrionis, H.A., Ortiz-Bernad, I., Resch, C.T., Long, P.E., Dayvault, R., et al., 2003. Stimulating the in situ activity of geobacter species to remove uranium from the groundwater of a uranium-contaminated aquifer. Appl. Environ. Microbiol. 69, 5884–5891.

Arai, Y., Marcus, M.A., Tamura, N., Davis, J.A., Zachara, J.M., 2007. Spectroscopic evidence for uranium bearing precipitates in vadose zone sediments at the Hanford 300-are site. Environ. Sci. Technol. 41, 4633–4639.

Argonne National Lab, 2007. Radiological and chemical fact sheets to support health risk analyses for contaminated areas. http://www.ead.anl.gov/pub/doc/ANL_ContaminantFactSheets_All_070418.pdf

Barnett, M.O., Jardine, P.M., Brooks, S.C., 2002. U(VI) adsorption to heterogeneous subsurface media: application of a surface complexation model. Environ. Sci. Technol. 36, 937–942.

Boukhalfa, H., Icopaini, G.A., Reilly, S.D., Neu, M.P., 2007. Plutonium(IV) reduction by the metal-reducing bacteria *Geobacter metallireducens* GS15 and *Shewanella oneidensis* MR1. Appl. Environ. Microbiol. 73, 5897–5903.

Brooks, S.C., Fredrickson, J.K., Carroll, S.L., Kennedy, D.W., Zachara, J.M., Plymale, A.E., et al., 2003. Inhibition of bacterial U(VI) reduction by calcium. Environ. Sci. Technol. 37, 1850–1858.

Burow, K., Jurgens, B., Dalgish, B., Shelton, J., 2005. Effects of well operation on quality of water from a public-supply well in Modesto, CA. Abstr. Programs Geol. Soc. Am. 37, 247.

Dodge, C.J., Francis, A.J., Gillow, J.B., Halada, G.P., Eng, C., Clayton, C.R., 2002. Association of uranium with iron oxides typically formed on corroding steel surfaces. Environ. Sci. Technol. 36, 3504–3511.

DOE, 2006. Rifle, Colorado, processing sites and disposal site. http://www.lm.doe.gov/documents/sites/co/rifle_d/fact_sheet/rifle.pdf

Dong, W., Brooks, S.C., 2006. Determination of the formation constants of ternary complexes of uranyl and carbonate with alkaline earth metals (Mg^{2+}, Ca^{2+}, Sr^{2+}, and Ba^{2+}) using anion exchange method. Environ. Sci. Technol. 40, 4689–4695.

Eaton, A., Franson, M.A., American Water Works Association, and Water Environment Federation., 2005. Standard methods for the examination of water and wastewater, twenty-first ed. American Public Health Association, Washington, DC. (Managing editor, Mary Ann Franson).

EPA, 1996. Fact sheet: environmental characteristics of EPA, NRC, and DOE sites contaminated with radioactive substances. http://www.epa.gov/radiation/docs/cleanup/540-f-94-023.pdf

EPA, 2003. National primary drinking water standards. 816-F-03-016.

EPA, 2007. Commonly encountered radionuclides. http://www.epa.gov/radiation/radionuclides/index.html

Finneran, K.T., Anderson, R.T., Nevin, K.P., Lovley, D.R., 2002. Potential for bioremediation of uranium-contaminated aquifers with microbial U(VI) reduction. J. Soils Sediments Contam. 11, 339–357.

Focazio, M., Szabo, Z., Kraemer, T., Mullin, A., Barringer, T., DePaul, V., 1998. Occurrence of selected radionuclides in groundwater used for drinking water in the United States: a reconnaissance survey. USGS water-resources investigations report 00-4273.

Fox, P.M., Davis, J.A., Zachara, J.M., 2006. The effect of calcium on aqueous uranium(VI) speciation and adsorption to ferrihydrite and quartz. Geochim. Cosmochim. Acta 70, 1379–1387.

Fredrickson, J.K., Zachara, J.M., Kennedy, D.W., Duff, M.C., Gorby, Y.A., Li, S.-W., et al., 2000. Reduction of U(VI) in goethite (α-FeOOH) suspensions by a dissimilatory metal-reducing bacterium. Geochim. Cosmochim. Acta 64, 3085–3098.

Ginder-Vogel, M., Criddle, C.S., Fendorf, S., 2006. Thermodynamic constraints on the oxidation of biogenic UO_2 by Fe(III) (hydr)oxides. Environ. Sci. Technol. 40, 3544–3550.

Gorby, Y.A., Lovley, D.R., 1992. Enzymatic uranium precipitation. Environ. Sci. Technol. 26, 205–207.

Gu, B., Chen, J., 2003. Enhanced microbial reduction of Cr(VI) and U(VI) by different natural organic matter fractions. Geochim. Cosmochim. Acta 67, 3575–3582.

Gu, B., Yan, H., Zhou, P., Watson, D.B., Park, M., Istok, J., 2005. Natural humics impact uranium bioreduction and oxidation. Environ. Sci. Technol. 39, 5268–5275.

Guillaumont, R., Fanghanel, T., Fuger, J., Grenthe, I., Neck, V., Palmer, D.A., et al., 2003. Update on the Chemical Thermodynamics of Uranium, Neptunium, Plutonium, Americium, and Technetium, OECD Nuclear Energy Agency. Elsevier, Amsterdam.

Hazen, T.C., Tabak, H.H., 2005. Developments in bioremediation of soils and sediments polluted with metals and radionuclides: 2. Field research on bioremediation of metals and radionuclides. Rev. Environ. Sci. Biotechnol. 4, 157–183.

Hirsch, R., Hamilton, P., Miller, T., Meyers, D. 2008. Water availability—the connection between water use and quality. USGS fact sheet 2008–3015. http://pubs.usgs.gov/fs/2008/3015/

Istok, J.D., Senko, J.M., Krumholz, L.R., Watson, D., Bogle, M.A., Peacock, A., et al., 2004. In situ bioreduction of technetium and uranium in a nitrate-contaminated aquifer. Environ. Sci. Technol. 38, 468–475.

Jeon, B.-H., Kelly, S.D., Kemner, K.M., Barnett, M.O., Burgos, W.D., Dempsey, B.A., 2004. Microbial reduction of U(VI) at the solid-water interface. Environ. Sci. Technol. 38, 5649–5655.

Jeon, B.-H., Dempsey, B.A., Burgos, W.D., Barnett, M.O., Roden, E.E., 2005. Chemical reduction of U(VI) by Fe(II) at the solid-water interface using natural and synthetic Fe(III) oxides. Environ. Sci. Technol. 39, 5642–5649.

Krepelova, A., Sachs, S., Bernhard, G., 2006. Uranium(VI) sorption onto kaolinite in the presence and absence of humic acid. Radiochim. Acta 94, 825–833.

Liger, E., Charlet, L., Van Cappellen, P., 1999. Surface catalysis of uranium(VI) reduction by iron(II). Geochim. Cosmochim. Acta 63, 2939–2955.

Liu, C., Zachara, J.M., Smith, S.C., McKinley, J.P., Ainsworth, C.C., 2003. Desorption kinetics of radiocesium from subsurface sediments at Hanford Site, USA. Geochim. Cosmochim. Acta 67, 2893–2912.

Liu, C., Jeon, B.-H., Zachara, J.M., Wang, Z., Dohnalkova, A., Fredrickson, J.K., 2006. Kinetics of microbial reduction of solid phase U(VI). Environ. Sci. Technol. 40, 6290–6296.

Lloyd, J.R., Renshaw, J.C., 2005. Bioremediation of radioactive waste: radionuclide-microbe interactions in laboratory and field-scale studies. Curr. Opin. Biotechnol. 16, 254–260.

Lovley, D.R., Phillips, E.J.P., 1992. Reduction of uranium by *Desulfovibrio desulfuricans*. Appl. Environ. Microbiol. 58, 850–856.

Lovley, D.R., Phillips, E.J.P., Gorby, Y.A., Landa, E.R., 1991. Microbial reduction of uranium. Nature 350, 413–416.

Marshall, M.J., Beliaev, A.S., Dohnalkova, A.C., Kennedy, D.W., Shi, L., Wang, Z., 2006. c-Type cytochrome-dependent formation of U(IV) nanoparticles by Shewanella oneidensis. PLoS Biol. 4.

McKinley, J.P., Zachara, J.M., Smith, S.C., Liu, C., 2007. Cation exchange reactions controlling desorption of $^{90}Sr^{2+}$ from coarse-grained contaminated sediments at the Hanford site, Washington. Geochim. Cosmochim. Acta 71, 305–325.

Missana, T., Maffiotte, C., García-Gutiérrez, M., 2003. Surface reactions kinetics between nanocrystalline magnetite and uranyl. J. Colloid Interface Sci. 261, 154–160.

Morris, K., Raiswell, R., 2002. Biogeochemical cycles and remobilization of the actinide elements. In: Keith-Roach, M.J., Livens, F.R. (Eds.) Interactions of Microorganisms with Radionuclides. Elsevier, Amsterdam.

N'Guessan, A.L., Vrionis, H.A., Resch, C.T., Long, P.E., Lovley, D.R., 2008. Sustained removal of uranium from contaminated groundwater following stimulation of dissimilatory metal reduction. Environ. Sci. Technol. 42, 2999–3004.

NRC, 2008. 10 CFR Part 40 Appendix A. http://www.nrc.gov/reading-rm/doc-collections/cfr/part040/part040-appa.html

NRC, 2004. Medical use of radioactive materials. http://www.nrc.gov/reading-rm/doc-collections/fact-sheets/med-use-radioisotopes-bg.html

NRC, 2006a. Tritium, radiation protection limits, and drinking water standards. http://www.nrc.gov/reading-rm/doc-collections/fact-sheets/tritium-radiation-fs.html

NRC, 2006b. Uranium mill tailings. http://www.nrc.gov/reading-rm/doc-collections/fact-sheets/mill-tailings.pdf

NRC, 2007. Radioactive waste. http://www.nrc.gov/reading-rm/doc-collections/fact-sheets/radwaste.html

O'Loughlin, E.J., Kelly, S.D., Cook, R.E., Csencsits, R., Kemner, K.M., 2003. Reduction of uranium(VI) by mixed iron(II)/iron(III) hydroxide (green rust): Formation of UO_2 nanoparticles. Environ. Sci. Technol. 37, 721–727.

Payne, T.E., Davis, J.A., Waite, T.D., 1996. Uranium adsorption on ferrihydrite-effects of phosphate and humic acid. Radiochim. Acta 74, 239–243.

Peper, S.M., Brodnax, L.F., Field, S.E., Zehnder, R.A., Valdez, S.N., Runde, W.H., 2004. Kinetic study of the oxidative dissolution of UO_2 in aqueous carbonate media. Ind. Eng. Chem. Res. 43, 8188–8193.

Peretyazhko, T., Zachara, J.M., Heald, S.M., Kukkadapu, R.K., Liu, C., Plymale, A.E., et al., 2008. Reduction of Tc(VII) by Fe(II) sorbed on Al (hydr)oxides. Environ. Sci. Technol. 42, 5499–5506.

Scott, T.B., Allen, G.C., Heard, P.J., Randell, M.G., 2005. Reduction of U(VI) to U(IV) on the surface of magnetite. Geochim. Cosmochim. Acta 69, 5639–5646.

Singer Jr., D.M., Farges, F., Brown, G.E., 2007. Biogenic UO2-characterization and surface reactivity. AIP Conf. Proc. 882, 277–279.

Stucki, J.W., Lee, K., Goodman, B.A., Kostka, J.E., 2007. Effects of in situ biostimulation on iron mineral speciation in a sub-surface soil. Geochim. Cosmochim. Acta 71, 835–843.

Suzuki, Y., Kelly, S.D., Kemner, K.M., Banfield, J.F., 2002. Nanometre-size products of uranium bioreduction. Nature 419, 134.

Suzuki, Y., Kelly, S.D., Kemner, K.M., Banfield, J.F., 2003. Microbial populations stimulated for hexavalent uranium reduction in uranium mine sediment. Appl. Environ. Microbiol. 69, 1337–1346.

Szabo, Z., DePaul, V., 1998. Radium-226 and Radium-228 in shallow ground water, Southern New Jersey: USGS fact sheet 062-98. http://www.pubs.er.usgs.gov/usgspubs/fs/fs06298

Tabak, H.H., Lens, P., Van Hullebusch, E.D., Dejonghe, W., 2005. Developments in bioremediation of soils and sediments polluted with metals and radionuclides-1. Microbial processes and mechanisms affecting bioremediation of metal contamination and influencing metal toxicity and transport. Rev. Environ. Sci. Biotechnol. 4, 115–156.

Tso, T.C., Hallden, N.A., Alexander, L.T., 1964. Radium-226 and polonium-210 in leaf tobacco and tobacco soil. Science 146, 1043–1045.

Ulrich, K., Singh, A., Schofield, E.J., Bargar, J.R., Veeramani, H., Sharp, J.O., Bernier-Latmani, R., Giammar, D.E., 2008. Dissolution of Biogenic and Synthetic UO2 under Varied Reducing Conditions. Environ. Sci. Technol. 42, 5600–5606.

Van Hullebusch, E.D., Lens, P.N.L., Tabak, H.H., 2005. Developments in bioremediation of soils and sediments polluted with metals and radionuclides. 3. Influence of chemical speciation and bioavailability on contaminants immobilization/mobilization bio-processes. Rev. Environ. Sci. Biotechnol. 4, 185–212.

Vrionis, H.A., Anderson, R.T., Ortiz-Bernad, I., O'Neill, K.R., Resch, C.T., Peacock, A.D., et al., 2005. Microbiological and geochemical heterogeneity in an in situ uranium bioremediation field site. Appl. Environ. Microbiol. 71, 6308–6318.

Waite, T.D., Davis, J.A., Payne, T.E., Waychunas, G.A., Xu, N., 1994. Uranium(VI) adsorption to ferrihydrite: application of a surface complexation model. Geochim. Cosmochim. Acta 58, 5465–5478.

Wersin Jr., P., Hochella, M.F., Persson, P., Redden, G., Leckie, J.O., Harris, D.W., 1994. Interaction between aqueous uranium (VI) and sulfide minerals: spectroscopic evidence for sorption and reduction. Geochim. Cosmochim. Acta 58, 2829–2843.

World Health Organization, 2005. Uranium in drinking water. WHO report WHO/SDE/WSH/ 03.04/118. http://www.who.int/water_sanitation_health/dwq/chemicals/uranium290605.pdf

Wu, W.-M., Carley, J., Gentry, T., Ginder-Vogel, M.A., Fienen, M., Mehlhorn, T., 2006. Pilot-scale in situ bioremedation of uranium in a highly contaminated aquifer. 2. Reduction of U(VI) and geochemical control of U(VI) bioavailability. Environ. Sci. Technol. 40, 3986–3995.

Wu, W.-M., Carley, J., Luo, J., Ginder-Vogel, M.A., Cardenas, E., Leigh, M.B., et al., 2007. In situ bioreduction of uranium (VI) to submicromolar levels and reoxidation by dissolved oxygen. Environ. Sci. Technol. 41, 5716–5723.

Chapter 11

Volatile and Semivolatile Contaminants

Satinder Ahuja

Ahuja Academy of Water Quality at UNCW, Calabash, NC, USA

INTRODUCTION

As mentioned in the previous chapters, a large number of contaminants can find their way into the water sources we use for drinking water. Of these contaminants, volatile and semivolatile contaminants can enter directly from various spills by improper disposal, or from the atmosphere in the form of rain, hail, and snow. Snow and rain are nature's way of providing freshwater; however, now they are generally contaminated with various pollutants that we release into the atmosphere, most of which are volatile contaminants. According to a USGS report (U.S. Geological Survey, 2006), volatile organic compounds (VOCs) are a group of organic compounds with inherent physical and chemical properties that allow these compounds to move between water and air. In general, VOCs have high vapor pressure, low-to-medium water solubilities, and low molecular weights. By contrast, semivolatile organic compounds (SOCs) have a higher boiling point than VOCs; however, they can be volatilized under various environmental conditions and can pollute water (see Section SEMIVOLATILE COMPOUNDS).

BOILING POINTS OF SOME VOCs

VOCs are also called purgeable organic compounds because their volatile nature allows them to be purged by aeration of water supplies. The boiling points of a

Handbook of Water Purity and Quality
Copyright © 2009 by Elsevier Inc. All rights of reproduction in any form reserved.

large number of volatile compounds are given in Table 1 to provide the reader a better understanding of the relative volatility of these compounds.

It may be noticed from the table that these compounds have a wide range of boiling points (from $-24°C$ to $157°C$). Some of these compounds, for example, chloromethane, vinyl chloride, bromomethane, and bromoethane are highly volatile, with boiling points below the ambient temperature; they are found in most inhabited parts of the world. However, a fairly large number of these compounds have boiling points over the boiling point of water ($100°C$). Table 1 also provides the detection limit in nanograms (ng) of these compounds.

CLASSIFICATION OF VOLATILE ORGANIC COMPOUNDS

Various solvents and a number of other VOCs have been used for more than 100 years. For example, chloroform and trihalomethanes (THMs) have been present in chlorinated drinking water since the first application of chlorination in 1908. According to the USGS, as many as 55 VOCs can be found in U.S. groundwater and drinking-water supply wells. VOCs have been detected in water in 90 out of 98 aquifer studies across the nation. The largest frequencies were found in the West, New England, and the Mid-Atlantic states. Twenty percent of the samples from aquifers contained one or more of these compounds at $0.2\,\mu g/L$ level. THMs, which originate as chlorination by-products and solvents, were the most frequently detected VOCs. A few compounds such as methyl *tert*-butyl ether (MTBE), ethylene dibromide, and dichloropropane had regional or local patterns. MTBE was the third most detected VOC. Chloroform was most frequently detected in wells. VOCs may be classified into seven groups as follows:

- Fumigants
- Gasoline hydrocarbons
- Gasoline oxygenates
- Organic synthetic compounds
- Refrigerants
- Solvents
- Trihalomethanes

For further information on some of the VOCs that belong to these groups see Tables 2 and 3.

Sources of Contamination of VOCs and Their Health Effects

According to the U.S. Environmental Protection Agency (EPA), the following volatile organic contaminants are the most likely ones to be found in drinking water:

- Benzene
- Carbon tetrachloride
- Chlorobenzene

TABLE 1 Boiling Points and Detection Limits for Some Volatile Organic Compounds

Compound	Detection Limit (ng)	Boiling Point (°C)
Chloromethane	58	−24
Bromomethane	26	4
Vinyl chloride	14	−13
Chloroethane	21	13
Methylene chloride	9	40
Acetone	35	56
Carbon disulfide	11	46
1,1-Dichloroethene	14	32
1,1-Dichloroethane	12	57
trans-1,2-Dichloroethene	11	48
Chloroform	11	62
1,2-Dichloroethane	13	83
1,1,1-Trichloroethane	8	74
Carbon tetrachloride	8	77
Bromodichloromethane	11	88
1,1,2,2-Tetrachloroethane	23	146
1,2-Dichloropropane	12	95
trans-1,3-Dichloropropene	17	112
Trichloroethene	11	87
Dibromochloromethane	21	122
1,1,2-Trichloroethane	26	114
Benzene	26	80
cis-1,3-Dichloropropene	27	112
Bromoform	26	150
Tetrachloroethene	11	121
Toluene	15	111
Chlorobenzene	15	132
Ethylbenzene	21	136
Styrene	46	145
Trichlorofluoromethane	17	24
Iodomethane	9	43
Acrylonitrile	13	78
Dibromomethane	14	97
1,2,3-Trichloropropane	37	157
Total xylenes	22	138–144

Source: Adapted from EPA (1996).

TABLE 2 VOCs Found in About 1% or More of Aquifier Samples

Compound Name	VOC Group
Chloroform	Trihalomethane
Perchloroethene	Solvent
Methyl *tert*-butyl ether	Gasoline oxygenate
Trichloroethene	Solvent
Toluene	Gasoline hydrocarbon
Dichlorodifluoromethane	Refrigerant
1,1,1-Trichloroethane	Solvent
Chloromethane	Solvent
Bromodichloromethane	Trihalomethane
Trichlorofluoromethane	Refrigerant
Bromoform	Trihalomethane
Dibromochloromethane	Trihalomethane
trans-1,2-Dichloroethene	Solvent
Methylene chloride	Solvent
1,1-Dichloroethane	Solvent

Source: USGS.

TABLE 3 VOCs Found at Concentration(s) of Potential Human Health Concern

Compound Name	VOC Group	Domestic Wells	Public Wells
Trichloroethene	Solvent	X	X
Dibromochloropropane	Fumigant	X	
Perchloroethene	Solvent	X	X
1,1-Dichloroethene	Organic synthesis compound	X	X
1,2-Dichloropropane	Fumigant	X	
Ethylene dibromide	Fumigant	X	
Methylene chloride	Solvent		X
Vinyl chloride	Organic synthesis compound		X

Source: USGS.

- *o*-Dichlorobenzene
- *p*-Dichlorobenzene
- 1,1-Dichloroethylene
- *cis*-1,1-Dichloroethylene
- *trans*-1,1-Dichloroethylene
- Dichloromethane
- 1,2-Dichloroethane
- 1,2-Dichloropropane
- Ethylbenzene
- Styrene
- Tetrachloroethylene
- 1,2,4-Trichlorobenzene
- 1,1,1,-Trichloroethane
- 1,1,2-Trichloroethane
- Trichloroethylene
- Toluene
- Vinyl chloride
- Xylenes

It should be noted that most of these compounds are hydrocarbons or halogenated hydrocarbons. A number of them have been found to be carcinogenic.

The USGS reports that a fairly large number of VOCs can be found in aquifers (Table 2). VOCs that were found in about 1% or more of aquifer samples at the assessment level of 0.2 µg/L are included in the table. Note that the compounds in the table are listed in the order of decreasing detection frequency.

VOCs that were found at concentrations of potential concern to human health in domestic and public wells are given in Table 3. The compounds are listed by decreasing order of concentrations of potential concern.

According to the USGS, the contamination caused by 10 frequently detected VOCs can occur from various sources (see Table 4). The source of contamination generally relates to gasoline or solvents. The most common variables are source and transport.

The sources of contamination of a number of VOCs, along with their potential health effects at maximum contamination level (MCL), are given in Table 5. The primary source of contamination from these chemicals is discharge from chemical, agricultural, pharmaceutical, petroleum, plastic, rubber, textile, and metal degreasing factories. Other sources are leaching from gas storage tanks and landfills and various other industrial activities. Dry-cleaning industries also add their share of pollutants.

SEMIVOLATILE COMPOUNDS

A number of semivolatile compounds can be detected in our water supplies. According to Usenko et al. (2007), SOCs are ubiquitous throughout the

TABLE 4 Positive Associations, in Order of Decreasing Importance, for 10 Frequently Detected VOCs in Aquifers

Compound	Occurrence Associated with	Type of Variable
Gasoline hydrocarbons		
1,2,4-Trimethylbenzene	Few septic systems	S
	Cool climates	T
	Dry climates	T
	Gasoline UST sites	S
Toluene	Cool climates	T
	Old construction	S
	Gasoline LUST sites	S
	Domestic wells	I
	Oxic water	F
	Hydric soils	T
Gasoline oxygenate		
MTBE	Wet climates	T
	MTBE-use areas	S
	Shallow depth to top of well screen	T
	Public wells	I
	Cool climates	T
	Oxic water	F
	Gasoline LUST sites	SI
Solvents		
Chloromethane	Anoxic water	F
	High silt in soil	T
	Undeveloped land	S
Methylene chloride	Domestic wells	I
	Shallow well depth	T
	Septic systems	S
	Sparse sand in soil	T
TCA	Oxic water	F
	Shallow depth to top of well screen	T
	Low soil organic content	T
	Cool climates	S
	Urban land	S
	RCRA facilities	S
	Old construction	S
TCE	Urban land	S
	Oxic water	F
	Wet climates	T

(Continued)

TABLE 4 (Continued)

Compound	Occurrence Associated with	Type of Variable
	Public wells	I
	Sparse hydric soils	T
	Septic systems	S
PCE	Shallow depth to top of well screen	T
	Oxic water	F
	Public wells	I
	Urban land	S
	Septic systems	S
Trihalomethanes		
Bromodichloromethane	Oxic water	F
	Sewer systems	S
	Low groundwater recharge	T
	Public wells	I
Chloroform	Urban land	S
	Oxic water	F
	Wet climates	T
	Public wells	I
	Sparse hydric soils	T
	Septic systems	S
	RCRA facilities	S

Source: USGS.
TCA, 1,1,1,-trichloroethane; TCE, trichloroethene; PCE, perchloroethene; MTBE, methyl tert-butyl ether; RCRA, Resource Conservation and Recovery Act; LUST, leaking underground storage tank; UST, underground storage tank; F, fate; S, source; T, transport; I, indeterminate.

environment because of anthropogenic activities, and they have the potential to accumulate in polar and mountainous regions (Wania and Mackay, 1993; Simonich and Hites, 1995; Blais et al., 1998; Grimalt et al., 2001; Daly and Wania, 2005; Usenko et al., 2005; Hageman et al., 2006; Mast et al., 2006). SOCs can undergo atmospheric long-range transport via large-scale winds (Wania and Mackay, 1993; Simonich and Hites, 1995; Killin et al., 2004; Daly and Wania, 2005; Usenko et al., 2005; Hageman et al., 2006). The EPA has also classified certain SOCs as persistent, bioaccumulative, and toxic (PBT) chemicals (U.S. Environmental Protection Agency, 2007).

In mountainous regions, diurnal winds have the potential to transport SOCs from lower elevation source regions to higher elevations (Daly and Wania, 2005). Due to the potential for transport and deposition of these PBT SOCs to sensitive remote ecosystems, the Western Airborne Contaminant Assessment Project (WACAP) was developed to study the atmospheric deposition of SOCs

TABLE 5 Sources of Volatile Contaminants and Potential Health Effects

Contaminant	MCLG[1] (mg/L)[3]	MCL[2] (mg/L)[3]	Potential Health Effects from Ingestion of Water	Sources of Contaminant in Drinking Water
Benzene	0	0.005	Anemia; decrease in blood platelets; increased risk of cancer	Discharge from factories; leaching from gas storage tanks and landfills
Carbon tetrachloride	0	0.005	Liver problems; increased risk of cancer	Discharge from chemical plants and other industrial activities
Chlorobenzene	0.1	0.1	Liver or kidney problems	Discharge from chemical and agricultural chemical factories
o-Dichlorobenzene	0.6	0.6	Liver, kidney, or circulatory system problems	Discharge from industrial chemical factories
p-Dichlorobenzene	0.075	0.075	Anemia; liver, kidney or spleen damage; changes in blood	Discharge from industrial chemical factories
1,2-Dichloroethane	0	0.005	Increased risk of cancer	Discharge from industrial chemical factories
1,1-Dichloroethylene	0.007	0.007	Liver problems	Discharge from industrial chemical factories
cis-1,2-Dichloroethylene	0.07	0.07	Liver problems	Discharge from industrial chemical factories
trans-1,2-Dichloroethylene	0.1	0.1	Liver problems	Discharge from industrial chemical factories
Dichloromethane	0	0.005	Liver problems; increased risk of cancer	Discharge from drug and chemical factories
1,2-Dichloropropane	0	0.005	Increased risk of cancer	Discharge from industrial chemical factories

Contaminant	MCLG[1][2]	MCL[2][3]	Potential health effects from ingestion of water	Sources of contaminant in drinking water
Ethylbenzene	0.7	0.7	Liver or kidney problems	Discharge from petroleum refineries
Styrene	0.1	0.1	Liver, kidney, or circulatory system problems	Discharge from rubber and plastics factories; leaching from landfills
Tetrachloroethylene	0	0.005	Liver problems; increased risk of cancer	Discharge from factories and dry cleaners
Toluene	1	1	Nervous system, kidney, or liver problems	Discharge from petroleum factories
1,2,4-Trichlorobenzene	0.07	0.07	Changes in adrenal glands	Discharge from textile finishing factories
1,1,1-Trichloroethane	0.20	0.2	Liver, nervous system, or circulatory problems	Discharge from metal degreasing sites and other factories
1,1,2-Trichloroethane	0.003	0.005	Liver, kidney, or immune system problems	Discharge from industrial chemical factories
Trichloroethylene	0	0.005	Liver problems; increased risk of cancer	Discharge from metal degreasing sites and other factories
Vinyl chloride	0	0.002	Increased risk of cancer	Leaching from PVC pipes; discharge from plastics factories
Xylenes (total)	10	10	Nervous system damage	Discharge from petroleum factories; discharge from chemical factories

Source: Adapted from EPA, National Primary Drinking Water Regulations.

[1]Maximum contaminant level goal (MCLG): the level of a contaminant in drinking water below which there is no known or expected risk to health. MCLGs allow for a margin of safety and are nonenforceable public health goals.

[2]Units are in milligrams per liter (mg/L).

[3]Maximum contaminant level (MCL): the highest level of a contaminant that is allowed in drinking water. MCLs are set as close to MCLGs as feasible using the best available treatment technology and taking cost into consideration. MCLs are enforceable standards.

to high-elevation and high-latitude lake catchments in eight national parks throughout the western United States from 2003 to 2005 (Landers et al., 2003). A study estimated that 50–98% of the pesticides in the Rocky Mountains in 2002–2003 snowpack were due to regional sources (Hageman et al., 2006). This was attributed to revolatilization of pesticides from soils, atmospheric transport, and deposition. Westerly winds predominate in the Colorado Rocky Mountains (Burns, 2003). However, atmospheric deposition of nitrogen, in the form of NO_x and NH_3, has been shown to originate from fossil-fuel combustion and agriculture sources in major metropolitan areas <150 km east of the Rocky Mountains (Denver, Boulder, and Fort Collins) and is linked to summer diurnal mountain winds.

Lake sediments preserve the chronology of SOC deposition to lake catchments, and radiometric dating of an undisturbed sediment core permits reconstruction of the historical deposition (Fernandez et al., 1999, 2000; Grimalt et al., 2004a, b). A study was conducted to quantify 98 SOCs in sediment from two high-elevation lake catchments located on either side of the Continental Divide in Rocky Mountain National Park, to reconstruct the history of SOC deposition and to identify possible SOC source regions to the two Rocky Mountain National Park lakes (Usenko et al., 2007). The sediment cores were dated using 210 Pb and 137 Cs and analyzed for polybrominated diphenyl ethers (PBDEs), organochlorine pesticides, phosphorothioate pesticides, thiocarbamate pesticides, amide herbicides, triazine herbicides, polychlorinated biphenyls (PCBs), and polycyclic aromatic hydrocarbons (PAHs) using the method described in Section *EPA Methods for SOCs* on page 252. SOC deposition profiles were reconstructed, and deposition half-lives and doubling times were calculated for U.S. historic-use pesticides (HUPs) and current-use pesticides (CUPs) as well as PBDEs, PCBs, and PAHs. Sediment records indicate that the deposition of CUPs has increased in recent years, whereas the deposition of HUPs has decreased as U.S. restrictions were imposed, but has not been eliminated. This is most likely due to the revolatilization of HUPs from regional soils, atmospheric transport, and deposition. The differences in the magnitude of SOC sediment fluxes, flux profiles, time trends within those profiles, and isomeric ratios suggest that SOC deposition in high-elevation ecosystems is dependent on regional upslope wind directions and site location with respect to regional sources and topographic barriers.

MONITORING VOCs AND SOCs

VOCs can confer some odor to water; drinking water is frequently monitored for any potential contaminants that are responsible for the odor; these are discussed in the next section. The monitoring methods utilized for the analysis of VOCs and SOCs are discussed in two separate sections. For the reader's convenience, some of the EPA methods commonly utilized for the analyses of these compounds have also been grouped in the separate subsections.

Odors in Drinking Water

At times, municipal water may have an odor, and frequently, that odor relates to the chlorination of water. However, it goes without saying that various compounds can impart odor to water. For example, the musty odor in drinking water may be the result of the by-products of blue–green algae. Some people can smell certain compounds at concentrations of 10 parts per trillion (ppt) or less. Therefore, many utilities and beverage companies analyze water for the presence of compounds such as isopropyl-methoxypyrazine (IPMP), isobutyl-methoxypyrazine (IBMP), methylisoborneol (MIB), and geosmine. All of these compounds can be analyzed by solid-phase microextraction (SPME) and gas chromatography (GC) on Supelco's SLB-5 ms column at 60°C (2 min) and then 8°C/min. gradient to 200°C. The detectabilities range from 10 to 20 ppt. Remediation of odors and removal of undesirable VOCs are discussed in Section REMEDIATION.

Analysis of VOCs

VOCs can be analyzed in various ways. However, the most commonly used method is GC with headspace analysis (see section *Chemistry of Static Headspace GC*). A large number of these compounds (see Table 1) were analyzed after sample preparation by EPA method 5041. The detection limits given in the table were determined according to 40 CFR, Part 136, Appendix B, using standards spiked onto clean tubes. As clean tubes were used, the values cited above represent the best ones that the methodology can achieve. The presence of an emission matrix can affect the ability of the methodology to perform at its optimum level. It should be noted that compounds with boiling points >130°C are not suitable for quantitative sampling by EPA method 0030.

Chemistry of Static Headspace GC

An exact volume or weight of sample is placed in a closed sealed vial. This creates two distinct phases in the vial—an aqueous phase containing volatile components and a gaseous phase, or headspace. Volatile components in the sample can partition into the headspace. An aliquot of the headspace can then be analyzed by GC. The equations for the headspace relationship are as follows:

$$A \propto C_G = \frac{C_0}{K + \beta}$$

$$\beta = V_G / V_S$$

$$K = C_S / C_G$$

$$K \propto \frac{1}{P_i^0 - \gamma_i}$$

A = area
β = phase ratio
V_G = volume of gas phase

V_S = volume of sample phase
V_V = total vial volume
C_0 = initial analyte concentration in sample
C_G = analyte concentration in gas phase
C_S = analyte concentration in sample phase
K = partition coefficient
P_i^0 = analyte vapor pressure
γ_i = activity coefficient

Partition coefficient (K) is defined as the equilibrium distribution of an analyte between the sample and the gas phase. Therefore, as per definition, the compounds that have lower K values tend to partition more readily into the gas phase. The relationship of K may also be described as a relationship between analyte vapor pressure and activity coefficient. High salt concentrations and other solvents can decrease the solubility of analytes in the sample phase.

The phase ratio (β) is defined as a volume of the headspace over the volume of the sample in the vial. Lower values for β will yield higher responses for the volatile components with low K values. It should be noted that decreasing β will not always increase the response, because when β is decreased by increasing sample size, compounds with high K values will partition less into headspace, compared to compounds with low K values.

Some of the main techniques utilized for the extraction of organic contaminants are given in Table 6. The table also shows whether a given technique is

TABLE 6 Main Techniques for Extraction/Concentration of Organic Contaminants from Water

Technique	Applicability	Detection Limit (MS)	Analysis Time	Solvent Used (mL)
Headspace	VOCs	ppb	30 min	0
P&T	VOCs	ppb	30 min	0
CLS	VOCs	ppt	2 h	0
LLE	SOCs	ppt	1 h	200
SPE	SOCs	ppt	30 min	50
SPME	VOCs, SOCs	ppt	5 min	0
MIMS	VOCs	ppb	Run time	0
T-MIMS	VOCs, SOCs	ppt	5–30 min	0

Source: Adapted from Kooester and Clement (1993).
ppb, parts per billion; ppt, parts per trillion; P&T, purge and trap; CLS, closed-loop stripping; LLE, liquid–liquid extraction; SPE, solid-phase extraction; SPME, solid-phase microextraction; MIMS, membrane introduction mass spectrometry; T-MIMS, trap MIMS.

applicable for VOCs or SOCs. The detection limit and analysis time are also included in this table.

A large number of compounds can be resolved and identified by GC. The combination of GC with mass spectrometry (MS) increases the versatility of this technique, making it the method of choice for detection and quantification of these contaminants. Of special interest is membrane introduction mass spectrometry (MIMS), as it is an efficient technique for the trace level detection of VOCs in water. Detection limits of a number of VOCs are given in Table 7 at the parts per billion (ppb) level with MIMS. These limits can be extended down to the ppt level with cryotrap membrane introduction mass spectrometry (CT-MIMS).

EPA method 624, commonly used for the SPB-624 column, was modified to fast GC column dimensions, and then sample size, linear velocity, and oven

TABLE 7 CT-MIMS Gains and Detection Limits for a Series of VOCs

VOCs	Monitored Ion (m/z)	MIMS Signal (kcounts)	CT-MIMS Signal (kcounts)	CT-MIMS Gain	MIMS Detection Limit (ppb)	CT-MIMS Detection Limit (ppt)
Benzene	78	11.4	1,099	96	1	10
Toluene	91	11.0	1,139	103	1	10
Xylene	106	10.4	987	95	1	10
Chlorobenzene	112	14.5	1,419	98	1	10
Benzaldehyde	106	1.8	170	95	15	150
Acetone	58	3.4	321	95	10	100
2-Butanone	43	7.9	836	106	5	50
Ethyl ether	59	10.9	1,136	104	1	10
Tetrahydrofuran	72	0.4	35.4	96	50	500
Carbon tetrachloride	117	3.4	311	92	5	50
Chloroform	83	6.0	583	98	2	20
Dichloromethane	49	1.6	193	118	20	200
1,1,2,2-Tetrachloroethane	83	3.0	275	93	5	50
Chlorodibromethane	127	5.3	579	110	2	20

Source: Adapted from Menendes et al. (1996).

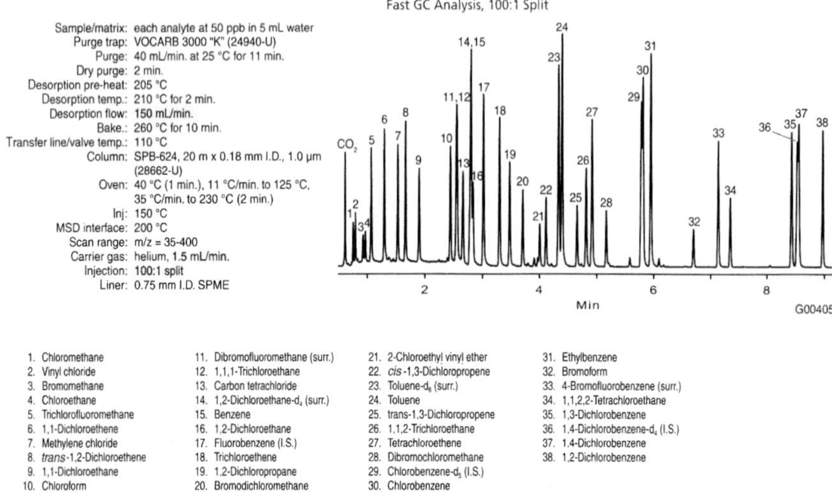

1. Chloromethane
2. Vinyl chloride
3. Bromomethane
4. Chloroethane
5. Trichlorofluoromethane
6. 1,1-Dichloroethene
7. Methylene chloride
8. *trans*-1,2-Dichloroethene
9. 1,1-Dichloroethane
10. Chloroform
11. Dibromofluoromethane (surr.)
12. 1,1,1-Trichloroethane
13. Carbon tetrachloride
14. 1,2-Dichloroethane-d$_4$ (surr.)
15. Benzene
16. 1,2-Dichloroethane
17. Fluorobenzene (I.S.)
18. Trichloroethene
19. 1,2-Dichloropropane
20. Bromodichloromethane
21. 2-Chloroethyl vinyl ether
22. *cis*-1,3-Dichloropropene
23. Toluene-d$_8$ (surr.)
24. Toluene
25. trans-1,3-Dichloropropene
26. 1,1,2-Trichloroethane
27. Tetrachloroethene
28. Dibromochloromethane
29. Chlorobenzene-d$_5$ (I.S.)
30. Chlorobenzene
31. Ethylbenzene
32. Bromoform
33. 4-Bromofluorobenzene (surr.)
34. 1,1,2,2-Tetrachloroethane
35. 1,3-Dichlorobenzene
36. 1,4-Dichlorobenzene-d$_4$ (I.S.)
37. 1,4-Dichlorobenzene
38. 1,2-Dichlorobenzene

FIGURE 1 Fast GC analysis of VOCs.
Source: Sigma-Aldrich.com/gc, vol. 26.1, p. 6

temperature ramp rates were optimized (see Figure 1). The analysis time was reduced from 17 to <10 min.

EPA Methods for VOCs

VOCs are among the most common chemical pollutants tested by the U.S. EPA method TO-14 for their analysis in soil, sludge, drinking water, and wastewater. The great number and chemical diversity of VOCs that can be present in these samples demands utilization of capillary GC columns that are capable of separating almost 100 compounds. This high separating power is usually accomplished with long (75 or 105 m) 0.53 mm ID columns; however, narrow diameter columns (0.25 mm ID) with shorter lengths (30 or 60 m) can also be effective. Columns with specially developed bonded stationary phases provide the polar, polarizable, and dispersive interactions needed to separate large numbers of VOCs. The columns with different bonded stationary phases can also elute VOCs in different orders or can separate pairs of VOCs that cannot be separated by the other columns.

EPA drinking water methods 524 and 8260 implement purge-and-trap (P&T) technology and are widely used for the analysis of VOCs. To ensure reliable analysis, a P&T concentrator must be able to effectively control the large amounts of water being transformed to the adsorbent trap prior to analysis by GC–MS. Minimizing the amount of water being transferred significantly improves water quality. Seventy-five compounds were analyzed by this approach at the 200-ppb level. This approach results in less carryover from high-level samples.

VOCs, such as those tested by EPA method 8260B, are usually determined with a P&T system connected to GC. The column used for the gas chromatograph should have a selective stationary phase to resolve the volatile pollutants

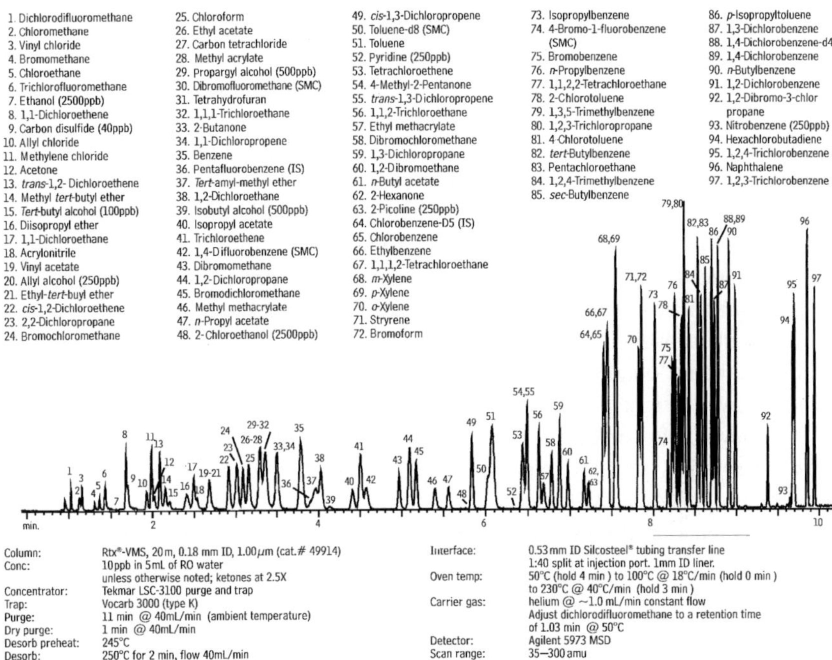

FIGURE 2 Resolution of bromomethane and chloromethane and some other isomeric compounds. *Source*: Restek Tech Guide lit. 59887A

and should have sufficient film thickness to retain and resolve the low-boiling volatile compounds (e.g., dichlordifluoromethane), and should be thermally stable to elute high-boiling volatile compounds such as hexachlorobutane and naphthalene. Figure 2 shows the resolution of a large number of volatile compounds on Rtx-VMS column in <10 min.

EPA method 8260C (SW-846). This analytical method can be utilized for the measurement of volatile organic pollutants in a wide range of matrices ranging from groundwater to aqueous sludge (EPA method 8260C, Rev 3, 2006 http://www.epa.gov/epaoswer/hazwaste/test/pdfs/8260c.pdf). The method lists 100 organic compounds with diverse physical properties that can present significant challenges to analysts. It is useful for analyzing VOCs by GC–MS. This method should be used for the analysis of solid, nonaqueous liquid/organic solid, aqueous liquid, and drinking water samples for the following contaminants: allyl alcohol, 2-chloroethanol (ethylene chlorohydrin), cyanogen chloride, ethylene oxide, propylene oxide, 1,4-dithiane (degradation product of HD), 1,4-thioxane (degradation product of HD). Appropriate sample preparation techniques should be used prior to analysis. *Note*: For carbon disulfide and 1,2-dichloroethane only, EPA method 524.2 (rather than 8260C) should be used for the analysis of drinking water samples.

Volatile compounds are introduced into a GC by P&T or other procedures (see Section 1.2 in Method 8260C). The analytes can be introduced directly to a wide-bore capillary column or cryofocused on a capillary precolumn before being flash-evaporated to a narrow-bore capillary for analysis. Alternatively, the effluent from the trap is sent to an injection port operating in the split mode for injection to a narrow-bore capillary column. The column is temperature-programed to separate the analytes, which are then detected with a mass spectrometer (MS interfaced to the GC). Analytes eluted from the capillary column are introduced into the MS via a jet separator or a direct connection. The estimated quantitation limit (EQL) for an individual analyte is dependent on the instrument as well as the choice of sample preparation/introduction method. Using standard quadrupole instrumentation and the P&T technique, EQLs are $5\,\mu g/kg$ (wet weight) for soil/sediment samples and $5\,\mu g/L$ for groundwater.

Somewhat lower limits may be achieved using an ion-trap MS or other instrumentation of improved design. No matter which instrument is used, EQLs will be proportionately higher for sample extracts and samples that require dilution or when a reduced sample size is used to avoid saturation of the detector.

Analysis of SOCs

As mentioned earlier, an analytical method was developed for trace analysis of 98 SOCs in remote, high-elevation lake sediment. Sediment cores from Lone Pine Lake (west of the Continental Divide) and Mills Lake (east of the Continental Divide) in Rocky Mountain National Park, Colorado. Sediment samples were allowed to thaw in sealed glass jars, ground with sodium sulfate, and extracted using a pressurized liquid extraction. Interferences were removed from the sediment extract using a 20-g silica solid-phase extraction cartridge and gel permeation chromatography. The sediment extracts were analyzed for target SOCs by GC–MS, using both electron impact ionization and electron capture negative ionization with selective ion monitoring as described in detail by Usenko et al. (2005) and Ackerman et al. (2005).

EPA Methods for SOCs

EPA method 8270D (SW-846). This is a challenging method that covers a wide range of compound classes: neutral, acidic, and basic compounds including anilines, phenols, PAHs, etc., that differ in both volatility and reactivity (EPA method 8270D (SW-846): Semivolatile Organic Compounds by Gas Chromatography/Mass Spectrometry (GC/MS), Revision 4, 1998. http://www.epa.gov/epaoswer/hazwaste/test/pdfs/8270d.pdf). A number of compounds were resolved on Rxi 5SilMs column (Figure 3).

Some of the semivolatile compounds that are frequently analyzed by GC–MS using method 8270D (Revision February 4, 2007) are shown in Table 8.

This method should be used for the analysis of solid, nonaqueous liquid/organic solid, aqueous liquid, and drinking water samples for the contaminants

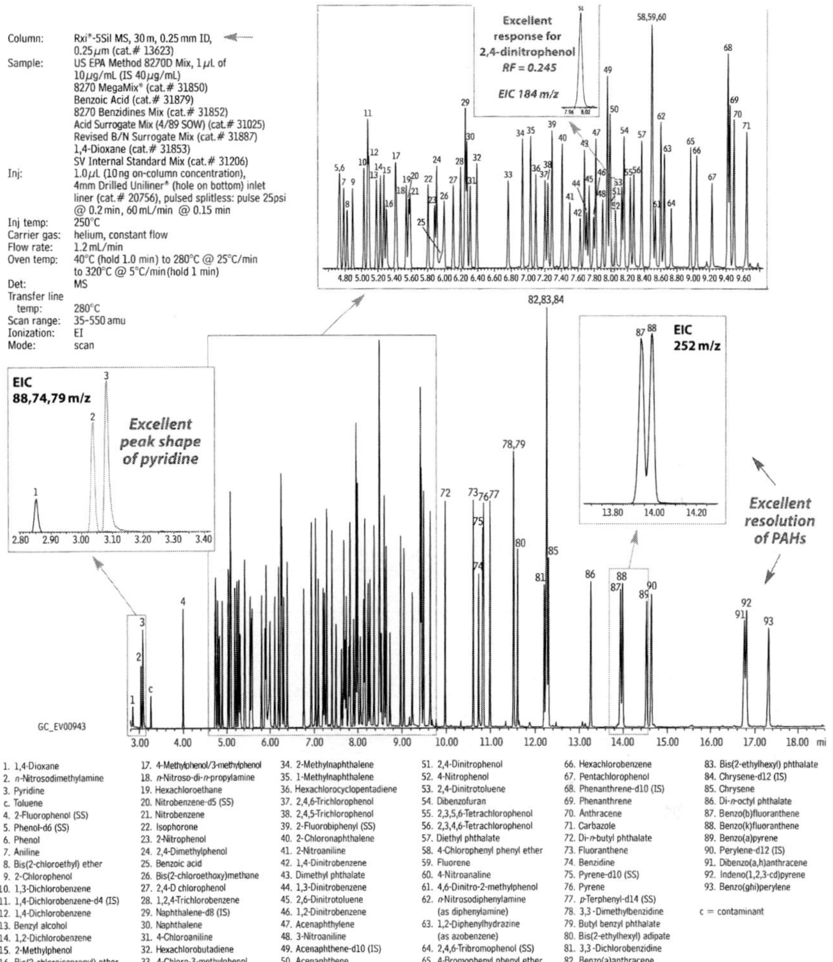

FIGURE 3 GC separation of SOCs.
Source: Restek Tech Guide.

identified below. Appropriate sample preparation techniques should be used prior to analysis. *Note.* For dichlorvos, fenamiphos, mevinphos, and SOCs only, EPA method 525.2 (rather than method 8270D) can be used for the analysis of drinking water samples. For organophosphate pesticides only, EPA methods 614 and 507 should be used for the analysis of aqueous liquid and drinking water samples, respectively. For chloropicrin only, EPA method 551.1 may be used for the analysis of aqueous liquid and drinking water samples.

Samples are prepared for analysis by GC–MS, using the appropriate sample preparation and, if necessary, sample cleanup procedures. SOCs are introduced into the GC–MS by injecting the sample extract into a GC with a narrow-bore

TABLE 8 Semivolatile Compounds Analyzed by Method 8270D

Analyte(s)
Chlorfenvinphos
3-Chloro-1,2-propanediol[1]
Chloropicrin[2]
Chlorosarin
Chlorosoman
Chlorpyrifos
Crimidine[3]
Cyclohexyl sarin (GF)
Dichlorvos
Dicrotophos
Diisopropyl methylphosphonate (DIMP) (degradation product of GB)
Dimethylphosphite
1,4-Dithiane (degradation product of HD)[4]
Ethyldichloroarsine (ED)
Fenamiphos
Methamidophos
Methyl hydrazine (monomethylhydrazine)
Methyl parathion
1-Methylethyl ester ethylphosphonofluoridic acid (GE)
Mevinphos
Mustard, nitrogen (HN-1) [*bis*(2-chloroethyl) ethylamine]
Mustard, nitrogen (HN-2) [2,2'- dichloro-N-methyldiethylamine N,N-*bis*(2-chloroethyl) methylamine]
Mustard, nitrogen (HN-3) [*tris*(2-chloroethyl)amine]
Mustard, sulfur (HD)/mustard gas (H)[5]
Nicotine sulfate
Organophosphate pesticides, NOS[6]
Parathion
Phencyclidine
Phenol
Phorate

(Continued)

TABLE 8 (Continued)

Phosphamidon
R 33 (VR) [methylphosphonothioic acid, S-[2-(diethylamino)ethyl] O-2-methylpropyl ester]
Sarin (GB)[5]
Semivolatile organic compounds, NOS[6]
Soman (GD)
Strychnine
Tabun (GA)
Tear gas (CS) [chlorobenzylidene malonitrile]
Tetraethyl pyrophosphate
Tetramethylene-disulfotetramine2,7,8
Thiodiglycol (TDG) (degradation product of HD)
1,4-Thioxane (degredation product of HD)[4]
Trimethyl phosphate
VE [phosphonothioic acid, ethyl-, S-(2-(diethylamino)ethyl) O-ethyl ester]
VG [phosphonothioic acid, S-(2-(diethylamino)ethyl) O,O-diethyl ester]
VM [phosphonothioic acid, methyl-,S-(2-(diethylamino)ethyl) O-ethyl ester]
VX [O-ethyl-S-(2-diisopropylaminoethyl) methyl-phosphonothiolate][5]
Dimethylphosphoramidic acid (degradation product of GA)[9]
EA2192 [diisopropylaminoethyl methylthiophosphonate] (hydrolysis product of VX)[9]
Ethyl methylphosphonic acid (EMPA) (degradation product of VX)[9]
Isopropyl methylphosphonic acid (IMPA) (degradation product of GB)[9]
Methylphosphonic acid (MPA) (degradation product of VX, GB, and GD)[9]
Pinacolyl methyl phosphonic acid (PMPA) (degradation product of GD)[9]

[1]*For this analyte, SW-846 method 8270D must be modified to include a derivatization step.*
[2]*This analyte requires determination using an injection port temperature of less than 200°C.*
[3]*If problems occur when using this method, it is recommended that SW-846 method 8321B be used.*
[4]*If problems occur when using this method, it is recommended that SW-846 method 8260C and appropriate corresponding sample preparation procedures (i.e., 5035A for solid samples, 3585 for nonaqueous liquid/organic solid samples, and 5030C for aqueous liquid and drinking water samples) be used.*
[5]*For this analyte, refer to EPA SW-846 method 8271 for GC–MS conditions.*
[6]*NOS, not otherwise specified.*
[7]*This analyte may require SIM analyses in order to be determined.*
[8]*When analyzing for tetramine, the injection temperature must not exceed 250°C (the decomposition temperature of tetramine).*
[9]*This analyte should be determined only by this method if LC–MS (electrospray) procedures are not available to the laboratory. This analyte should be analyzed with GC–MS procedures using derivatization based on SW-846 method 8270D. Sample preparation methods should remain the same. Both electrospray LC–MS and GC–MS derivatization procedures are currently under development.*

fused-silica capillary column. The GC column is temperature-programmed to separate the analytes, which are then detected with an MS connected to the GC. Analytes eluted from the capillary column are introduced into the MS. For the determination of 3-chloro-1,2-propanediol, dimethylphosphoramidic acid, EA2192, EMPA, IMPA, MPA, and pinacolyl methyl phosphonic acid, a derivatization step is required prior to injection into the GC–MS. The phosphonic acids require derivatization with a trimethylsilyl agent, and 3-chloro-1,2-propanediol requires derivatization with a heptafluorobutyryl agent. The estimated detection limits vary with each analyte and range between 10 and 1000 µg/L for aqueous liquid samples and 660 and 3300 µg/kg for soil samples. The analytical range depends on the target analyte(s) and the instrument used.

REMEDIATION

In a field study, the performance of Ambersorb 563 adsorbent was evaluated in treating groundwater contaminated with VOCs. This adsorbent yielded a significant recovery of VOCs, and the treated groundwater consistently met drinking water standards. Treatment of groundwater contaminated with low concentrations of chlorinated organics may be costly to remediate. Ambersorb 563 carbonaceous adsorbent offers an alternative to granular activated carbon (GAC), at a significantly reduced operating cost.

A field demonstration study was performed by Roy F. Weston, Inc., to evaluate Ambersorb 563 adsorbent for remediating groundwater contamination by VOCs (U.S. Geological Survey, 2006). The study was conducted at Site 32/36 at Pease Air Force Base, Newington, New Hampshire, where groundwater was found to be contaminated with vinyl chloride, 1,1-dichloroethene, *cis*-1,2-dichloroethene, *trans*-1,2-dichloroethene, and trichloroethene. Concentrations ranged from ppb to low ppm for trichloroethene. The study included four service cycles, three steam regenerations, and one superloading in which the aqueous condensate from steam regeneration of an Ambersorb 563 service column was treated using a smaller column containing Ambersorb 563 adsorbent. (After superloading treatment, the aqueous condensate was discharged as part of the treated water stream.) A one-gallon-per-minute continuous pilot system was used, consisting of two adsorbent columns operating in parallel or in series. In the first service cycle, the columns were operated in parallel for direct comparison of Ambersorb 563 adsorbent with Filtrasorb® 400 GAC. In other cycles, two Ambersorb 563 columns were used in series to investigate the effect of multiple service cycles and steam regeneration on adsorbent performance.

CONCLUSIONS

Many compounds fit the categories of volatile and semivolatile compounds. A fairly long list of these compounds, their source of contamination, and negative health effects are given in this chapter. A number of methods recommended by EPA and the improvements offered by various scientists are also included in this chapter.

REFERENCES

Ackerman, L.K., Wilson, G.R., Simonich, S.L., 2005. Quantitative analysis of 39 polybrominated diphenyl ethers by isotope dilution GC/low-resolution MS. Anal. Chem. 77, 1979–1987.

Blais, J.M., Schindler, D.W., Muir, D.C.G., Kimpe, L.E., Donald, D.B., Rosenberg, B., 1998. Accumulation of persistent organochlorine compounds in mountains of western Canada. Nature 395, 585–588.

Burns, D.A., 2003. Atmospheric nitrogen deposition in the Rocky Mountains of Colorado and southern Wyoming: a review and new analysis of past study results. Atmos. Environ. 37, 921–932.

Daly, G.L., Wania, F., 2005. Organic contaminants in mountains. Environ. Sci. Technol. 39, 385–398.

EPA method 8260B-78, 1996. CD-Rom, Revision 2.

Fernandez, P., Vilanova, R.M., Grimalt, J.O., 1999. Sediment fluxes of polycyclic aromatic hydrocarbons in European high-altitude mountain lakes. Environ. Sci. Technol. 33, 3716–3722.

Fernandez, P., Vilanova, R.M., Martinez, C., Appleby, P., Grimalt, J.O., 2000. The historical record of atmospheric pyrolytic pollution over Europe registered in the sedimentary PAH from remote mountain lakes. Environ. Sci. Technol. 34, 1906–1913.

Grimalt, J.O., Fernandez, P., Berdie, L., Vilanova, R.M., Catalan, J., Psenner, R., 2001. Selective trapping of organochlorine compounds in mountain lakes of temperate areas. Environ. Sci. Technol. 35, 2690–2697.

Grimalt, J.O., Drooge, B.L.V., Ribes, A., Fernandez, A.E., Appleby, P., 2004a. Polycyclic aromatic hydrocarbon composition in soils and sediments of high altitude lakes. Environ. Pollut. 131, 13–24.

Grimalt, J.O., van Drooge, B.L., Ribes, A., Vilanova, R.M., Fernandez, P., Appleby, P., 2004b. Persistent organochlorine compounds in soils and sediments of European high altitude mountain lakes. Chemosphere 54, 1549–1561.

Hageman, K.J., Simonich, S.L., Campbell, D.H., Wilson, G.R., Landers, D.H., 2006. Atmospheric deposition of current-use and historic-use pesticides in snow at national parks in the western United States. Environ. Sci. Technol. 40, 3174–3180.

Killin, R.K., Simonich, S.L., Jaffe, D.A., DeForest, C.L., Wilson, G.R., 2004. Transpacific and regional atmospheric transport of anthropogenic semivolatile organic compounds to Cheeka Peak Observatory during the spring of 2002. J. Geophys. Res., 109.

Kooester, C.J., Clement, R.E., 1993. Crit. Rev. Anal. Chem. 24, 263.

Landers, D.H., Simonich, S.L., Campbell, D.H., Erway, M.M., Geiser, L., Jaffe, D., et al., 2003. Western Airborne Contaminants Assessment Project Research Plan, EPA/600/R-03/035. U.S. Environmental Protection Agency, Office of Research and Development, NHEERL, Western Ecology Division, Corvallis, OR. www2.nature.nps.gov/air/Studies/air_toxics/wacap.htm

Mast, M.A., Foreman, W.T., Skaates, S., 2006. Organochlorine compounds and current-use pesticides in snow and lake sediment in Rocky Mountain National Park, Colorado, and Glacier National Park, Montana, 2002–03. U.S. geological survey scientific investigations report 2006–5119, U.S. Geological Survey, Washington, D.C.

Menendes, M.A., Pimpim, R.S., Kotiaho, T., Eberlin, M.N., 1996. Anal. Chem. 68, 3502.

Simonich, S.L., Hites, R.A., 1995. Global distribution of persistent organochlorine compounds. Science 269, 1851–1854.

Usenko, S., Hageman, K.J., Schmedding, D.W., Wilson, G.R., Simonich, S.L., 2005. Trace analysis of semivolatile organic compounds in large volume samples of snow, lake water, and groundwater. Environ. Sci. Technol. 39, 6006–6015.

Usenko, S., Landers, D.H., Appleby, P.G., Simonich, S., 2007. Current and historical deposition of PBDEs, pesticides, PCBs, and PAHs to Rocky Mountain National Park. Environ. Sci. Technol. 41, 7235.

U.S. Environmental Protection Agency, 2007. Persistent Bioaccumulative and Toxic (PBT) Chemical Program. U.S. Environmental Protection Agency, Washington, D.C. http://www.epa.gov/pbt/index.htm

U.S. Geological Survey, 2006. Volatile organic compounds in the nation's groundwater and drinking-water supply wells, circular 1292.

Wania, F., Mackay, D., 1993. Global fractionation and cold condensation of low volatility organochlorine compounds in polar regions. Ambio 22, 10–12.

Chapter 12

Monitoring Disinfectants

Taha F. Marhaba

Department of Civil and Environmental Engineering, New Jersey Institute of Technology, 323 MLK Blvd., Newark, NJ 07102, USA

INTRODUCTION

Disinfection is a process that deliberately reduces the number of pathogenic microorganisms in water to achieve the principal objective of drinking water treatment: protection of public health. Chemical disinfection has been an integral part of the drinking water treatment process in the United States since the introduction of chlorine as a disinfectant in the early 1900s. Chlorine, along with some other disinfectants such as ozone and chlorine dioxide, was found to provide additional benefits including color, taste, and odor reduction. Therefore, the use of chemical disinfectants was made in as large quantities as required to achieve the desired quality. Although chlorine was known to react with organic

Handbook of Water Purity and Quality
Copyright © 2009 by Elsevier Inc. All rights of reproduction in any form reserved.

material in water, it was only in the early 1970s that scientists were able to identify the formation of chloroform ($CHCl_3$) and other volatile halogen-substituted organics in drinking water (Rook, 1971, 1974). These compounds were related to chlorine and were termed "by-products" of chlorination. These findings led to a large number of studies to learn about the formation of by-products and their effects. As more became known about the potential by-products, it was also found that alternative chemical disinfectants (such as ozone and chlorine dioxide) form by-products of their own. The United States Environmental Protection Agency (USEPA) regulates disinfection by-products (DBPs) and also the amount of disinfectants that can be used in drinking water. The balance between the risk of microbial contamination and DBP formation is still a challenge. Currently, there are several options for the disinfection of drinking water, each with its own merits as a disinfectant and the presence of by-products that must be minimized.

THEORY OF DISINFECTION

In waterworks practice, the disinfection process is expected to satisfy the following three requirements: (1) inactivation of the pathogenic and other harmful microorganisms in water (primary disinfection), (2) disinfectant residual maintenance in the distribution system (secondary disinfection), and (3) keeping the amount of by-products to a minimum. Different disinfectants offer different performances toward the achievement of these three requirements. This is mainly because the characteristics of a disinfectant that make it suitable for each of the three requirements are not the same. Today, in drinking water treatment, the following five disinfection agents are commonly used:

1. *Free chlorine* is a strong oxidant that rapidly kills most of the microorganisms. It is also by far the most commonly used disinfection agent. To lower the cost and, more importantly, to avoid the danger of the release of toxic chlorine gas, a relatively inexpensive sodium hypochlorite solution, which releases free chlorine upon dissolution in water is used.
2. *Combined chlorine (chloramines)* is not a very strong oxidant, but it is used for its ability to provide longer-lasting free chlorine residual after disinfection.
3. *Ozone* is the strongest oxidant in the list, and it also provides control over taste- and odor-producing compounds such as methyl isoborneol and geosmin. Use of ozone as a disinfection agent is becoming increasingly common.
4. *Chlorine dioxide* is also a fast-acting disinfection agent, but it is not often used because of the possibility of the production of excessive amounts of chlorite, which is regulated by USEPA under stage-2 disinfectant/disinfection by-products (D/DBP) rule.
5. *UV light* disinfection uses electromagnetic radiation. This method has two disadvantages: water must have a low level of color for it to work efficiently, and it does not leave any residual disinfectant.

TABLE 1 Characteristics of Commonly Used Disinfectants

Consideration	Disinfection Agent				
	Free Chlorine	**Combined Chlorine**	**Ozone**	**Chlorine Dioxide**	**Ultraviolet Light**
Effectiveness	Excellent	Fair	Excellent	Excellent	Good
Cost	Low	Moderate	High	Moderate	High
Size of plant	All sizes	All sizes	Medium to large	Small to medium	Small to medium
Safety concern	High	High	Moderate	High	Low
Toxicity at operating temperatures	High	High	High	High	High
Residual	Long	Long	None	Moderate	None
Odor removal	Moderate to high	Moderate	High	High	N/A
Contact time	Moderate	Moderate	Short	Moderate	Short
pH dependency	High	High	Low	Low	None
Regulatory limit on residuals (USEPA, 2007)	4 mg/L	4 mg/L	0.8 mg/L	N/A	N/A
Solubility	Moderate	High	High	High	Moderate
Frequency of use as primary disinfectant	High	Moderate	Moderate	Low	Low, but increasing
Stability	High	Moderate	Low	Low	N/A

N/A, not applicable.

Table 1 summarizes information on each of these common disinfection agents.

Mechanisms of Disinfectants

Chlorine compounds and ozone are used for disinfection processes because for they are strong oxidizing agents. However, oxidation is not the only process that governs disinfection, but more interactions happen simultaneously and contribute to inactivation of microorganisms in water. Action of a disinfection agent can be explained through the following five mechanisms:

1. *Cell wall destruction* that results in cell lysis and death. Ozone causes direct oxidation/destruction of cell walls. Chlorine hydrolyzes the cell, causing mechanical disruption.
2. *Cell permeability alteration* that can result in loss of selective permeability of cytoplasmic membrane, allowing nitrogen, phosphorous, and other vital nutrients to flow out. Chlorine alters the cell wall permeability, resulting in lethal damage to the cell. Ozone can break the carbon–nitrogen bond, causing nitrogen to escape.
3. *Alteration of the colloidal nature of the protoplasm* by heat, radiation, and highly alkaline or acidic agents that causes coagulation or denaturing of the cell protein, that in turn produces permanent cell damage. Chlorine compounds and UV radiation can alter the colloidal nature of the protoplasm.
4. *Deoxyribonucleic acid (DNA) or ribonucleic acid (RNA) alteration* that results in disruption of the replication process (because DNA and RNA carry genetic information for reproduction) and inactivation of the organism. Ozone can damage the constituents of nucleic acids, but UV radiation alters the DNA or RNA by causing formation of double bonds within the cells of microorganisms as well as rupturing DNA. DNA or RNA absorb the UV photon, causing formation of a double bond.
5. *Inhibition of enzyme activity* by strong oxidizing agents, altering the enzyme's chemical properties. Chlorine and ozone are capable of causing inactivation of enzymes.

Factors Influencing the Action of Disinfectants

As discussed earlier, different disinfectants have different characteristics that give them certain advantages or disadvantages over one another. However, there are many other influencing factors that govern the choice of the right disinfectant. The following are some of the important factors that must be considered:

1. *Contact time* is perhaps the most important of all the factors that influence the disinfection process. Chick's law governs the effect of contact time on the action of disinfectant (Chick, 1908). In simple words, Chick's law states that for a given concentration of disinfectant, the degree of disinfection achieved is directly proportional to the contact time (Chick, 1908).

$$\frac{\mathrm{d}N_t}{\mathrm{d}t} = -kN_t \quad \text{or} \quad \ln\frac{N_t}{N_0} = -kt \tag{1}$$

where $\mathrm{d}N_t/\mathrm{d}t$ = rate of change in organism concentration with time
t = time
k = inactivation rate constant (1/time units), which can be obtained by plotting $-\ln(N_t/N_0)$ against the contact time t

N_t = number of microorganisms at time t
N_0 = number of microorganisms at time $t = 0$

2. *Concentration of disinfectant* governs the inactivation rate constant. This relationship is explained by Watson equation (Watson, 1908).

$$k = k'C^n \qquad (2)$$

where k = inactivation rate constant (1/time units)
k' = die-off constant
C = disinfectant concentration
n = dilution coefficient

3. *Temperature* is another important factor that influences the degree of *disinfection* achieved. A form of van't Hoff–Arrhenius equation shown below can explain the effect of temperature (IUPAC Goldbook Definition, 1997).

$$\ln \frac{t_1}{t_2} = \frac{E(T_2 - T_1)}{RT_1T_2} \qquad (3)$$

where T_1, T_2 = temperatures (unit, K)
t_1, t_2 = time to achieve given percentage kill at T_1 and T_2, respectively
E = energy of activation (J/mole or cal/mole)
R = universal gas constant (8.3114 J/mole K or 1.99 cal/mole K)

Some typical values for the activation energy for aqueous chlorine and chloramines at normal temperatures and at different pH values are given in Table 2 (Fair et al., 1948; Metcalf & Eddy et al., 2003).

TABLE 2 Typical Activation Energy Values for Chlorine Compounds[1]

Compound	pH	E (cal/mole)
Aqueous chlorine	7.0	8,200
	8.5	6,400
	9.8	12,000
	10.7	15,000
Chloramines	7.0	12,000
	8.5	14,000
	9.5	20,000

For E values in J/mol use E (J/mol) = E (cal/mole) × (4.1781).
[1]*(Fair et al., 1948; Metcalf and Eddy, 2003)*

4. *Types of organisms* present in water influence the action of disinfectant and also the choice of the best disinfectant. The nature and condition of microorganisms will also decide how effectively the disinfectant will work. For example, older bacterial cells that have developed a slime coating are killed at a much slower rate compared with the rate of killing of growing, viable cells. Viruses and protozoa may have considerably different killing rate with each disinfectants. Therefore, sometimes a combination of chemical disinfectant and heat or UV radiation is used for better efficiency.

METHODS OF DISINFECTION

Disinfection with Chlorine (Free and Combined)

The chemical form of chlorine in water decides its effectiveness as a disinfectant. Chlorine as an under-pressure compressed gas or in a water solution may be used as disinfectant. Sodium hypochlorite and solid calcium hypochlorite solutions are widely used to avoid the risk or release of toxic chlorine gas. Temperature, pH, and organic content in water are some of the factors that influence the effectiveness of chlorine as a disinfectant. The concentration of hypochlorite or any other oxidizing disinfectant is often expressed as *available chlorine*, which refers to the relative amount of chlorine present in chlorine (under pressure liquid form) or hypochlorite. Upon application, chlorine gas quickly hydrolyzes to form hypochlorous acid (HOCl) and hydrochloric acid (HCl).

$$Cl_2 + H_2O \leftrightarrow H^+ + Cl^- + HOCl$$

For this reaction, the equilibrium constant is as follows:

$$K_H = \frac{[H^+][Cl^-][HOCl]}{[Cl_{2(aq)}]} \tag{4}$$

where K_H = equilibrium constant (mole/L^2)
$K_H = 4.48 \times 10^4$ at 25°C (White, 1972)

Henry's law describes the dissolution of gaseous chlorine to form dissolved molecular chlorine (Downs and Adams, 1973).

$$Cl_{2(g)} = \frac{Cl_{2(aq)}}{H} = \frac{[Cl_{2(aq)}]}{P_{Cl_2}} \tag{5}$$

where $[Cl_{2(aq)}]$ = molar concentration of Cl_2
P_{Cl_2} = partial pressure of chlorine in atmosphere
H = Henry's law constant (mole/L atm)
H = 4.805×10^{-6} exp (2818.48/T)
T = temperature (K)

HOCl is a weak acid and it dissociates further to produce hypochlorite ions (OCl^-) and hydrogen ions.

$$HOCl \leftrightarrow OCl^- + H^+$$

For this reaction, the acid dissociation constant is as follows.

$$K_a = \frac{[OCl^-][H^+]}{[HOCl]} \tag{6}$$

where K_a = acid dissociation constant (mole/L)
K_a = 3.7×10^{-8} at 25°C (Morris, 1966; Lin, 2001)
K_a = 2.61×10^{-8} at 20°C (Morris, 1966; Lin, 2001)

The correlating equation of K_a as a function of temperature (T) in Kelvin is as follows (Morris, 1966):

$$\ln K_a = 23.184 - 0.058T - 6908/T \tag{7}$$

In the presence of certain constituents in water, chlorine can react and transform to less effective chemical forms, referred to as combined chlorine. In the presence of ammonium ions, free chlorine undergoes the following step-wise reactions to form monochloramine (NH_2Cl), dichloramine ($NHCl_2$), and trichloramine (NCl_3). These formations are also pH dependent.

$$NH_4^+ + HOCl \leftrightarrow NH_2Cl + H_2O + H^+$$
$$\text{or}$$
$$NH_{3(aq)} + HOCl \leftrightarrow NH_2Cl + H_2O \text{ (pH > 7)}$$
$$NH_2Cl + HOCl \leftrightarrow NHCl_2 + H_2O \text{ (not favored at high pH)}$$
$$NHCl_2 + HOCL \leftrightarrow NCl_3 + H_2O$$

Free available chlorine is the sum of [HOCl] and [OCl^-].
Combined chlorine is the sum of the three chloramines formed in the reactions above.
Total available chlorine is the sum of free chlorine and combined chlorine.

As mentioned earlier, at different pH values chlorine takes different forms in water. These different forms of chlorine have different disinfection capacity. For example, the disinfection capability of HOCl is generally a higher disinfection capability than that of OCl⁻ (White, 1972, 1978). It is therefore a common practice to add an alkaline agent to raise the pH of the solution to maintain higher HOCl concentration. This also helps to reduce the risk of chlorine exposure due to accidental spills or leakage. Distribution of free chlorine with respect to the variation in pH of water at 20°C is shown in Figure 1 (AWWA and Pontius, 1990).

Breakpoint chlorination is a phenomenon in which all the ammonium ions disappear and the solution possesses free chlorine residue. It occurs when the molar ratio of chlorine to ammonia is greater than 1.0. Under ideal conditions, at breakpoint chlorination, the reduction of chlorine and oxidation of ammonia occurs at a 2:1 ratio. Further addition of chlorine results in more and more free available chlorine. This phenomenon is very important in calculating the chlorine dosage to maintain the chlorine residue in the distribution system in order to comply with regulations. Also, the breakpoint chlorination concept

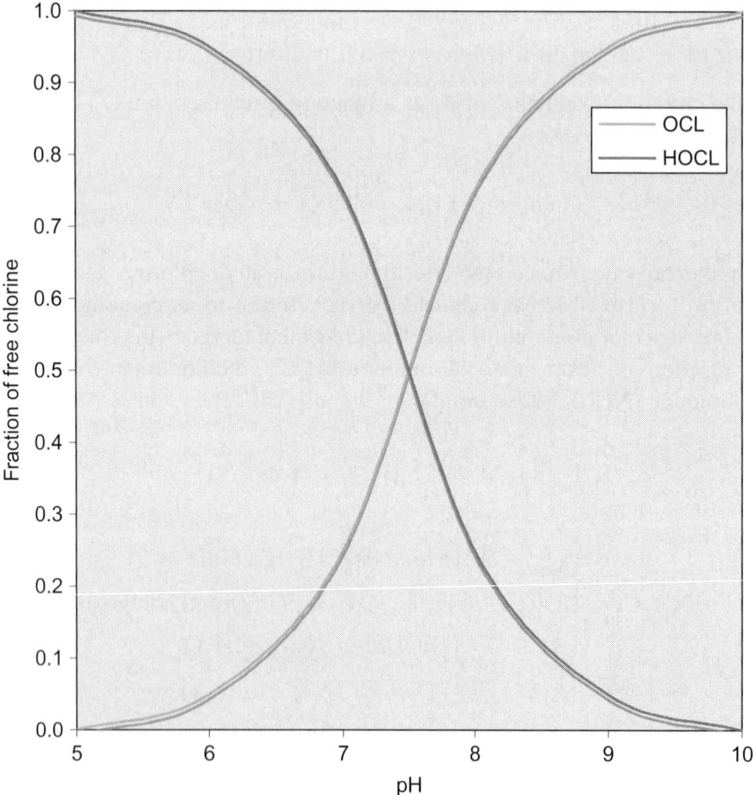

FIGURE 1 Effect of pH on free chlorine distribution at 20°C (AWWA, 1990).

can be used for ammonia removal from water. Figure 2 shows the curve for chlorine dosage vs. chlorine residue to explain the phenomenon of breakpoint chlorination.

1. Chlorine is reduced to chlorides by metallic ions and compounds that are oxidized easily (Fe^{2+}, H_2S, etc.).
2. Chlorine reacts with ammonia to form chloramines, which are weak disinfectants.
3. The nitrogen trichloride formation reaction is favored, and the chloramines are consumed in the reaction with free chlorine. In this zone, nitrogen gas is formed, which leaves the system, and breakpoint chlorination is reached.
4. Free chlorine residue is observed in water, and further addition of chlorine only increases the residue concentration.

Some alternative disinfectants such as ozone and UV radiation do not leave a residual disinfectant in water. Therefore, to meet the regulations for residual disinfectant concentration, it is sometimes necessary to add a residual disinfectant after they (ozone or UV radiation) are used. This is often done through the addition of chloramines. Chloramines, when used in this manner, provide a long–lasting residual disinfectant without most of the negative aspects of chlorination.

Disinfection with Chlorine Dioxide

Chlorine dioxide (ClO_2) is a neutral and unstable (at atmospheric conditions) gas that (at high concentrations) reacts rapidly with reducing agents. It is a compound of chlorine in $+IV$ oxidation state and has 1.4 times oxidation power than that of chlorine. The chemistry of ClO_2 in water is complex and pH dependent (Aieta and Berg, 1986).

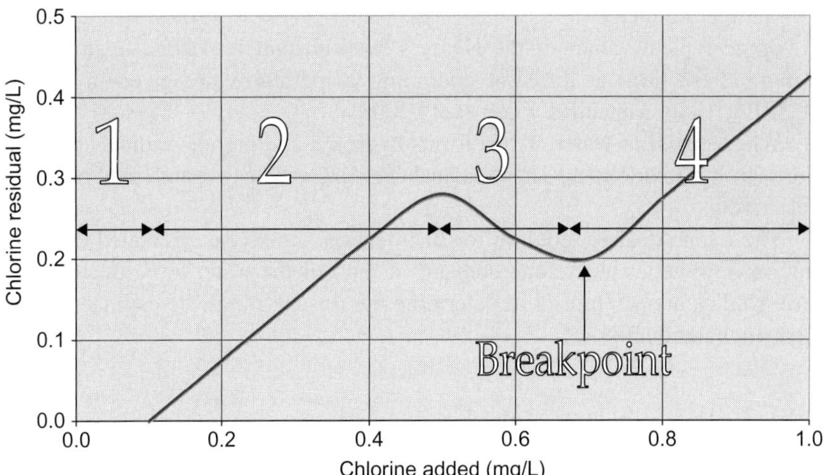

FIGURE 2   Breakpoint chlorination.

$$ClO_2 + 5e^- + 4H^+ \leftrightarrow Cl^- + 2H_2O \text{ (low pH)}$$
$$ClO_2 + e^- \leftrightarrow ClO_2^- \text{ (neutral pH)}$$
$$2ClO_2 + 2OH^- \leftrightarrow ClO_2^- + ClO_3^- + H_2O \text{ (high pH)}$$

Chlorine dioxide may be explosive in the presence of organic substances, on exposure to light, and at elevated temperatures. To minimize such risks, it is generated on-site just prior to usage.

Disinfection with Ozone

Ozone (O_3) is a blue or colorless gas having a pungent odor. It is unstable at ambient temperature, it is one of the most powerful oxidizing agents. However, the half-life of ozone in water is very short because of various decomposition processes, and as a result ozone does not leave free residue in the system. This is the main disadvantage in the use of ozone as a disinfectant. Most commonly, secondary disinfection with chloramines is used to overcome this problem.

Ozone is generated by the action of electric fields on oxygen.

$$3O_2 + \text{energy} \leftrightarrow 2O_3$$

Henry's law constant (temperature and pH dependent) can be used to describe the solubility of ozone in water (AWWA, 1990). Temperature dependence of the solubility constant H is (AWWA and Pontius, 1990) as follows:

$$H = (1.29 \times 10^6 / T) - 3721 \tag{8}$$

where H = Henry's law constant (mole/L atm)
T = temperature (K)

Some typical values of the Henry's law constant for ozone (expressed as atm/mole fraction) at different operating temperatures are given in Table 3 (USEPA, 1986; Metcalf & Eddy et al., 2003).

When mixed in water, ozone forms hydroxyl and organic radicals by reacting with hydroxide ions. These radicals oxidize various organic materials non-selectively.

The ozone dosage required for disinfection should be calculated considering ozone kinetics and ozone demand (if any) of the water to be treated. Pilot case studies are often used to determine the dosage range for estimated variations in water quality.

Disinfection with UV Light

UV radiation is a very effective viricide and bactericide and reasonably effective at inactivating cysts, as long as the UV can pass through the water

TABLE 3 Values of Henry's Law Constant for Ozone (USEPA, 1986; Metcalf and Eddy, 2003)

Temperature (K)	Henry's Constant (atm/mole fraction)
273.15	1,940
278.15	2,180
283.15	2,480
288.15	2,880
293.15	3,760
298.15	4,570
303.15	5,980

without being absorbed. It can also be used for treating *Giardia lamblia* and *Cryptosporidium* with some effect (Ware et al., 2003). As the use of UV radiation does not require the addition of any chemicals to water being treated, no by-products are formed. Usually, a low-level color is added to the water to minimize the amount of UV absorbed. A complete mixing in all directions is also necessary to make sure that all microorganisms will come equally close to the UV source. Although UV disinfection is currently practiced more in wastewater treatment plants, its use in drinking water treatment is spreading fast.

The UV dose can be defined by the following equation:

$$D = I \times t$$

(9)

where D = UV dose (mJ/cm^2 and J = Ws)
I = UV intensity (mW/cm^2)
t = contact time (s)

In equation 9, D is analogous to the Ct value used for chemical disinfection.

The main disadvantage of the use of UV radiation, like ozone treatment, is the lack of a disinfectant residual to ensure that no microbial recontamination occurs. Hence, it can be used only as a primary disinfectant in combination with another secondary disinfectant (chlorine based). Otherwise, an increase in microorganisms after treatment may occur because of properties of the pipeline system that transports the water or the presence of substances in the pipe that form part of the food chain for bacteria. Despite this concern relating to the lack of residual effects, many communities do not experience the hygiene problems in their pipeline systems that would lead to recontamination.

MONITORING DISINFECTANTS

Disinfection By-products

DBPs that form when disinfectants are added to water are potentially toxic and/or are carcinogenic substances. Depending on the disinfectant used and the precursor materials present in the water, several classes of DBPs may form, including trihalomethanes (THMs), haloacetic acids (HAAs), chlorate, chlorite, bromate, and haloacetonitriles (HANs). The most common precursors of DBPs is natural organic matter (NOM) such as organic debris and leaves that find their way into surface waters. While serving the purpose of killing pathogens in raw water, the disinfectant may also react with precursor material, mainly dissolved organic matter, to form DBPs. Several factors are responsible for the concentration of DBPs formed. Most influential among these factors are total organic carbon (TOC) concentration, chlorine dose and contact time, pH, and temperature (Krasner et al., 1989).

Regulations Governing Disinfectants and Disinfection By-products Monitoring

Chemical disinfection has been an integral part of drinking water treatment processes in the United States for over a century. However, it was only in the early 1970s, that the Dutch scientist, J. Rook, identified chloro- and bromo- THMs, the first class of halogenated DBPs in chlorinated drinking water (Rook 1971, 1974). The USEPA conducted a survey in 1975 that identified chloroform as being dominant in most chlorinated drinking waters. In cases where bromide was present in the source water, the addition of chlorine formed brominated THMs. In addition, the concentrations of THMs in the finished water were correlated to the TOC concentrations in the raw water. NOM, which is the major constituent of TOC, was found to be a primary component of the precursors that react with chlorine to form THMs. Chloroform was identified in 1976 by the National Cancer Institute as a suspected human carcinogen, leading the USEPA to set a maximum contaminant level (MCL) for total THMs (TTHMs) of 0.100 mg/L. This standard applied to systems serving over 10,000 people. The decision to regulate TTHMs was made because the health effects of individual THMs were not well-known.

In the 1980s, dichloroacetic acid (DCAA), trichloroacetic acid (TCAA), HANs, haloketones, chloropicrin, cyanogen chloride, and chloral hydrate, to name a few, were found in chlorinated drinking water. Several of these halogenated DBPs were found by the National Academy of Sciences in 1987 to have adverse health effects. More concerns about waterborne viral diseases escalated, leading the USEPA to promulgate the Surface Water Treatment Rule (SWTR) and the Total Coliform Rule (TCR) in 1989 (USEPA 1989a, b). The SWTR required three logs, or 99.9%, inactivation of *Giardia* cysts and four logs, or 99.99%, inactivation of viruses. Log credits were given to utilities, depending on the level of treatment. The *Ct* concept was introduced, requiring that water be in contact with a sufficient concentration of disinfectant (*C*) for a sufficient contact time (t) in order to provide adequate disinfection under different water-quality conditions. Hence Ct values for different disinfectants were developed. The SWTR and TCR also required a residual disinfectant level and disinfection practices in the distribution system for continued microbial protection.

Cryptosporidium is a waterborne parasite that has been responsible for major disease outbreaks in three U.S. cities, the most recent of which infected 370,000 people in Milwaukee, Wisconsin, in April of 1993. The USEPA proposed mandatory testing for *Cryptosporidium* as a result of these outbreaks.

Another vital piece of data is the public identification of more halogenated DBPs, given that the hundreds of halogenated DBPs found to date account for only about 50% of total organic halide (TOX). Because of the complex issues that USEPA faced, it had to draw on the expertise of others to prepare a rule on their own. Hence in 1992, the USEPA initiated a negotiated rule-making process, termed RegNeg, which consisted of an advisory committee of utilities, state drinking water agencies, environmental groups, consumer advocates, public health officials, and individuals who had interests in the regulation.

Disinfectants/Disinfection By-products Rules

The 1996 Safe Drinking Water Act (SDWA) amendments require the USEPA to develop rules and standards for DBPs in drinking water. USEPA has attempted to confront this challenge of balancing the risk associated with DBPs against the risk associated with microbial disease, but several factors have made this difficult, including (a) the lack of occurrence data for viruses, *Giardia*, *Cryptosporidium*, and halogenated DBPs, (b) uncertainty about the health effects of various DBPs, and (c) the lack of data on DBPs of alternative disinfectants such as ozone and chlorine dioxide, which were to become widely adopted (Means and Krasner, 1993). Because of the complexity of the issues, the regulations were to be proposed in two stages. Stage 1 of the D/DBPs rule was proposed in 1994 and became effective in 1998. It lowered the TTHM MCLs to 0.80 mg/L and provided MCLs for three other classes of DBPs. The D/DBPs rule required water systems to use treatment methods to reduce the formation of DBPs and to meet the standards (USEPA, 2007a). The rule also provided maximum residual disinfectant level goals (MRDLG) for three disinfectants. Table 4 (USEPA, 2007b) lists the three disinfectants along with their potential health effects that may be caused by overdose.

The stage-2 DBPR (published in the *Federal Register* on January 4, 2006, USEPA, 2007c) strengthens public health protection for customers of systems that deliver disinfected water by requiring such systems to meet MCLs. On average, each monitoring location needs to meet the standard (instead of as a system-wide average as in previous rules) for two groups of DBPs, TTHM, and five haloacetic acids (HAA5). TTHM is defined as the sum of four individual THMs: chloroform, bromoform, dibromochloromethane, and bromodichloromethane. HAA5 is defined as the sum of five HAAs: monochloroacetic acid (MCAA), DCAA, TCAA, monobromoacetic acid (MBAA), dibromoacetic Acid (DBAA).

TABLE 4 Maximum Residual Disinfectant-Level Goals (USEPA, 2007b)

Disinfectant Residual	MRDLG (mg/L)	MRDL (mg/L)	Potential Health Effects
Chlorine (as Cl_2)	4	4	Eye/nose irritation, stomach discomfort
Chloramines (as Cl_2)	4	4	Eye/nose irritation, stomach discomfort, anemia
Chlorine dioxide (as ClO_2)	0.8	0.8	Anemia, nervous system effects in infants and young children.

TABLE 5 MCL, MCLG, and Potential Health Effects of DBPs (USEPA, 2007b, c, d, e)

Disinfection By-product	MCL (mg/L)	MCLG (mg/L)	Potential Health Effects
Total trihalomethanes (TTHM)	0.080	0	Liver, kidney, or central nervous system problems; increased risk of cancer
Haloacetic acids (HAA5)	0.060	N/A	Increased risk of cancer
Bromate	0.010	0	Increased risk of cancer
Chlorite	1.0	0.8	Anemia, nervous system effects in infants and young children

The stage-2 regulation reduces DBP exposure and related potential health risks and provides more equitable public health protection. It has set new standards for MCLG for each DBP and MRDLG for each disinfectant listed in Tables 5 (USEPA, 2007b, c, d, e) and 6 (USEPA, 2007a, b, c, d). Table 5 also gives the potential health effects of the DBPs.

The information collection rule (ICR) was proposed to provide necessary data for stage 2 of the D/DBP regulation. The ICR requires utilities serving over 10,000 people to begin monitoring for microbial contaminants and DBPs. Granular activated carbon (GAC) and membrane pilot testing are required for surface

TABLE 6 Maximum Contaminant-Level Goals for DBPs (USEPA, 2007a, b, c, d)

Disinfection By-product	MCLG (mg/L)
Chloroform	0.07
Bromodichloromethane	0
Bromoform	0
Bromate	0
Monochloroacetic acid	0.07
Dichloroacetic acid	0
Trichloroacetic acid	0.02
Chlorite	0.8
Dibromochloromethane	0.06

water utilities with raw water TOCs greater than 4 mg/L and serving over 100,000 people, as well as for groundwater with finished water TOCs greater than 2 mg/L and serving over 50,000 people.

Sampling and Monitoring for D/DBPs

Regularly monitoring D/DBPs is the responsibility of the public water system. A certified drinking water laboratory must analyze collected samples. Federal and/or state regulations specify monitoring frequency for each contaminant or contaminant group. Monitoring frequency and number of compliance monitoring sites is different for different types of water system and population served. Population-based monitoring is believed to provide better risk-targeting and ease of implementation. All the data collected by monitoring are stored in a database that is managed by Bureau of Safe Drinking Water. In addition to monitoring data for regulated contaminants, some state agencies collect data for certain unregulated parameters as well.

The stage-2 D/DBPs rule focuses on reducing concentrations of and monitoring for TTHM and HAA5. These two groups are considered indicators for various DBPs that may be present in water that is treated with either chlorine or chloramines. Similarly, a reduction in concentrations of TTHM and HAA5 generally indicates an overall reduction in concentration of DBPs.

It is required that disinfectant residual be measured monthly at the end of each treatment process that uses chlorine. If free chlorine is used, free and total chlorine residuals must be reported; whereas, if ammonia is present in source water, total chlorine residual must be reported. This data, along with the contact time and applied disinfectant concentration, would show a better picture of the formation of DBPs.

The D/DBPs rule provides several routine monitoring requirements for the water treatment and distribution systems. The rule requires that all systems develop and implement a monitoring plan. At a minimum, each monitoring plan must include

- Specific schedules and locations for collecting DBPs or disinfectant residual samples
- Calculations for determining compliance with the MCLs and MCLGs
- Distribution sampling locations that are reflective of the entire distribution system involved in the sampling

Table 7 describes the routine monitoring requirements for D/DBPs (USEPA, 2007f).

Analytical Methods

The SWTR and stage-1 D/DBPs rule specified methods for analysis of regulated D/DBPs. Analytical methods generally include information on the collection, transport, and storage of samples; define procedures to concentrate, separate, identify, and quantify components contained in samples; specify quality control criteria the analytical data must meet; and designate how to report the results of the analyses (USEPA, 2008). In 2002, the EPA updated the approved methods for analysis of D/DBPs (USEPA, 2008). These updates specified that in order to comply with SWTR, public water systems must measure disinfectant concentrations using EPA-approved methods. The EPA has approved five methods for the measurement of free chlorine, four methods for combined chlorine, six methods for total chlorine, two methods for chlorine dioxide, three methods for HAA5, three methods for TTHMs, three methods for TOC/DOC, two methods for monthly measurement of chlorite, and one method for daily monitoring of chlorite, two methods for bromide, one method for bromate, and one method for measurement of UV (USEPA, 2004a, b, c, d, 2007a, b, c, d). Tables 8–10 list the EPA-approved methods for the analysis of regulated D/DBPs and the precursors of DBPs, respectively. Most of the methods specified are based on the *Standard Methods for the Examination of Water and Wastewater* (APHA et al., 1999) and *USEPA Drinking Water Methods for Chemical Contaminants* (USEPA, 2004d).

Using *Ct* Values for Monitoring Disinfectants

Ct value refers to the product of residual of disinfectant (*C*, mg/L) and contact time (*t*, min). *Ct* value is generally dependent on pH and temperature. The Final Surface Water Treatment Regulations published by USEPA (1989a, b) made it mandatory for public surface water (or groundwater under the influence of surface water) treatment plants to determine and apply *Ct* values daily. USEPA promotes *Ct* values as indicators of the effectiveness of inactivation of *G. lamblia* and viruses. Thus, monitoring and maintaining *Ct* values directly monitors efficient running of disinfection process and disinfectants. More recently, long-term 2

TABLE 7 Routine Monitoring Requirements (USEPA, 2007f)

Parameter	Coverage	Monitoring Frequency	Compliance
TTHM HAA5	Surface water and groundwater under the direct influence of surface water serving three 10,000	4/plant/quarter	Running annual average
	Surface water and measure under the direct influence of surface water serving 500–9,999	1/plant/quarter	Running annual average
	Surface water and measure under the direct influence of surface water serving <500	1/plant/year in month of warmest water temperature[*]	Running annual average of increased monitoring
	Measure serving > = 10,000	1/plant/quarter	Running annual average
	Measure serving <10,000	1/plant/year in month of warmest water temperature[*]	Running annual average of increased monitoring
Bromate	Ozone plants	Monthly	Running annual average
Chlorite	Chlorine dioxide plants	Daily at entrance to distribution system; monthly in distribution system	Daily/follow-up monitoring
Chlorine dioxide	Chlorine dioxide plants	Daily at entrance to distribution system	Daily/follow-up monitoring
Chlorine/ chloramines	All systems	Same location and frequency as TCR sampling	Running annual average
DBP precursors	Conventional filtration	Monthly for total organic carbon and alkalinity	Running annual average

[*]*System must increase monitoring to one sample per plant per quarter if an MCL is exceeded.*

enhanced surface water treatment rule (LT2) (released simultaneously with the stage-2 DBPR) was published by USEPA to address concerns about risk trade-offs between pathogens and DBPs (USEPA, 2007d). The purpose of the LT2 rule was to reduce illness linked with the contaminant *Cryptosporidium* and other

TABLE 8 EPA-Approved Methods for Analysis of Regulated Disinfectants and Disinfectant Residual (APHA et al., 1999; USEPA, 2008)

Residual	Standard Method Number	Methodology	Recommended Source
Free chlorine	4500-Cl D	Chloride by potentiometric method	*Standard Methods for the Examination of Water and Wastewater*, 18th, 19th, and 20th editions
	4500-Cl F[*]	Chlorine residual by DPD ferrous titration	
	4500-Cl G[*]	Chlorine residual by DPD colorimetric method	
	4500-Cl H[*]	Chlorine residual by syringaldazine (FACTS) method	
Combined chlorine	4500-Cl D	Chloride by potentiometric method	
	4500-Cl F[*]	Chlorine residual by DPD ferrous titration	
	4500-Cl G[*]	Chlorine residual by DPD colorimetric method	
Total chlorine	4500-Cl D	Chloride by potentiometric method	
	4500-Cl E[*]	Chlorine residual by low–level amperometric titration	
	4500-Cl F[*]	Chlorine residual by DPD ferrous titration	
	4500-Cl G[*]	Chlorine residual by DPD colorimetric method	
	4500-Cl I[*]	Chlorine residual by iodometric electrode technique	
Chlorine dioxide	4500-ClO$_2$ C	Chlorine dioxide by the amperometric method I	
	4500-ClO$_2$ D[*]	Chlorine dioxide by the DPD method	
	4500-ClO$_2$ E[*]	Chlorine dioxide by the amperometric method II	
Ozone	4500-O3 B	Ozone residual by indigo colorimetric method	

[*]*To comply with Stage 1 D/DBP rule, use SM 19th ed. only.*

disease-causing microorganisms in drinking water. This rule applies to all public water systems that use surface water or groundwater under the direct influence of surface water to treat reservoir discharge to inactivate four-log (99.99%) viruses, three-log (99.9%) *G.* and two-log (99%) *Cryptosporidium* cysts. These requirements are necessary to protect against the risk of contamination.

TABLE 9 EPA-Approved Methods for Analysis of Regulated Disinfectant By-products (USEPA, 2004a, b, c, d, 2008)

DBP	EPA Method Number	Methodology	Recommended Source
Total trihalomethanes (TTHM)	502.2	VOCs by purge and trap capillary GC with photoionization and electrolytic conductivity detectors in series	USEPA, (2004a,b)
	524.2	Purgeable organic compounds by capillary column GC/mass spectrometry	
	551.1	Chlorinated disinfection by-products and chlorinated solvents by liquid–liquid extraction and GC with an electron capture detector	
Haloacetic acids (five) (HAA5)	552.1	Haloacetic acids and dalapon by ion exchange liquid–solid extraction and GC with electron capture detector	
	552.2	Haloacetic acids and dalapon by liquid–liquid extraction, derivatization, and GC with electron capture detector	
Bromate	300.1	Determination of inorganic anions in drinking water by ion chromatography	USEPA, (2004c,d)
Chlorite (daily monitoring)	300.0	Inorganic anions by ion chromatography	
	300.1	Determination of inorganic anions in drinking water by ion chromatography	
Chlorite (distribution system monitoring)	300.0	Inorganic anions by ion chromatography	
	300.1	Determination of inorganic anions in drinking water by ion chromatography	

TABLE 10 EPA-Approved Methods for Analysis of Precursors of Disinfectant By-products (APHA et al., 1999; USEPA, 2008)

DBP Precursor	Standard Method Number	Methodology	Recommended Source
Total organic carbon	5310 B	Total organic carbon by combustion method	*Standard Methods for the Examination of Water and Wastewater*, 18th, 19th, and 20th editions
	5310 C	Total organic carbon by persulfate–UV or heated persulfate method	
	5310 D	Total organic carbon by wet-oxidation method	
Alkalinity	2320 B	Alkalinity by titration	

Disinfectant dosage control is critical to complying with Ct regulations. Normally, frequent measurements of disinfectant residual are made and the dosage is adjusted to obtain desired residual (e.g., 0.5 mg/L for chlorine). Measurements of disinfectant residual and dosage adjustments can be done manually or using an automatic analyzer and/or a flow-measurement device.

Ct values are dependent on the configuration of the entire treatment system, residual concentration of disinfectant, and the type and number of point applications (USEPA, 2007g). Detailed instructions on Ct value calculation are given in the USEPA guidance manual (USEPA, 1999a, 2006, 2007g).

Table 11 (USEPA, 1986, 1999a, b, 2006, 2007g; WEF, 1996; Metcalf & Eddy et al., 2003, MWH, 2005) gives the estimated range of Ct values for various levels of inactivation of bacteria, viruses, and protozoan cysts (pH~7 and T~20°C)

DISINFECTION BY-PRODUCTS CONTROL

The USEPA started taking steps towards the control of THMs in drinking water in the late 1970s and published a guidance manual in 1981 for controlling THMs in drinking water (USEPA, 1979; Symons et al., 1975, 1981). Several options were considered effective for the removal of DBP precursors, including

- Aeration and air stripping
- Enhanced coagulation
- Activated carbon adsorption

TABLE 11 Estimated Range of Ct Values for Various Levels of Inactivation of Bacteria, Viruses, and Protozoan Cysts (pH~7 and T~20°C)

Disinfectant	Unit	Inactivation Level				Microorganisms
		1 Log	2 Log	3 Log	4 Log	
Chlorine	mg min/L	0.1–0.2	0.4–0.8	1.5–3.0	10–12	Bacteria
			2.5–3.5	4–5	6–7	Viruses
		20–30	35–45	70–80		Protozoan cysts
Chloramine	mg min/L	4–6	12–20	30–75	200–250	Bacteria
			300–400	500–800	200–1,200	Viruses
		400–650	700–1,000	1,100–2,000		Protozoan cysts
Chlorine dioxide	mg min/L	2–4	8–10	20–30	50–70	Bacteria
			2–4	6–12	12–20	Viruses
		7–9	14–16	20–25		Protozoan cysts
Ozone	mg min/L		3–4			Bacteria
			0.3–0.5	0.5–0.9	0.6–1.0	Viruses
		0.2–0.4	0.5–0.9	0.7–1.4		Protozoan cysts
UV radiation*	mJ/cm²		30–60	60–80	80–100	Bacteria
			20–30	50–60	70–90	Viruses
		5–10	10–15	15–25		Protozoan cysts

Source: Adapted from Metcalf and Eddy, 2003; MWH, 2005; USEPA, 1986; WEF, 1996; USEPA, 1999a, b, 2006, 2007g.
*UV dose = UV intensity × time.

- Oxidation by ozone or chlorine dioxide
- Clarification by coagulation, precipitate softening, or direct filtration
- Oxidation with potassium permanganate and lowering pH
- Moving the point of chlorination downstream in the treatment process to allow for the removal of precursors to DBPs prior to disinfection.

Although aeration and activated carbon adsorption were considered effective removal technologies, many water utilities opted to move the point of chlorination downstream from raw water application to sedimentation basin effluent, decreasing chlorine dosages, and using chloramines as an alternative primary or a secondary disinfectant in place of free chlorine (Chen and Rest, 1996; Singer, 1994, 1999). Many utilities were able to meet the TTHMs MCL with such modifications, but questions were raised regarding the integrity of the finished water in terms of microbial inactivation. Some utilities found that they could meet either the TTHMs MCL or the SWTR/TCR regulations, but not both.

Further complicating the issues were the findings that, in addition to chlorine DBPs having adverse health effects, alternative disinfectants formed by-products with adverse health effects as well (USEPA, 2007e). In 1983, Haag and Hoigne found that bromate ion and brominated organics were formed in bromide containing ozonated water (Haag and Hoigne, 1983). Bromate was later identified as a suspected human carcinogen. Chlorite, which causes hepatotoxicity in animals, was found in chlorine dioxide disinfected water. In addition to the findings of other DBPs, the emergence of *Cryptosporidium* in water supplies added more worries to the issues. The main challenge was, and still is, to effectively disinfect water while minimizing the formation of harmful by-products.

When DBPs are a concern in a water treatment process, there are two main approaches to solving the problem: (1) to control the precursors that react with the disinfectant to form the unwanted DBP and (2) to allow the DBPs to form and then use a separate removal process for the DBPs.

Precursor control and removal strategies are mainly focused on NOM present in water. NOM is considered to be the major precursor for DBP formation. NOM is very site-specific, and the different components of NOM (e.g., humic and fulvic acids) are removed with varying degrees of effectiveness by different strategies. Research into this area has focused on characterizing the behavior of NOM according to apparent molecular weight (AMW). Precursor management is often grouped into the following three categories:

1. Control at the source by managing inputs into the watershed to lower precursor concentrations.
2. Physical/chemical removal, which involves the removal of precursors by processes such as coagulation, adsorption, and membrane separation.
3. Oxidation/transformation, which involves processes that change the form of precursors.

After DBPs have formed, it is possible to remove them with a subsequent treatment process. The USEPA has specified air stripping and GAC adsorption as techniques for the removal of THMs.

Using Chlorine

Free chlorine (HOCl) is the most widely used of all the oxidative disinfectants because of its low cost and proven effectiveness. It is an excellent bactericide, viricide, and cysticide. As chlorine has been the disinfectant of choice for nearly 100 years and is used by the majority of water treatment systems, its DBPs are usually considered to be of the greatest concern. Chlorine DBPs form when free chlorine (HOCl) is added to water and reacts with the NOM present. The generalized equation describing the formation of the halogenated DBPs is

$$\text{HOCl} + \text{NOM} + \text{Br}^- \rightarrow \text{THMs and other halogenated DBPs}$$

The major halogenated DBPs that result from the addition of chlorine to drinking water are THMs, HAAs, HANs, cyanogen halides, halopicrins, haloketones, haloaldehydes, and halophenols. In the absence of bromide ions (Br^-), only the chlorinated by-products are formed. In the presence of bromide, free chlorine (HOCl) rapidly oxidizes bromide to hypobromous acid (HOBr), which then reacts, along with the remaining HOCl, with NOM to produce the mixed chloro-bromo DBPs.

It has been found that THMs and HAAs are the most common DBPs found in the treatment process. USEPA has set an MCL of 0.080 mg/L for TTHMs and an MCL for HAA5 of 0.060 mg/L. Some of the major types of these DBPs are listed in Table 12 (Marhaba and Washington, 1998).

TABLE 12 DBPs of Chlorine (Marhaba and Washington, 1998)

Generic Name	Chemical Compounds
Trihalomethanes (THMs)	Chloroform Bromodichloromethane Bromoform Dibromochloromethane
Haloacetic acids (HAAs)	Monochloroacetic acid (MCAA) Dichloroacetic acid (DCAA) Trichloroacetic acid (TCAA) Monobromoacetic acid (MBAA) Dibromoacetic acid (DBAA)
Haloacetonitriles (HANs)	Dichloroacetonitrile Trichloroacetonitrile Dibromoacetonitrile Tribromoacetonitrile
Cyanogen halides	Cyanogen chloride Cyanogen bromide

TABLE 13 THM Precursors Removal (Marhaba and Washington, 1998)

Removal Process	Description
Source control	Increasing the adsorption or exchange capacity of soil in the watershed may reduce the amount of DOC transport in water. The exchange capacity may be increased by the addition of absorbents such as alum sludge from the treatment plant, lime, and gypsum.
Enhanced coagulation	Enhanced coagulation can reduce THM precursors. This technique is very useful for utilities already using conventional coagulation. One or more of the following means may serve to achieve enhanced coagulation: pH adjustment, an increase in coagulant dose, and alternate coagulants. However, there are concerns associated with enhanced coagulation, including turbidity removal, corrosion, and increases in contaminant concentrations, such as aluminum, in the finished water.
Anion exchange	Anion exchange can remove much of NOM (organic precursors) present in water, which exists as anions at alkaline conditions.
Slow sand filtration	Slow sand filters (granular media such as anionic resins and GAC) may achieve significant removal (75–90%) of organic carbon and THM formation potential.
Reverse osmosis	The reverse osmosis (RO) process may be used prior to chlorination to reduce concentrations of nonvolatile organics and hence THMs in treated water. However, this method is cost-effective only at high concentrations of THM precursors, or if the plant is already using RO.
Adsorption	Granular activated carbon (GAC), powdered activated carbon (PAC), and other adsorbing materials may absorb NOM.
High-energy electron beam irradiation	Chloroform may be controlled at pilot scale by the use of innovative high-energy electron beam irradiation. However, this method is expensive and requires a high amount of energy.
Advanced oxidation	The use of ozone (in combination if free or combined chlorine to satisfy residual requirements) as a disinfectant may significantly reduce the formation of THMs.

NOM is the predominant precursor for the formation of chlorinated DBPs, and because of the fact that THMs were the first DBPs to be regulated, most methods of precursor removal deal with lowering the concentration of NOM in water. Table 13 (Marhaba and Washington, 1998) lists several processes that have been found useful in the removal of THMs.

Using Chloramines

Chloramine has been used as a primary disinfectant in some treatment plants since the early 1920s. Chloramines are formed by adding chlorine and ammonia to water at certain ratios of chlorine to ammonia. It is not as effective as free chlorine in disinfection or oxidation, and it may take 100 times longer to achieve the same bacteriological kill. Chloramines require significantly greater Ct values than free chlorine, and when chloramines are used, it is often in combination with additional disinfectants.

The use of chloramines can greatly reduce the formation of THMs and HAAs, but it may instead form chloral hydrate. Although chloral hydrate is currently not regulated, it is being considered for future legislation for classification as a DBP. In water containing cyanide, chloramines will form cyanogen chloride and cyanogen bromide to a greater degree than free chlorine. If the chloramines used in the disinfection process are formed by chlorination, followed by the addition of ammonia, THMs may form. Chloramination may also result in nitrate and nitrite formation as the chloramines decompose. The major chloramine DBPs are listed in Table 14 (Marhaba and Washington, 1998).

Little research has been done on technologies for the removal of chloramine-specific DBPs. However, since it forms DBPs similar to chlorine DBPs, many of the processes applied to chlorination DBPs will be effective for chloramine DBPs.

Using Chlorine Dioxide

Chlorine dioxide is primarily used as an oxidant, although recently it has been used as a primary disinfectant as well. It requires lower Ct values and inactivates

TABLE 14 DBPs of Chloramines (Marhaba and Washington, 1998)

Chlorine dioxide DBPs	Chlorate
	Chlorite
	1,1-Dichloropropanone
Chloramine DBPs	Cyanogen bromide
	Cyanogen chloride
	Haloacetic acids (HAAs)
	Nitrate/nitrite
	THMs

TABLE 15 Chlorite Removal (Marhaba and Washington, 1998)

Removal Process	Description
Reduction by ferrous sulfate (FeSO$_4$)	Sulfur compounds, such as ferrous sulfate (FeSO$_4$), react quickly (5–15 s) to remove chlorites. However, the reaction of sulfur dioxide (SO$_2$) and chlorite forms significant amounts of chlorate in waters that have high levels of dissolved oxygen.
GAC adsorption	GAC is of limited effectiveness for ClO$_2^-$ removal. Also, chlorate (ClO$_3^-$) formation will occur across the GAC medium.

Giardia, a disinfectant-resistant pathogen, five times faster than free chlorine. As an oxidant, it is highly effective for taste and odor control and iron and manganese oxidation. Chlorine dioxide is an unstable gas, which requires that it be manufactured on-site.

Chlorine dioxide will form very few, if any, halogenated DBPs. What few by-products it does form, however, are very undesirable. The DBPs of greatest concern with chlorine dioxide are chlorate and chlorite, both of which are toxic and carcinogenic (Cordie, 1986; Karpel Vel Leitner et al, 1996).

Although chlorine dioxide is a very effective disinfectant and forms no THMs, there is a concern that 50–70% of the applied ClO$_2$ dosage will remain as residual chlorite (ClO$_2^-$). Table 15 (Marhaba and Washington, 1998) lists several processes that have been found useful in the removal of chlorite.

Using Ozone

Ozone is formed by the passage of air or oxygen through an electrical discharge. The resultant air–ozone stream can be bubbled through water in a contact chamber. Ozone is considered the most effective oxidant and disinfectant used in the water treatment process and utilities, but it is unstable and does not maintain a residual in the water supply system. As a result, when ozone is used as a primary disinfectant, a secondary disinfectant must also be used so that the residual can be maintained (Ferguson et al, 1991). Like chlorine dioxide, ozone is an unstable gas and must be manufactured on-site.

The use of ozone for disinfection will produce no chlorinated THMs, HAAs, or other chlorinated by-products. It will, however, form various oxidation products in the presence of NOM, following the reaction

$$O_3 + NOM \rightarrow \text{Oxidation by-products}$$

TABLE 16 DBPs of Ozone (Glaze et al., 1993; Marhaba, 1994; Marhaba and Washington, 1998)

Generic Name	Chemical Compounds
Aldehydes	Formaldehyde Acetaldehyde Glyoxal Methyl glyoxal
Aldoacids and ketoacids	Pyruvic acid
Carboxylic acids	Oxalic acid Succinic acid Formic acid Acetic acid
Peroxides	Hydrogen peroxide
Brominated by-products	Bromate Bromoform Brominated acetic acids Bromopicrin Brominated acetonitriles

Oxidation by-products include aldehydes, aldo- and ketoacids, acids, and hydrogen peroxide. These are listed in detail in Table 10 (Dore et al, 1988; Marhaba and Washington, 1998). Although ozone itself does not produce halogenated DBPs, it can produce brominated DBPs if bromide-containing waters are ozonated, following the reaction:

$$O_3 + NOM + Br^- \rightarrow \text{Brominated by-products}$$

Ozone will oxidize the bromide (Br^-) to hypobromous acid (HOBr), which will react with NOM to produce the fully brominated analogs of the chlorination by-products (Table 11) shown in Table 16 (Glaza et al, 1993; Marhaba, 1994; Marhaba and Washington, 1998). Bromate ion is the by-product of greatest concern (Siddiqui et al, 1996a, b). It has been classified by USEPA as a B2 carcinogen (a probable human carcinogen.) Table 17 (Marhaba 1994; Marhaba and Washington, 1998) lists several processes that have been found useful in the removal of bromate ion.

CONCLUSIONS

Disinfection of water is often needed to assure its drinkability. This chapter discusses various methods of disinfection. It is important to monitor disinfectants and their by-products.

TABLE 17 Bromate Ion Removal (Marhaba, 1994; Marhaba and Washington 1998))

Removal Process	Description
UV radiation	UV radiation may reduce bromate ion to bromide ion.
Reduction by ferrous sulfate ($FeSO_4$)	Addition of ferrous sulfate will reduce the by-product of concern to a less harmful form (e.g., bromate ion is reduced to bromide ion).
Biologically active filtration	Several ozonation DBPs are biodegradable and may be reduced by using biologically active filtration.

ACKNOWLEDGMENTS

The author thanks Mr. Ashish Borgaonkar, doctoral student at the New Jersey Institute of Technology, for his assistance in the preparation of this chapter.

REFERENCES

Aieta, E.M., Berg, J.D., 1986. A review of chlorine dioxide in drinking water treatment. J. Am. Water Works Assoc. 78 (6), 62.

APHA, AWWA, WEF, 1999. Standard Methods for the Examination of Water and Wastewater, Twentieth ed. American Public Health Association, Library of Congress Catalog, Washington DC.

Pontius, F.W.AWWA, 1990. Water Quality and Treatment—A Handbook of Community Water Supplies, Fourth ed. McGraw-Hill, Inc., American Water Works Association, New York, NY.

Singer, P.C.AWWA, 1999. Formation and Control of Disinfection By-products in Drinking Water. American Water Works Association, Denver, CO.

Chen, P.P., Rest, G.B., 1996. Disinfection By-products: The Techniques of Control. Public Works, 36–38.

Chick, H., 1908. Investigation of the laws of disinfection. J Hyg (Br) 8, 92.

Condie, L.W., 1986. Toxicological problems associated with chlorine dioxide. J. Am. Water Works Assoc. 78 (6), 73.

Dore, M., Merlet, N., Legube, B., Croue, J.P., 1988. Interactions between ozone, halogens, and organic compounds. Ozone Sci. Eng. 10, 153.

Downs, A.J., Adams, C.J., 1973. The Chemistry of Chlorine, Bromine, Iodine, and Astaline. Pergamon Publication, Oxford, UK.

Fair, G.M., Morris, J.C., Chang, S.L., Weil, I., Burden, R.P., 1948. The behavior of chlorine as a water disinfectant. J. Am. Water Works Assoc. 40, 1051.

Ferguson, D.W., Grainith, J.T., McGuire, M.J., 1991. Applying ozone for organics control and disinfection: a utility perspective. J. Am. Water Works Assoc. 83 (5), 32.

Glaze, W.H., Weinberg, H.S., Cavanagh, J.E., 1993. Evaluating the formation of brominated DBPs during ozonation. J. Am. Water Works Assoc. 85 (1), 96.

Haag, W.R., Hoigne, J., 1983. Ozonation of bromide-containing waters: kinetics of formation of hypobromous acid and bromate. Environ. Sci. Technol. 23 (7), 838.

IUPAC Goldbook Definition, 1997. Arrhenius Equation, IUPAC Compendium of Chemical Technology, second ed, Blackwell Scientific Publications, Oxford, UK.

Karpel Vel Leitner, N., De Laat, J., Dore, M., Suty, H., 1996. The use of ClO_2 in drinking water treatment: formation and control of inorganic by-products (ClO_2^-, ClO_3^-). In: Disinfection By-products in Water Treatment: The Chemistry of Their Formation and Control. Edited by Minear, R.A., Amy. G.A., CRC Press, Inc, Boca Raton, FL, pp. 393–407.

Krasner, S.W., McGuire, M.J., Jacangelo, J.G., Patania, N.L., Reagan, K.M., Aieta, E.M., 1989. The occurrence of disinfection byproducts in US drinking water. J. Am. Water Works Assoc. 81 (8), 41.

Lin, S.D., 2001. Water and Wastewater Calculations Manual. McGraw-Hill, Inc, New York, NY.

Marhaba, T.F., 1994. Treatment/Removal of Bromate Ion and Its Precursor by Granular Activated Carbon and Reverse Osmosis from Drinking Water. Bell & Howell, Ann Arbor, MI p. 280.

Marhaba, T.F., Washington, M.B., 1998. Drinking water disinfection and by-products—history and current practice. Adv. Environ. Res. 02 (01), 103–115.

Means III, E.G., Krasner, S.W., 1993. D-DBP regulation: issues and ramifications. J. Am. Water Works Assoc. 85 (2), 68.

Metcalf and Eddy, Tchobanoglous, G., Burton, F.L., Stensel, H.D. 2003. Wastewater Engineering—Treatment and Reuse, fourth ed. Metcalf & Eddy, Inc., Tata McGraw-Hill Publishing Company Limited, New Delhi, India.

Morris, J.C., 1966. The acid ionization constant of HOCL from 5 °C to 15 °C. J. Phys. Chem. 70 (12), 3789.

MWH, Crittenden, J.C., Trussell, R.R., Hand, D.W., Howe, K.J., Tchobanoglous, G.M, 2005. Water Treatment—Principles and Design, second ed, MWH, John Willy and Sons, Inc, Hoboken, NJ.

Rook, J.J., 1971. Headspace analysis in water. H_2O 4 (17), 385–387 (translated).

Rook, J.J., 1974. Formation of haloforms during chlorination of natural waters. Water Treat. Exam. 23 (2), 234–243.

Siddiqui, M., Amy, G., Zhai, W., McCollum, L., 1996a. Removal of bromate after ozonation during drinking water treatment. In: Disinfection By-products in Water Treatment: The Chemistry of Their Formation and Control, Edited by Minear, R.A., Amy, G.A., CRC Press, Inc, Boca Raton, FL, pp. 207–234.

Siddiqui, M.S., Amy, G.L., McCollum, L.J., 1996b. Bromate destruction by UV irradiation and electric arc discharge. Ozone Sci. Eng. 18, 271.

Singer, P.C., 1994. Control of disinfection by-products in drinking water. J. Environ. Eng. 120 (4), 727–744.

Symons, J.M., Bellar, T.A., Carswell, J.K., De Marco, J., Kropp, K.L., Robeck, G.G., 1975. National organics reconnaissance survey for halogenated organics in drinking water. J. Am. Water Works Assoc. 67, 634.

Symons Jr., J.M., Stevens, A.A., Clark, R.M., Geldreichn, E.E., Love, O.T., DeMarco, J., 1981. Treatment Techniques for Controlling Trihalomethanes in Drinking Water EPA-600/2-81-156. USEPA, Municipal Environmental Research Laboratory, Cincinnati, OH.

USEPA, 1979. National interim primary drinking water regulations: control of trihalomethanes in drinking water. Federal Register 44, 68624.

USEPA, 1986, Design Manual for Municipal Wastewater Disinfection, EPA/625/I-86/021, USEPA, Cincinnati, OH.

USEPA, 1989a. National primary drinking water regulations: filtration and disinfection: turbidity, *Giardia Lamblia*, viruses, *Legionella*, and heterotrophic bacteria. Federal Register 54, 27544.

USEPA, 1989b. Total coliforms; final rule. Federal Register 54, 27544.

USEPA, 1999a. Microbial and Disinfection By-product Rule Simultaneous Compliance Guidance Manual, August 1999 ed.. USEPA, Washington, DC.

USEPA, 1999b. Alternative Disinfectants and Oxidants Guidance Manual, EPA/815/R-99-014, USEPA, Cincinnati, OH.

USEPA, 2004a. Methods for the Determination of Organic Compounds in Drinking Water—Supplement II, EPA/600/R-92/129, USEPA, Cincinnati, OH.

USEPA, 2004b. Methods for the Determination of Organic Compounds in Drinking Water—Supplement III, EPA/600/R-95/131, USEPA, Cincinnati, OH.

USEPA, 2004c. Methods for the Determination of Inorganic Substances in Environmental Samples, EPA/600/R-93/100, USEPA, Cincinnati, OH.

USEPA, 2004d. Methods for the Determination of Organic and Inorganic Compounds in Drinking Water, Volume 1, EPA/815/R-00/014, USEPA, Cincinnati, OH.

USEPA, 2006. UV Disinfection Guidance Manual for the Final Long Term 2 Enhanced Surface Water Treatment Rule, November 2006 ed. USEPA, Washington, DC.

USEPA, 2007a. Interim enhanced surface water treatment rule. http://www.epa.gov/OGWDW/mdbp/ieswtr.html (Retrieved on August, 2007)

USEPA, 2007b. National Primary Drinking Water Regulations: Disinfection and Disinfection By-products, Federal Register 40, 69389

USEPA, 2007c. Stage 2 Disinfectants and Disinfection By-product Rule, (Stage 2 DBP rule) http://www.epa.gov/safewater/disinfection/stage2/regulations.html (Retrieved on August, 2007)

USEPA, 2007d. Long Term 2 Enhanced Surface Water Treatment Rule, (LT2). http:www.epa.gov/safewater/disinfection/lt2/index.html (Retrieved on August, 2007)

USEPA, 2007e. Health and Environmental Effects Research, http://www.epa.gov/nheerl/research/drinking_water.html (Retrieved on August, 2007).

USEPA, 2007f. National Primary Drinking Water Regulations: Monitoring Requirements for Public Drinking Water Supplies; Final Rule, Federal Register 61, 24353.

USEPA, 2007g. Disinfection By-products: A Reference Resource, http://www.epa.gov/enviro/html/icr/gloss_dbp.html (Retrieved on August, 2007)

USEPA, 2008. Analytical methods approved for drinking water compliance monitoring-approved methods for D/DBPs. http://www.epa.gov/safewater/methods/rules_dbp.html (Retrieved on April, 2008)

Ware, M.W., Schaefer III, F.W., Hayes, S.L., Rice, E.W., 2003. Inactivation of *Giardia muris* by Low Pressure Ultraviolet Light. 2003 USEPA Science Forum.

Watson, H.E., 1908. A note on the variation of the rate of disinfection with change in the concentration of the disinfectant. J. Hyg. (Br.) 8.

White, G.C., 1972. Handbook of Chlorination. Van Nostrand Reinhold Company.

White, G.C., 1978. Disinfection of Wastewater and Water for Reuse. Van Nostrand Reinhold Company, New York, NY.

Chapter 13

The Evolution of Analytical Technology and Its Impact on Water-Quality Studies for Selected Herbicides and Their Degradation Products in Water

Michael T. Meyer and Elisabeth A. Scribner (Retired)

U.S. Geological Survey, Lawrence, KS, USA

INTRODUCTION

The purpose of this chapter is to describe advances in analytical instrumentation and methods for the analyses of herbicides and their degradation products and to assess their impact on major findings of broad surveys of herbicides in water conducted by the U.S. Geological Survey (USGS) over the last two decades. The connection of clean water has historically been linked to the health and longer life of the human body (Bottled Water Blues, 2002). In many parts of the world,

Handbook of Water Purity and Quality
Copyright © 2009 by Elsevier Inc. All rights of reproduction in any form reserved.

hunger and disease have been overcome with clean water and considered a gift of life to many people (Global Water, 2008). As early as 1908, the disinfection of urban water supplies with chlorination was introduced and quickly adopted by many American cities. The results were a steep decline in typhoid deaths as well as the absence of cholera and dysentery within the American population during the early 1900s (Greatest Achievements, 2008).

Since that time, standards for water purity have been set and continually revised by governments as new contaminants that may impact human health are identified. These water-purity standards have brought continued improvement in water quality of existing water sources by reducing the amount of pollution in drinking water, treating wastewater, diverting wastewater discharge from drinking-water supplies, implementing new filtration practices, and other innovative techniques (Greatest Achievements, 2008). The demand of governmental higher standards has led to a new way of managing rivers by not only looking at the content of the water and its quality, but also researching the whole life of the river and its collection structure (Brown, 2004).

The global population has tripled since 1938, causing water use to increase (United Nation Population Fund, 2001) and is expected to exceed 8 billion people by 2030 (FAO, 2000). In the United States alone, the population grew from 76 million in 1900 to 273 million in 1999 (Demographia, 2001). At the present time, the population exceeds 305 million, a growth of 33 million additional people since the 2000 U.S. census (U.S. Census Bureau, 2008). Increasing population is putting further pressure on the world's water and food supply; thus, the question arises as to not only is there enough quality water, but also whether food production can exceed demands from population growth expected in the next three decades?

Advances in farm equipment have modified the management and production of food, bringing a new efficiency to agriculture (Kusel, 2008). In the early 1900s, a farmer could harvest about 2.5 metric tons (100 bushels) of corn in a 9-h day, whereas today a farmer can combine 22.5 metric tons (900 bushels) of corn per hour, or 2.5 metric tons in 6 min. With these modern production methods, 0.4047 hectare, or 1 acre, of land can produce as much as 19 metric tons (42,000 lbs) of strawberries, 11,000 heads of lettuce, 11.3 metric tons (25,000 lbs) of potatoes, and 4 metric tons (8,800 lbs) of sweet corn (Field Crop News, 2008). The changes in farm equipment, along with the incorporation of fertilizers and crop protection chemicals, have further advanced the production and lowered the cost of food enjoyed by the American population (Field Crop News, 2008). Today, herbicides are used routinely on more than 90% of the acreage in which most U.S. crops are grown (Gianessi and Reigner, 2007).

In conjunction with the use of advanced farm equipment, farmers, advisors, and researchers should know which herbicides are best to combat certain resistant weeds (IA Weed Science, 1998; UW Weed Science, 1999; HRGW, 2004; Mennes, 2005:p.159). Guidelines for the management of crop resistance and a better understanding of mode-of-action classification is important for a more effective crop production (Nevill et al., 1998). A summary of resistant weeds by mode-of-action classification may be found at Weed Science (2008).

The importance of advanced analytical methods and environmental knowledge of organic contaminants has been very significant to maintain a position of "cutting edge" science in the study of herbicides during the past 20 years. Many marketplace companies not only sell state-of-the-art technology, equipment, and software for updating an environmental laboratory, but they also offer training and consulting by salespeople who understand the new technology and how it applies to the development of methods as the equipment becomes more accepted by, in particular, the regulatory community.

It is vital that state-of-the-art instrumentation for analyzing organic contaminants continually be introduced into the marketplace the advancement of analytical instrumentation has given scientists the capability to continually broaden their studies of the fate of herbicides and their degradation products over the last two decades. It is important to note that in the various studies summarized in this chapter, pieces of information are continually added to the puzzle of understanding science in a fast-changing world.

HERBICIDE USE

Herbicides are not new, but usage has changed over the centuries. During the time of the Roman Empire, insects were controlled by burning sulfur (pesticide) and weeds were controlled with salt (herbicide). By the end of the nineteenth century, farmers in the United States were using various arsenates, sulfates, and sulfurs to control insects in field crops. At the end of World War II, herbicides such as aldrin, dieldrin, endrin, and 2,4-dichlorophenoxyacetic (2,4-D) were introduced and found to be effective (Delaplane, 2000).

Herbicide use in the United States increased greatly from the mid-1960s through the mid-1980s, with a dip between 1982 and 1987. The usage increased in the 1990s with 1997 being the highest-use year of herbicides in terms of millions of pounds of active ingredients applied to crops (Oregon State, 2008). According to Gianessi and Reigner (2006), herbicide use then declined between 1997 and 2002 because of the substitution of lower application rate compounds for previously used higher rate compounds. The reduced usage rate was particularly noticeable for corn, where use decreased by 48 million lbs between 1997 and 2002. Herbicides are regulated by the U.S. Environmental Protection Agency (USEPA) in cooperation with the U.S. Department of Agriculture (USDA), and the Federal Drug Administration (FDA) (Fishel, 2007).

COMMONLY USED HERBICIDES

Five of the most commonly used agricultural herbicides (glyphosate, atrazine, 2,4-D, acetochlor, and S-metolachlor) are described in this section by the weight of active ingredient per year. Glyphosate, often referred to as Roundup, does not belong to a herbicide class. It is a broad-spectrum herbicide widely used to kill unwanted plants in agriculture and nonagricultural landscapes. It has been registered for use in the United States since 1974 (Cox, 2004). Agricultural uses of glyphosate include maize, cotton, soybean, and sugar beet acres (Gianessi, 2005). Glyphosate use rose from 15,900 metric tons (35 million lbs) in 1997 to 46,000 metric tons (102 million pounds) in 2002. Glyphosate's increase resulted from the rapid adoption of genetically engineered crops and no-till farming practices, both of which incorporate glyphosate for weed control (Gianessi and Reigner, 2006).

The first triazine herbicide, atrazine, was discovered by J.R. Geigy, Ltd., in Switzerland (LeBaron et al., 2008). It was first registered in the United States in 1958. Atrazine has a range of trade names such as Marksman, Coyote, Atrazina, Atrazol, and Vectal (PAN, 2002) and is estimated to be the second most heavily used herbicide in the United States at 35,000 metric tons (77 million lbs) (Gianessi and Reigner, 2006). Its top agricultural use areas, in terms of use per acre, include the Midwestern United States, particularly Iowa, Illinois, Indiana, Ohio, and Nebraska, and also Delaware. Approximately 75% of the field corn acreage grown in the United States is treated with atrazine. It is also used heavily on lawns in Florida and throughout the Southeastern United States. Atrazine and its degradation products are frequently detected in rivers, groundwater, and reservoirs, which are related directly to the volume of usage and their tendency to persist in soil and move with water (USEPA, 2008).

Of the herbicides in use today, the world's most widely used herbicide and third most commonly used in the United States is 2,4-D, a member of the chlorophenoxy family. Agricultural use in the United States is 18,000 metric tons (40 million lbs; Gianessi and Reigner, 2006). Common trade names for 2,4-D include Aqua-Kleen, Barrage, Plantgard, and Salvage (Extoxnet, 1996a). It was developed during World War II to increase crop yields during rationing and became the first successful selective broadleaf plant herbicide, which allowed for weed control in wheat, corn, and rice (Dinnage, 2007).

Acetochlor ranked fourth of the most used agricultural herbicide active ingredients per year at 16,000 metric tons (35 million lbs) during 2002 (Gianessi and Reigner, 2006). Various trade names for acetochlor include Guardian, Harness, Relay, and Surpass. It was first registered in 1994 and is a chloroacetanilide herbicide. Acetochlor is used for the control of most annual grasses and certain broad-leaf weeds. Crops on which acetochlor is applied include cabbage, citrus, coffee, corn, cotton, green peas, maize, onion, soybeans, sugar beets, and vineyards. Acetochlor is applied preemergence, preplant incorporated, and is compatible with most other pesticides and fluid fertilizers when used at recommended rates (Extoxnet, 1996b).

S-metolachlor is an herbicide from the chloroacetanilide family and is the fifth most widely used herbicide in the United States at 11 metric tons (24 million lbs; Gianessi and Reigner, 2006). *S*-metolachlor, which is the resolved isomer of metolachlor, was registered in 1997. New formulations based primarily on the *S*-metolachlor isomer are more active on a gram-for-gram basis than metolachlor formulations composed of a 50:50 a racemic mixture of the R and S isomers (Shaner et al., 2006). It is effective at application rates around 35% lower than original metolachlor. While metolachlor is heavily used in the United States, it is in the process of being phased out in Europe (Kiely et al., 2004).

Other commonly used herbicides in the United States to kill unwanted vegetation include pendimethalin, trifluralin, alachlor, propanil, dimethenamid, mancozeb, and dicamba (Kiely et al., 2004). Additional triazine herbicides presented in several water-quality studies by the USGS include cyanazine, introduced on the market in 1972 and voluntarily withdrawn from the market by the manufacturer in 2000 (Scribner et al., 2005; Thurman and Scribner, 2008); prometryn (Coupe et al., 1998); and simazine (Coupe et al., 2005).

Isoxaflutole, a new herbicide to the market, is a member of the benzoyl isoxazole family (Pallet et al., 2001). Isoxaflutole was first used in 1999 in the United States. Usage fluctuated from 97 metric tons (214,000 lb) in 1999, peaked at 199 metric tons (439,000 lb) in 2001, then decreased to 145 metric tons (320,000 lb) in 2003 (USDA, 1991–2004).

PERSPECTIVE ON ROLE OF ANALYTICAL METHODS DEVELOPMENT

The importance of the development of analytical methods and environmental knowledge of organic contaminants has been crucial to developing continued information on the wide range of herbicides and their degradation products in our water resources for the past 20 years. Since the early 1900s, the primary goal of many research laboratories studying herbicides has been to develop analytical methods to understand the occurrence of organic contaminants and their fate and degradation pathways in the environment.

Mass Spectrometry

To assess the effects of agricultural nonpoint-source pollution in surface water and groundwater and the environmental fate and effects of emerging organic contaminants, it is important that robust analytical methods be developed for these contaminants in soil and water. In the early 1950s, the fragmentation of small organic molecules was beginning to be understood, but the mass spectrometer was very limited in sensitivity and resolution. This early instrument was the forerunner of today's reasonably priced bench top instruments seen in most chemical laboratories in the world (Mass Spectrometry Resource, 2005).

Gas Chromatography/Mass Spectrometry

The next major development was gas chromatography (GC)—a type of chromatography in which the mobile phase is a carrier gas and the stationary phase is a glass or metal tubing column. In 1956, the first biologically important molecules were successfully analyzed. Coupling the GC to MS provided "data rich" mass spectra for more definitive compound identification and in many cases increased sensitivity. New ionization techniques developed over the last 25 years have expanded the world of biological chemistry to MS. The development of GC/MS was the trigger for the development of MS for organic compounds (Mass Spectrometry Resource, 2005).

GC/MS provided a powerful tool to clearly identify compounds on the basis of their mass fragmentation spectra. During the 1980's and early 1990s, GC/MS was mostly used in research laboratories to understand the regional transport and degradation of herbicides and to interpret the hydrogeologic processes governing their occurrence, fate, and transport, as most USEPA contract laboratory methods were based on GC with nitrogen–phosphorous and electron capture detection. However, over the past two decades, the GC/MS has also become a standard instrument in environmental contract laboratories as the USEPA has developed approved methods for a wide range of contaminants.

GC/MS is limited to the analysis of a small range of organic molecules often characterized as nonionic, semivolatile. As molecules increase in polarity, they are difficult to be analyzed by GC/MS without derivatization. Liquid chromatography (LC) is much more amenable to the separation and analysis of compounds with a wide range of polarities, but adding mass spectrometers to LCs was problematic until relatively recently.

Liquid Chromatography/Mass Spectrometry

LC/MS is an analytical chemistry technique that combines the physical separation capabilities of LC, also known as high performance liquid chromatography (HPLC), with the mass analysis capabilities of MS. It is a powerful technique used for many applications and can have very high sensitivity and specificity. Generally, its application is oriented toward the specific detection and potential identification of chemicals in the presence of other chemicals. A major difference between HPLC using the diode array detection vs. LC/MS is the use of smaller diameter and particle-size LC columns and lower mobile-phase flow rates in LC/MS. For example, a typical HPLC method may use a 4.6 × 150–250 mm column with 5 μm particle size, whereas in LC/MS a 2.1–3 × 100–150 testing mm column with 3 μm particles is more common.

Coupling an HPLC with a mass spectrometer has proved to be a difficult task, requiring a great deal of research to overcome the challenge. The advancement of MS/ionization interfaces for LC revolutionized the ability of environmental researchers to determine the occurrence of a much wider variety of compounds than previously could be accomplished with GC/MS. The introduction of the thermospray ionization interface, which is capable of producing ions from an aqueous solution to spray directly into the MS, provided a significant advancement in coupling LC to MS. The primary problem was the inability of the particle beam and thermospray interfaces to dissipate the mobile phase before it entered the low vacuum portion of the mass spectrometer. Thus, the majority (90% or more) of the mobile phase is needed to be split before entering the source of the mass spectrometer. Thus, the sensitivity required for environmental analysis was problematic.

The introduction of the atmospheric pressure ionization (API) interfaces, electrospray ionization (ESI), and atmospheric pressure chemical ionization (APCI) overcame the shortcoming of earlier interfaces by evaporating the mobile phase during the ionization process. This, in addition to the orthogonal spray interface, also provided a means to remove potentially interfering nonvolatile molecules (salts, buffers, and detergents) from entering the mass spectrometer. In 2002, John Fenn was a cowinner of the Nobel Prize in Chemistry for the development of a soft desorption ionization method for mass spectrometric analyses of biological macromolecules—invention of the API interface. This new interface also provided the robustness and sensitivity needed to develop trace-level methods for the analysis of a wide variety of small polar molecules and their degradation products that were not readily amendable to analysis by GC/MS. Of the two API interfaces, the ESI has been more widely used than the APCI. These methods allow ionization at atmospheric pressure and can rapidly separate complex mixtures and readily identify its components.

There are several types of mass analyzers that can be used in LC/MS. Examples include single quadrupole, triple quadrupole, ion trap, time of flight (TOF), quadrupole–time of flight (Q-TOF), and a Q-trap LC/MS/MS system (Thurman et al., 2001, 2003). While several robust methods have been developed on single quadrupole instruments for the analysis of herbicides and other compounds, the triple quadrupole mass spectrometers have become the instrument of choice for quantitative environmental analysis. The TOF, Q-TOF, and ion-trap mass spectrometers generally are used for structural elucidation and the identification of unknown compounds.

Enzyme-Linked Immunosorbent Assay

Enzyme-linked immunosorbent assay (ELISA) tests were developed independently by research groups in Sweden and the Netherlands in 1971 (Lequin, 2005). In the late 1980s, ELISA began to be more widely used in the environmental field to screen for herbicides such as atrazine, with detection levels of 0.1–0.2 µg/L. The advantage of the ELISA kits was that a large number of samples could be screened inexpensively. This solid-phase assay simply works because of the fact that proteins (antibodies) can be positively attached to plastics (96-well microtiter plate, test tube, or magnetic particles) with relative positive values. Currently, ELISA is commercially available for a wide variety of organic contaminants.

WATER-QUALITY STUDIES

There have been many herbicide studies conducted by scientific researchers in government and state agencies, universities, corporations, and other groups in the United States and worldwide. The outcome is that a huge amount of data have been collected to investigate and understand the occurrence, concentrations, fate, and transport of several classes of herbicides and their degradation products in groundwater, surface water, and precipitation. Much of these data hve been shared in scientific meetings and published in many notable journals.

The USGS has implemented many of the large herbicide surveys conducted in the United States since the late 1980s, mostly through the Toxic Substances Hydrology and National Water-Quality Assessment (NAWQA) programs. The USGS has spent many years monitoring and collecting data of chemicals and sediments in the large rivers of the United States (USGS, 2006), investigating the status and trends in the quality of the nation's groundwater and surface-water resources (Gilliom et al., 2006), and presenting scientific information on the occurrence, movement, flux, fate, and effects of agricultural chemicals in the nation's hydrologic environments (Buxton, 2000).

The development of analytical methods for herbicides and other organic contaminants has been an important component of the USGS National Water Quality Laboratory Methods Research Development Program, Denver, Colorado, and of the USGS Organic Geochemistry Research Laboratory (OGRL) in Lawrence, Kansas. Many research studies on the fate and transport of herbicides and their degradation products in surface water and groundwater has been conducted by scientists at the OGRL since the late 1980s.

Analytical methods developed (by OGRL for various herbicides and their degradation products include solid-phase extraction SPE) GC/MS (Thurman et al., 1990; Meyer et al., 1993; Thurman and Mills, 1998; Kish et al., 2000), HPLC (Hostetler and Thurman, 2000), LC/MS (Thurman et al., 2001; Lee et al., 2001, 2002a, b; Lee and Strahan, 2003), and LC/MS/MS (Meyer et al., 2007a). The OGRL continues to develop robust methods to measure new and understudied herbicides, antibiotics, algal toxins, and cyanotoxins, and their degradation products.

Much of the understanding of the occurrence, fate, and transport of the wide variety of herbicides and their degradation products has resulted from large regional and national water-quality surveys that have been conducted from the 1980s to the present. The advancement of analytical equipment, ELISA tests, and solid-phase extraction technologies has been the backbone for the development of these studies.

The summarized surveys of (1) surface water, (2) groundwater, and (3) precipitation, which have had significant impacts on the understanding of herbicides in our nation's water resources, are given in the subsequent sections. In some cases, these surveys were repeated over multiple years. New methods were developed and incorporated into those studied to increase knowledge on the changing use of herbicides and, also, to better understand the wide variety of degradation products that are transported into the hydrogeological system.

Surface Water

Midcontinent Herbicide Reconnaissance, 1989–1990; 1994–1995; 1998; 2000

GC/MS results, 1989–1990. Although the rapid increase in herbicide concentrations with the first rainfall, after the application of preemergent herbicides, had been demonstrated in individual watershed studies, it had never been documented on a regional scale. During 1989, a reconnaissance survey of 147 streams in 10 Midwestern States, within the Corn Belt, was conducted to determine the geographic and seasonal distribution of acetanilide and triazine herbicides. The SPE-GC/MS method (Thurman et al., 1990; Meyer et al., 1993) developed for this study measured the most commonly used triazine and acetanilide corn and soybean herbicides, and also two dealkylated triazine degradation products, deethylatrazine (DEA) and deisopropylatrazine (DIA). Samples for this study were later reanalyzed for three cyanazine degradation products, cyanazine amide (CAM), deethylcyanazine (DEC), and deethylcyanazine amide (DCAM). The streams were sampled before application of herbicides (preplanting), during the first major runoff event after application of herbicides (post-planting), and during a low-flow period in the fall when most of the streamflow was derived from groundwater (harvest).

The major results of this study were (1) the high concentration pulses of pesticides flushed into streams after application of herbicides were a regional phenomenon in the Corn Belt; (2) DEA, DIA, and CAM were commonly detected in surface water during the post-planting event spring-flush; and (3) the ratio of DEA-to-atrazine (DAR) was indicative of seasonal transport and surface water–groundwater interaction (Thurman et al., 1991, 1992; Meyer et al., 1993, 2001; Meyer and Thurman, 1996).

Atrazine occurred in 98% of the post-planting samples followed by alachlor (86%), metolachlor (83%), and cyanazine (63%). Atrazine was the most frequently detected herbicide occurring in 91% of the preplanting samples and 76% of the harvest samples, whereas alachlor, metolachlor, and cyanazine were only detected in 18%, 34%, and 5% of the preplanting samples and 12%, 44%, and 0% of the harvest samples, respectively. These data indicated that cyanazine and alachlor degraded more rapidly and, thus, were less persistent than atrazine and metolachlor. DEA, a degradation product of atrazine, occurred frequently in preplanting, post-planting, and harvest samples, whereas DIA, CAM, and DEC mostly occurred in the post-planting samples. These findings were significant because they indicated that some of the parent herbicides persist from year to year in soil and water, and degradation products, such as DEA, also persist and are mobile. Results of studies of water samples collected from streams in eastern

Iowa during the 1990s further substantiated these findings (Kalkhoff et al., 2003; Schnoebelen et al., 2003).

A follow-up sampling was conducted in 1990 because of increased concern about the findings of high post-application concentrations of herbicides in 1989. The distribution of major herbicide concentrations detected in these streams was essentially the same in 1989 and 1990 for both the pre- and post-application samples and reaffirmed that the flush of herbicides following application is an annual occurrence (Goolsby et al., 1991a, b; Goolsby and Battaglin, 1995; Scribner et al., 1993, 1994, 1998, 2000, 2005; Thurman et al., 1991, 1992, 1994; Battaglin and Goolsby, 1999).

GC/MS results, 1994–1995, 1998, 2002. In 1994, 1995, 1998, post-application runoff samples were collected at 50 of the sites that were sampled in 1989–1990 to help determine if changes in herbicide usage were reflected in their occurrence and transport. For example, in 1990 and 1992, label changes decreased the recommended application rate of atrazine, acetochlor was introduced to replace alachlor in 1994, glyphosate-tolerant (GT) soybeans were introduced in 1997 and GT corn in 2000, the production of cyanazine was stopped in 1999 and its use discontinued on December 31, 2002, and metolachlor was replaced with *S*-metolachlor in 1998. Metolachlor was a 50–50 ratio of the *R* and *S* isomers of which the *S* isomer is herbicidally active. To lower the amount of metolachlor per acre that needed to be applied, *S*-metolachlor was introduced in 1997 (Scribner et al., 1998).

Comparison of the trends in median concentrations of the 1989 to 2002 reconnaissance water-quality studies showed that changes in herbicide uses were reflected in their transport in stream water. The median concentration of alachlor decreased from approximately 1.5 to <0.1 µg/L between 1989 and 1998. The median concentration of acetochlor increased from <0.1 µg/L in 1984 when it was first measured to approximately 1 µg/L in 1998. Median concentration of metolachlor was similar from 1989 to 1998 indicating that the use of *S*-metolachlor may not have been prevalent in 1998. However, between 1998 and 2002, the median concentration of metolachlor decreased from 1.4 to 0.75 µg/L. The median concentration of cyanazine decreased from 2 to <0.05 µg/L between 1998 and 2002, indicating a substantial decrease in usage that resulted from its removal from the market. The detection frequency of atrazine was >98% in all of the surveys conducted between 1989 and 2002, but the median concentration decreased from 11 to 4 µg/L, indicating a potential effect from the voluntary reduction in the application rate of atrazine that occurred in 1990 and 1992 and also from the increase in the use of GT corn. Although these surveys showed that the herbicide concentrations in midwestern streams are quite variable during post-application runoff, changes in herbicide use affecting herbicide concentrations in streams was indicated.

LC/MS results, 2002. The addition of three new LC/MS methods added 14 acetanilide herbicide degradation products, 11 triazine herbicide degradation products, and glufosinate and glyphosate and its degradation product aminomethylphosphonic acid (AMPA) to the 2002 stream-water reconnaissance survey, expanding the knowledge of the wide variety of herbicides and herbicide degradation products transported to surface water. These data are presented in Battaglin et al. (2005) and Scribner et al. (2003, 2004).

The major findings from this study include the following: (1) many of the acetanilide ethane sulfonic acid (ESA) and oxanilic acid (OXA) degradation products were detected in stream water year round, (2) didealkylatrazine (DDA) and hydroxyatrazine (HA) had similar detection frequencies to DEA, (3) CAM was not detected indicating cyanazine was not being applied, and (4) AMPA was detected more frequently and often at higher concentrations than glyphosate.

For example, for the triazine degradation products, DEA was the most frequently detected followed by HA, DDA, and DIA, respectively. For the acetanilide degradation products, metolachlor ESA and OXA were detected most frequently, followed by acetochlor ESA and OXA, and alachlor ESA, respectively.

In the case of glyphosate, it was detected at or above $0.10\,\mu g/L$ in 35% of preemergence, 40% of postemergence, and 31% of harvest season samples, with a maximum concentration of $8.7\,\mu g/L$. AMPA was detected at or above $0.10\,\mu g/L$ in 53% of preemergence, 83% of postemergence, and 73% of harvest season samples, with a maximum concentration of $3.6\,\mu g/L$. It is probable that glyphosate is not as mobile and is transformed more rapidly in the environment than the other herbicides (Scribner et al., 2003, 2007; Battaglin et al., 2005).

The data from these methods showed that herbicide degradation products accounted for a substantial portion of the total herbicide transport in streams that previously had not been recognized. These stream-water reconnaissance surveys in the Midwestern United States confirmed that herbicide degradation products were found to occur as frequently or more frequently and at concentrations that were often higher than the parent herbicides (Battaglin et al., 2003, 2005; Scribner et al., 2003, 2007).

Reservoirs, 1992–1993

The initial results from the 1989 and 1990 stream-water reconnaissance surveys led to the question of how herbicides are transported through the reservoir impoundments, which are prevalent throughout the Midwestern United States. Many reservoirs used for drinking water have drainage basins whose primary land use is crop production. Reservoirs were screened and selected from the reservoir database compiled by Ruddy and Hitt (1990) and are described in Scribner et al. (1996).

The important findings from this study were as follows: (1) reservoirs are repositories for contaminants that are introduced into midwestern streams, (2) herbicides and degradation products are detected more frequently throughout the year in reservoirs than in streams, and (3) long-term storage and mixing of water in reservoirs that originate as spring and summer storm runoff from cropland dampens and lengthens the pulse of herbicides transported and released through the reservoirs.

Analytical results from samples collected during 1992 indicate that a number of herbicides and degradation products were present and detected in 82–92% of the selected reservoirs during eight sampling periods. One of the most notable differences between the occurrence of herbicides in reservoirs and streams was a much higher frequency of detection of cyanazine and DIA in reservoirs. A possible explanation for this observation is that these two compounds are much more stable in the water of lakes and streams than in soil where organic matter and microorganisms promote rapid biodegradation. Consequently, late spring and summer runoff can flush large amounts of these two compounds into reservoirs, where they can persist for long periods of time. Neither cyanazine nor DIA was detected in streams during the fall because these compounds are no longer present in substantial amounts on the agricultural fields where they were applied (Scribner et al., 1996; Battaglin and Goolsby, 1998). Thus, herbicide concentrations in reservoir outflows behave differently than those in unregulated streams (Stamer and Zelt, 1992; Fallon and Thurman 1996; Thurman and Fallon, 1996; Stamer et al., 1998a). The mean of the individual concentrations in midwestern reservoirs of atrazine and its degradation products was 1.9 μg/L. Similarly, the mean sum of the individual concentrations of cyanazine and its degradation product, CAM, was 1.0 μg/L, which is also consistent with the fact that cyanazine usage in the study area was about half that of atrazine.

Perry Lake, 1992–1993

To understand in more detail how herbicides are transported through reservoirs, a three-dimensional survey was conducted in Perry Lake in northeastern Kansas during 1992 and 1993 (Fallon et al., 2002) using ELISA keyed to atrazine. A subset of samples were analyzed by GC/MS for 11 herbicides and 2 triazine and 3 cyanazine degradation products. In addition, the degradation product alachlor ESA was isolated by SPE and analyzed by ELISA using the method of Aga et al. (1994). The sampling strategy consisted of two components, five seasonal surveys with samples collected at randomly selected sites throughout the lake and at multiple depths at each site, and sampling of the inflow and outflow from the reservoir. Water samples were collected monthly throughout the year and during runoff events from April thru August.

Atrazine concentrations in Perry Lake increased 48% after application to croplands (from 2.7 to 4.0 μg/L). Three-dimensional computer images of atrazine concentrations and DAR values showed that recently applied atrazine mixed with atrazine applied the previous year as water moved sequentially through the reservoir. Changes in atrazine concentrations resulted from several factors including herbicide application, precipitation, and reservoir-residence time. Precipitation after atrazine application drove the system by flushing atrazine into the reservoir. The timing of the precipitation and runoff affected how much atrazine flushed into the reservoir. The volume of precipitation and runoff affected how long atrazine remained in the reservoir. Precipitation shortened reservoir-residence time by increasing inflow and outflow during wet periods. Below-normal precipitation in May and June 1992, combined with above-normal precipitation during the last

9 months of the study period, produced lower atrazine concentrations in the reservoir outflow than those found in previous years. Atrazine concentrations at the outflow were decreased, and were dampened (the pulse of water entering the reservoir) as water containing higher atrazine concentrations was temporarily stored and mixed with water having lower concentrations (Fallon and Thurman, 1996; Scribner et al.,1996; Thurman et al., 1996; Thurman and Fallon, 1996; Fallon et al., 2002; Thurman and Scribner, 2008).

The Quality of Our Nation's Waters—Pesticides in the Nation's Streams and Groundwater, 1992–2001

A USGS NAWQA Program report (Gilliom and Hamilton, 2006; Gilliom et al., 2006) showed that pesticides are widespread in streams and groundwater and occur in geographic and seasonal patterns along with land and pesticide usage. During the 1992–2001 sample-collection period, more than 95% of the samples collected from streams and almost 50% of the samples collected from wells contained at least one pesticide. Seventy-four of the 83 pesticide compounds analyzed were detected at least once in streams or groundwater. Major rivers, agricultural, and urban streams had relatively similar high frequencies of detection.

Pesticides most commonly detected in streams draining watersheds with mixed land use reflected multiple sources from agricultural and urban applications. The overall frequency of pesticide occurrence in mixed land-use streams was comparable to those monitored in agricultural and urban streams. Similarly, pesticides detected in major aquifers indicate the influence of agricultural and urban sources, but overall detection frequencies were lower in major aquifers than in shallow groundwater in agricultural and urban areas (Barbash et al., 2001; Gilliom and Hamilton, 2006; Gilliom et al., 2006; Rosen and Lapham, 2008).

Comparison of Fate and Transport of Isoxaflutole to Atrazine and Metolachlor in 10 Iowa Rivers, 2004

As more water soluble and lower application rate herbicides are marketed, it is important to develop methods on instruments capable of detecting these compounds at the concentrations that occur in the environment. In 1998, a new restricted use, preplanting, low-application rate herbicide, isoxaflutole, was registered for use on corn. Research had shown that isoxaflutole rapidly degraded to a herbicidally active degradation product diketonitrile (DKN), which is more stable and, which, in turn, degrades to a nonbiologically active benzoic acid analog (BA). However, no studies had been published on the transport of isoxaflutole or its degradation products in surface water. To assess whether isoxaflutole and its sequential degradation products occur in stream water, a SPE-LC/MS/MS method was developed to detect isoxaflutole, DKN, and BA at 0.002 μg/L (Meyer et al., 2007a). In addition, two LC/MS methods were used to analyze for a suite of triazine and acetanilide herbicides and their degradation products.

In 2004, samples were collected monthly from March through September from 10 major rivers in Iowa that drain to the Missouri and Mississippi Rivers. The purpose of the study was to (1) determine the seasonal transport of isoxaflutole and its degradation products and (2) compare its transport to the more commonly measured herbicides such as acetochlor, atrazine, and metolachlor.

The major findings of the study were: (1) DKN, the herbicidally active degradate for isoxaflutole, was frequently detected instead of isoxaflutole; (2) previous research was supported that isoxaflutole degrades rapidly to DKN; (3) seasonal transport of DKN and its BA degradate product is similar to that of atrazine and DEA but at significantly lower concentrations; and (4) the difference in the median concentration of atrazine and DKN detected in the post-application (May–June samples) was the same as the usage difference between atrazine and isoxaflutole.

Analytical results of 75 water samples show that isoxaflutole was detected in only four of the samples collected during the post-application (May–June) period, whereas DKN was detected in 53 water samples and BA was detected in 41 water samples collected from all three sampling periods. Metolachlor was the most frequently detected chloroacetanilide parent (59 of 60 samples) in all three sampling periods, followed by acetochlor (41/60) and alachlor (3/60). The ESA and OXA degradation products of acetochlor and metolachlor were present 100% of the samples respectively, whereas alachlor was detected in 56 of 60 ESA samples and 53 of 60 OXA samples. The degradation products were detected as frequently as or more frequently than their parent compounds. Atrazine was the most detected triazine parent compound with 65 detections of 69 samples during all three sampling periods, whereas the triazine degradation products of HA and DEA were detected in 66 and 65 samples, respectively, which is similar to other reported conclusions of the detection frequency of the atrazine parent and its degradation products (Scribner et al., 2006).

Findings of the isoxaflutole study include an improved understanding of the occurrence of the herbicide isoxaflutole and its degradation products in the hydrologic environment. Analytical results of the chloroacetanilide and triazine herbicides are consistent with previous studies, which show that large amounts of herbicides and their degradation products are flushed into streams with run-off. Also, the study confirmed prior findings that these herbicides occur as frequently as or more frequently than their parent herbicide (Scribner et al., 2006; Meyer et al., 2007a,b).

Groundwater

Midwestern United States Groundwater Reconnaissance Network, 1991–1994; 1995–1998

During the 1990s, the Midwestern United States was the focus of research on agricultural chemical contamination in groundwater because it was an area of intense application of herbicides. A regional monitoring network was designed that was geographically and hydrogeologically representative of near-surface aquifers in the corn and soybean producing areas of the Midwestern United States. A series of papers (Kolpin and Burkhardt,1991; Burkhardt and Kolpin, 1993; Kolpin et al., 1993, 1994, 1995, 1996a–c, 1997, 1998, 2000, 2001, 2004; Kolpin and Thurman, 1995; Kolpin, 1997; Burkhardt et al., 1999; Mills et al., 2005) highlighted the major findings of these groundwater studies. This study determined the relationship between soils, land use, groundwater age, and concentration and occurrence of herbicides and their degradation products in groundwater. The major findings of these studies were that (1) groundwater that was dated to be older than 1953 (predating the first use of herbicides) was found to have a much lower frequency of herbicide detection than water younger than 1953 and (2) inclusion of the herbicide degradation is important to assess the total contribution of herbicides to groundwater.

The initial phase of the groundwater research (1991–1994) focused on regional-scale studies. A total of 837 water-quality samples from 303 wells across 12 states were collected. Atrazine was the most frequently detected parent compound, being detected in 22.4% of the samples collected. Two atrazine degradation products, DEA and DIA, also were frequently detected compounds in these studies, supporting the relative stability of these compounds. Cyanazine production ceased in December 1999, although it was being used extensively during the early 1990s. It was detected in only 2.3% of the wells; however, CAM, a degradation product of cyanazine, was detected in samples from 11% of the wells. This greater frequency of detection suggests an increase in degradation product mobility to groundwater after transformation from cyanazine. As with atrazine and cyanazine, simazine also can be transformed to DIA but at a much faster rate than atrazine (Mills and Thurman, 1994). Simazine was detected in samples from 2.6% of the wells, and its dealkylation to DIA probably contributed little to the amount of DIA in groundwater.

A major finding for this study was documenting the importance of including herbicide degradates in water-quality studies investigating herbicides. Many such degradates were found much more frequently and at higher concentrations than their parent compounds. In addition, herbicides having a long half-life (such as atrazine) generally were detected more frequently in groundwater than those having shorter half-lives.

The second phase of the groundwater research (1995–1998) focused on statewide research on the occurrence of parent compounds and a wider range of herbicide degradates products. Samples were collected from 131 municipal wells covering all the major aquifer types in Iowa. An important finding of this study was the high frequency with which degradation products were detected in groundwater. Atrazine was the only herbicide for which the parent compound was at a similar level compared to its degradates. These results documented that aquifer types with the most rapid recharge rates were those most likely to contain detections of herbicide compounds. Eighty percent of the wells in which herbicides were not detected had older aged water. Overall, the frequency of detection increased from 17%, when only the parent compounds were considered to 53% when the parent compounds and degradates were considered. Thus, the transport of herbicide compounds to groundwater is substantially underestimated when herbicide degradates are not considered (Kolpin et al., 1998, 2000, 2001, 2004).

Chloroacetanilide Herbicides and Their Degradation Products in Groundwater of the United States, 1993–2003

During 1993–2003, a USGS study was conducted to investigate and document the occurrence, fate, and transport of chloroacetanilide herbicides and their degradation products in groundwater (Scribner et al., 2004). About 2,420 samples were collected to analyze for the chloroacetanilide parent herbicides acetochlor, alachlor, dimethenamid, flufenacet, and metolachlor and their ESA, OXA, and sulfinyl acetic acid (SAA) degradation products.

The major finding of this study was that the ESA and OXA degradates of metolachlor and alachlor that occurred more frequently and at higher concentrations than their parent herbicides. Metolachlor was the most frequently detected chloroacetanilide parent compound (12%), followed by acetochlor and alachlor (2% each), and dimethenamid (0.18%). The chloroacetanilide degradation products were detected more frequently than their parent compounds. Metolachlor ESA, the most frequently detected degradation product, was present in 45% of the samples analyzed, followed by alachlor ESA (35%) and metolachlor OXA (26%). Overall, the median concentrations of detections for the acetochlor, alachlor, and metolachlor parent compounds were less than their ESA or OXA degradation products. For example, the median concentration for metolachlor detections in groundwater (0.17 μg/L) was less than the median concentrations of detections for its degradation products, metolachlor ESA (0.97 μg/L) and metolachlor OXA (0.70 μg/L). However, the median concentration for the dimethenamid parent compound was greater than its ESA or OXA degradation product. Flufenacet and its degradation products of ESA, OXA, and the SAAs were not detected in any of the groundwater samples.

Analytical results showed that the methods were valuable for acquiring information about the fate and transport of the parent chloroacetanilide herbicides in water. Degradation products of chloroacetanilide herbicides in groundwater were detected more frequently and occurred at similar or higher concentrations than their parent compounds. Once again, this study confirmed that it is important to include both parent compounds and their degradation products in herbicide studies (Lee and Strahan, 2003; Scribner et al., 2004).

Precipitation

Herbicides in Rainfall Across the Midwestern and Northeastern United States, 1990–1991

During the late spring and summer of 1990 and 1991, a study by Goolsby et al. (1997) was focused on herbicides that could be transported into the atmosphere by various processes. This study was conducted prior to significant label rate reductions for atrazine-containing products. Once in the atmosphere, these compounds can be dispersed by air currents and redeposited by precipitation, snow, and dry deposition on the land surface, lakes, and streams.

The overall objective of the precipitation study was to (1) determine the occurrence and temporal distribution of herbicides and their degradation products in precipitation, (2) estimate the amounts of atrazine deposited by precipitation annually in individual States and over a large part of the United States, (3) relate annual deposition of atrazine to amounts applied annually, and (4) compare annual herbicide deposition by precipitation within the Mississippi River basin to the estimated annual amount transported out of the basin in streamflow.

Herbicide concentrations exhibited distinct geographic and seasonal patterns. The highest concentrations occurred in Midwestern Corn Belt States following herbicide application to cropland. Occurrence and concentrations of triazine herbicides were detected by ELISA in 5,297 samples collected during the study period. Herbicides having significant cross-reactivity in the ELISA methods other than alachlor and atrazine were rarely detected in this study and probably had little or no effect on the ELISA analysis. GC/MS analysis was performed on 2,085 of the precipitation samples (Goolsby et al., 1995; Pomes et al., 1998). The most frequently detected herbicide was atrazine, which was present in 30.2% of the samples analyzed. DEA was present in more than one-half of the samples that contained atrazine and was the third most frequently detected compound. Trace concentrations of DEA were detected in 12 samples that contained no detectable atrazine. Cyanazine was detected in 7.2% of the samples. Although herbicides were detected in a significant number of samples, concentrations were relatively low. Atrazine was also detected in low concentrations at sites in Maine and on Isle Royale in northern Lake Superior (Goolsby et al., 1997; Stamer et al., 1998b; Thurman and Cromwell, 2000).

Because of the large temporal and spatial variations in the amount of precipitation, it is difficult to make meaningful comparisons of herbicide concentrations among sites or over time on the basis of individual weekly samples. Therefore, comparisons were made with precipitation-weighted concentrations. The spatial distribution of precipitation-weighted concentrations of atrazine were calculated for a 13-week period from mid-April through mid-July 1990 and 1991 when concentrations were the highest. Precipitation-weighted concentrations of 0.2–0.4 μg/ L for atrazine were typical throughout the Midwest for this 13-week period, and weighted concentrations of 0.4–0.9 μg/L were recorded at sites in Iowa, Illinois, and Indiana. Overall, the spatial patterns of the weighted atrazine concentrations in 1990 and 1991 were similar and generally reflect atrazine use.

Nearly all of the deposition of atrazine and alachlor in rainfall occurred during April through July when concentrations were highest. Consequently, results should closely represent the total annual wet deposition of atrazine during the 2 years. Atrazine deposition rates ranged from more than 100 μg/m^2/year in the Midwestern States to <10 μg/m^2/year in the Northeastern States. Deposition rates throughout most of the Corn Belt ranged from $50\,\mu g/m^2/yr$ to more than $100\,\mu g/m^2/yr$ for atrazine.

One of the sampling sites was located on Isle Royale in the northwestern part of Lake Superior near the Canadian border and far from the U.S. Corn Belt. Atrazine, presumably from the Midwestern Corn Belt, was detected and verified by GC/MS analysis in samples from several rain events at this site during June 1990. These data prompted the collection of water samples from Lake Superior and from four small lakes on Isle Royale in late September 1990. The atrazine concentration in these samples, determined by isotope dilution methods, was 6.5 ng/L for Lake Superior and ranged from 2.5 to 20 ng/L for the four lakes sampled in Isle Royal National Park. The most likely source of atrazine to Lake Superior and Isle Royale is atmospheric deposition (Goolsby et al., 1995, 1997).

CONCLUSIONS

Continued advancements in research on water quality are necessary to provide environmental engineers, toxicologists, and Federal and State government regulatory agencies with new knowledge needed to assess the need for new or advanced water treatment and assessment of best management practices and regulation. Weeds are a huge problem in agriculture, not only in cost incurred by the farmer but also in assessing potential deleterious effects on water quality from the practices used to control the weeds.

In the large herbicide water-quality reconnaissance studies conducted in the United States over the last two decades, pesticides were found in almost every stream sample collected, with concentrations greatest in areas of the Nation with the greatest agricultural use. These findings, along with the other studies, are relevant to the water quality of source drinking-water supplies. While the streams and wells sampled in most of these studies are not directly used for drinking water, these water-quality studies show the need to consider criteria for contaminant levels of degradation products as well as the parent herbicides in treatment processes. The evolution of analytical equipment has provided scientists with the means to more fully assess the complex mixtures of organic contaminants that are transported into our Nation's water resources. The results of these water-quality surveys have shown that herbicides applied in row crop agriculture are transported into surface water as pulses in response to rainfall and that they also are transported to groundwater. The results of these studies have, in part, resulted in cases where manufacturers voluntarily reduce the recommended application rates, in removing herbicides from the market, and prompt the USEPA into introducing selected acetanilide degradates onto the contaminant candidate list.

Studies by many scientists have continually expanded our knowledge of the occurrence, persistence, fate, and transport of herbicides and their degradation products in the hydrologic environment. The results of large water-quality studies also provide important data sets for Federal, State, and local agencies, and utilities to develop effective regulatory and management strategies and provide a historical perspective on changing usage.

Any use of trade, product, or firm names is for descriptive purposed only and does not imply endorsement by the U.S. government. Support for this project was provided by the U.S. Geological Survey Toxic Substances Hydrology Program.

REFERENCES

Aga, D.S., Thurman, E.M., Pomes, M.L., 1994. Determination of alachlor and its sulfonic acid metabolite in water by solid-phase extraction and enzyme-linked immunosorbent assay. Anal. Chem. 66, 1495–1499.

Barbash, J.E., Thelin, G.P., Kolpin, D.W., Gilliom, R.J., 2001. Major herbicides in water. J. Environ. Qual. 30, 831–845.

Battaglin, W.A., Goolsby, D.A., 1998. Regression models of herbicide concentrations in out-flow from reservoirs in the midwestern USA, 1992–1993. J. Am. Water Resour. Assoc. 34, 1369–1390.

Battaglin, W.A., Goolsby, D.A., 1999. Are shifts in herbicide use reflected in concentration changes in midwestern rivers? Environ. Sci. Technol. 33, 2917–2925.

Battaglin, W.A., Thurman, E.M., Kalkhoff, S.J., Porter, S.D., 2003. Herbicides and transformation products in surface waters of the midwestern United States. J. Am. Water Res. Assoc. 39, 743–756.

Battaglin, W.A., Kolpin, D.W., Scribner, E.A., Kuivila, K.M., Sandstrom, M.W., 2005. Glyphosate, other herbicides, and transformation products in midwestern streams, 2002. Am. Water Res. Assoc. 41, 323–332.

Bottled Water Blues, 2002. Pure water—water purity. http://www.bottledwaterblues.com/Pure_Water.cfm

Brown, P., 2004. Most British rivers will fail new EU water purity rules. UK News, October 6, 2004. http://www.guardian.co.uk/uk/2004/oct/06/society.water

Burkhardt, M.R., Kolpin, D.W., 1993. Hydrologic and land-use factors associated with herbicides and nitrate in near-surface aquifers. J. Environ. Qual. 22, 646–656.

Burkhardt, M.R., Kolpin, D.W., Jaquis, R.J., Cole, K.J., 1999. Agrichemicals in groundwater of the midwestern United States: relations to soil characteristics. J. Environ. Qual. 28, 1908–1915.

Buxton, H.T., 2000. USGS toxic substances hydrology program, 2000: USGS Fact Sheet 062-00.

Coupe, R.H., Thurman, E.M., Zimmerman, L.R., 1998. Relation of usage to the occurrence of cotton and rice herbicides in three streams of the Mississippi Delta. Environ. Sci. Technol. 32, 3673–3680.

Coupe, R.N., Welch, H.L., Pell, A.B., Thurman, E.M., 2005. Herbicide and degradate flux in Yazoo River Basin. Int. J. Environ. Anal. Chem. 85, 1127–1140.

Cox, C., 2004. Glyphosate herbicide fact sheet. J. Pestic. Reform 24, 10–15.

Delaplane, K.S., 2000. Pesticide usage in the United States: history, benefits, risks, and trends. Cooperative Extension Service. The University of Georgia College of Agricultural and Environmental Sciences. http://www.pubs.caes.uga.edu/caespubs/pubs/PDF/B1121.pdf

Demographia, 2001. US population from 1900. Demographia. http://www.demographia.com/db=uspop1900.htm

Dinnage, R.J., 2007. Most widely used herbicide not a carcinogen—EPA, EarthNews. http://www.earthportal.org/news/?p=396

Extoxnet, 1996a. Extension toxicology network—2,4-D. http://extoxnet.orst.edu/pips/2,4-D.htm

Extoxnet, 1996b. Extension toxicology network—acetochlor. http://www.extoxnet.orst.edu/pips/acetochl.htm

Fallon, J.D., Tierney, D.P., Thurman, E.M., 2002. Movement of atrazine and deethylatrazine through a Midwestern reservoir. J. Am. Water Res. Assoc. 94, 54–65.

Fallon, J.D., Thurman, E.M., 1996. Determining the age, transport, and three-dimensional distribution of atrazine in a reservoir using immunoassay. In: Morganwalp, D.W., Aronson, D.A. (Eds.), USGS Toxic Substances Hydrology Program – Proceedings of the Technical Meeting, vol. 1, 20–24 September 1993, Colorado Springs, CO, USGS water-resources investigations report 94-4015, pp. 499–504.

FAO, 2000. Food and population: FAO looks ahead. Food and Agriculture Organization of the United Nations. http://www.fao.org/News/2000/000704-e.htm

Field Crop News, 2008. National Ag day, March 20, 2008, Paul H. Craig, Dauphin County Extension. http://www.fcn.agronomy.psu.edu/2008/fcn0803.cfm

Fishel, F.M., 2007. Pesticide use trends in the U.S.: global comparison. University of Florida, Institute of Food and Agricultural Sciences Extension. http://www.edis.ifas.ufl.edu/Pl180#TABLE_2

Gianessi, L.P., 2005. Economic and herbicide use impacts of glyphosate-resistant crops. Pest Manag Sci. 61, 241–245.

Gianessi, L.P., Reigner, N.P., 2006. National Pesticide Use Database: 2002. National Center for Food and Agricultural Policy available at http://www.croplifefoundation.org.

Gianessi, L.P., Reigner, N.P., 2007. The value of herbicides in US crop production. Weed Technol. 21, 559–566.

Gilliom, R.J., Hamilton, P.A., 2006. Pesticides in the nation's streams and ground water, 1992–2001—a summary: USGS fact sheet 2006–3028.

Gilliom, R.J., Barbash, J.E., Crawford, C.G., Hamilton, P.A., Martin, J.D., Nakagaki, N., et al., 2006. Pesticides in the nation's streams and ground water, 1992–2001: USGS Circular 1291.

Global Water, 2008. Overcoming hunger, disease, and poverty…with water. http://www.globalwater.org

Goolsby, D.A., Battaglin, W.A., 1995. Occurrence and distribution of pesticides in streams of the midwestern US. In: Leng, M.L. (Ed.), Agrochemical Environmental Fate – State of the Art. Lewis–CRC Press, Boca Raton, FL, pp. 159–173.

Goolsby, D.A., Thurman, E.M., Pomes, M.L., Meyer, M.T., Battaglin, W.A., 1997. Herbicides and their metabolites in rainfall: origin, transport, and deposition patterns across the Midwestern and Northeastern US. Environ. Sci. Technol. 31, 1325–1333.

Goolsby, D.A., Coupe, R.H., Markovchick, D.J., 1991a. Distribution of selected herbicides and nitrate in the Mississippi River and its major tributaries, April through June 1991: USGS water-resources investigations report 91-4163.

Goolsby, D.A., Thurman, E.M., Kolpin, D.W., 1991b. Geographic and temporal distribution of herbicides in surface waters of the upper Midwestern United States, 1989–90: USGS water-resources investigations report 91-4034.

Goolsby, D.A., Scribner, E.A., Thurman, E.M., Meyer, M.T., Pomes, M.L., 1995. Data on selected herbicides and two triazine metabolites in precipitation in the midwestern and northeastern U.S., 1990–91: USGS open-file report 95-469.

Greatest Achievements, 2008. Water supply & distribution history II—early years. http://www.greatachievements.org/?id=3614/

Hostetler, K.A., Thurman, E.M., 2000. Determination of chloroacetanilide herbicide metabolites in water using high-performance liquid chromatography-diode array detection and high-performance liquid chromatography/mass spectrometry. Sci. Total Environ. 248, 147–156.

HRGW, 2004. Part 1. Herbicide resistance in weeds. Herbicide resistance in grass weeds. http://www.aun.edu.eg/distance/agriculture/HRGW/1%20herb%20resist....htm

IA Weed Science, 1998. Herbicide Site of Action. Iowa State University, Ames, IA. http://www.weeds.iastate.edu/reference/siteofaction.htm

Kalkhoff, S.J., Lee, K.E., Porter, S.D., Terrio, P.J., Thurman, E.M., 2003. Herbicides and herbicide degradation products in upper midwest agricultural streams during August base-flow conditions. J. Environ. Qual. 32, 1025–1035.

Kiely, T., Donaldson, D., Gruber, A., 2004. Pesticide Industry Sales and Usage: 2000 & 2001 Market Estimates. USEPA, Office of Pesticides Program, Washington, D.C. http://www.epa.gov/oppbead1/pestsales/01pestsales/usage2001.htm

Kish, J.L., Thurman, E.M., Scribner, E.A., Zimmerman, L.R., 2000. Method of analysis by the U.S. Geological Survey Organic Geochemistry Research Group—determination of selected herbicides and their degradation products in water using solid-phase extraction and gas chromatography/mass spectrometry: USGS open-file report 00-385.

Kolpin, D.W., 1997. Agricultural chemicals in groundwater of the midwestern United States: relations to land use. J. Environ. Qual. 6, 1025–1037.

Kolpin, D.W., Burkhardt, M.R., 1991. Work plan for regional reconnaissance for selected herbicides and nitrate in ground water of the mid-continental United States, 1991: USGS open-file report 91-59.

Kolpin, D.W., Thurman, E.M., 1995. Post-flood occurrence of selected agricultural chemicals and volatile organic compounds in near-surface unconsolidated aquifers in the upper Mississippi River Basin, 1993: USGS circular 1120-G.

Kolpin, D.W., Burkhardt, M.R., Thurman, E.M., 1993. Hydrogeologic, water-quality, and land-use data for the reconnaissance of herbicides and nitrate in near-surface aquifers of the midcontinental US, 1991: USGS Open-file report 93-114.

Kolpin, D.W., Burkhardt, M.R., Thurman, E.M., 1994 Herbicides and nitrate in near-surface aquifers of the midcontinental United States, 1991: U.S. geological survey water-supply paper 2413.

Kolpin, D.W., Goolsby, D.A., Thurman, E.M., 1995. Pesticides in near-surface aquifers: an assessment using highly sensitive analytical methods and tritium. J. Environ. Qual. 24, 1125–1132.

Kolpin, D.W., Nations, B.K., Goolsby, D.A., Thurman, E.M., 1996a. Acetochlor in the hydrologic system in the midwestern United States. Environ. Sci. Technol. 30, 1459–1464.

Kolpin, D.W., Thurman, E.M., Goolsby, D.A., 1996b. Occurrence of selected pesticides and their metabolites in near-surface aquifers of the midwestern US. Environ. Sci. Technol. 30, 335–340.

Kolpin, D.W., Zichelle, K.E., Thurman, E.M., 1996c. Water-quality data for nutrients, pesticides, and volatile organic compounds in near-surface aquifers of the midcontinental United States, 1992–1994: U.S. geological survey open-file report 96-435.

Kolpin, D.W., Kalkhoff, S.J., Goolsby, D.A., Sneck-Fahrer, D.A., Thurman, E.M., 1997. Occurrence of selected herbicides and herbicide degradation products in Iowa's groundwater, 1995. Ground Water 35, 679–688.

Kolpin, D.W., Thurman, E.M., Linhart, S.M., 1998. The environmental occurrence of herbicides: the importance of degradates in groundwater. Arch. Environ. Contam. Toxicol. 35, 1–6.

Kolpin, D.W., Thurman, E.M., Linhart, S.M., 2000. Finding minimal herbicide concentrations in groundwater? Try looking for their degradates. Sci. Total Environ. 248, 115–122.

Kolpin, D.W., Thurman, E.M., Linhart, S.M., 2001. Occurrence of cyanazine compounds in groundwater: degradates more prevalent than the parent compound. Environ. Sci. Technol. 35, 1217–1222.

Kolpin, D.W., Schnoebelen, D.J., Thurman, E.M., 2004. Degradates provide insight to spatial and temporal distribution of herbicides in ground water. Ground Water 42, 601–608.

Kusel, D., 2008. Farming methods and practices. http://www.davidkusel.com/centennial/237farming.htm

LeBaron, H.M., McFarland, J.E., Burnside, O.C., 2008. The triazine herbicides: a milestone in the development of weed control technology. In: LeBaron, H.M., McFarland, J.E., Burnside, O.C. (Eds.), The Triazine Herbicides. Elsevier, San Diego, CA, pp. 1–12.

Lee, E.A., Strahan, A.P., 2003. Methods of analysis by the U.S. Geological Survey Organic Geochemistry Research Group—Determination of acetamide herbicides and their degradation products in water using online solid-phase extraction and high-performance liquid chromatography/mass spectrometry: USGS Open-File Report 03-173.

Lee, E.A., Kish, J.L., Zimmerman, L.R., Thurman, E.M., 2001. Methods of analysis by the U.S. Geological Survey Organic Geochemistry Research Group—update and additions to the determination of chloroacetanilide herbicide degradation compounds in water using high-performance liquid chromatography/mass spectrometry: USGS open-file report 01-10.

Lee, E.A., Strahan, A.P., Thurman, E.M., 2002a. Methods of analysis by the U.S. Geological Survey Organic Geochemistry Research Group—determination of glyphosate, aminomethylphosphonic acid, and glufosinate in water using online solid-phase extraction and high-performance liquid chromatography/mass spectrometry: USGS open-file report 01-454.

Lee, E.A., Strahan, A.P., Thurman, E.M., 2002b. Methods of analysis by the U.S. Geological Survey Organic Geochemistry Research Group—determination of triazine and phenylurea herbicides and their degradation products in water using solid-phase extraction and liquid chromatography/mass spectrometry: USGS open-file report 02-436.

Lequin, R.M., 2005. Enzyme immunoassay (EI)/enzyme-linked immunosorbent assay (ELISA). Clin. Chem. 51, 2415–2418.

Mass Spectrometry Resource, 2005. The history of mass spectrometry. University of Bristol, School of Chemistry. http://www.chm.bris.as.uk/ms/history.html

Mennes, H., 2005. Classification of herbicide mode of action. Herbicide Resistance Action Committee. http://www.hracglobal.com/Publications/ClassificationofModeofAction/

Meyer, M.T., Thurman, E.M., 1996. The transport of cyanazine metabolites in surface water of the Midwestern US. In: Morganwalp, D.W., Aronson, D.A. (Eds.), USGS Toxic Substances Hydrology Program—Proceedings of the Technical Meeting, Colorado Springs, CO, September 20–24, 1993, vol. 1. USGS Water-Resources Investigations Report 94-4015, pp. 463–470.

Meyer, M.T., Mills, M.S., Thurman, E.M., 1993. Automated solid-phase extraction of herbicides from water for gas chromatographic/mass spectrometric analysis. Chromatography 629, 55–59.

Meyer, M.T., Thurman, E.M., Goolsby, D.A., 2001. Differentiating nonpoint sources of deisopropylatrazine in surface water using discrimination diagrams. J. Environ. Qual. 30, 1836–1843.

Meyer, M.T., Lee, E.A., Scribner, E.A., 2007a. Methods of analysis by the U.S. Geological Survey Organic Geochemistry Research Group—Determination of dissolved isoxaflutole and its sequential degradation products, diketonitrile and benzoic acid, in water using solid-phase extraction and liquid chromatography/tandem mass spectrometry: USGS Techniques and Methods, book 5, chap. A9.

Meyer, M.T., Scribner, E.A., Kalkhoff, S.J., 2007b. Comparison of fate and transport of isoxaflutole to atrazine and metolachlor in 10 Iowa rivers. Environ. Sci. Technol. 41, 6933–6939.

Mills, M.S., Thurman, E.M., 1994. Preferential dealkylation reactions of s-triazine herbicides in the unsaturated zone. Environ. Sci. Technol. 28, 600–605.

Mills, P.C., Kolpin, D.W., Scribner, E.A., Thurman, E.M., 2005. Herbicides and degradates in shallow aquifers of Illinois: spatial and temporal trends. J. Am. Water Res. Assoc. June, 537–547.

Nevill, D., Cornes, D., Howard, S., 1998. The role of HRAC in the management of weed resistance. Weed resistance. Pesticide Outlook, 9(4), 17–22.

Oregon State, 2008. Use in the US. Pesticide use in the United States. http://www.oregonstate.edu/~muirp/uspestic.htm

Pallet, K.E., Cramp, S.M., Little, J.P., Veerasekaran, P., Crudace, A.J., Slater, A.E., 2001. Isoxaflutole – the background to its discovery and the basis of its herbicidal properties. Pest Manag. Sci. 57, 133.

PAN (Pesticide Action Network, U.K.), 2002. Pesticide news No. 56, pp. 20–21. http://www.pan-uk.org/pestnews/Actives/atrazine.htm (accessed 27.12.08).

Pomes, M.L., Thurman, E.M., Aga, D.S., Goolsby, D.A., 1998. Evaluation of microtiter-plate enzyme-linked immunosorbent assay for the analysis of triazine and chloroacetanilide herbicides in rainfall. Environ. Sci. Technol. 32, 163–168.

Rosen, M.R., Lapham, W.W., 2008. Introduction to the U.S. Geological Survey National Water-Quality Assessment (NAWQA) of ground-water quality trends and comparison to other national programs. J. Environ. Qual. 37, 190–198.

Ruddy, B.C., Hitt, K.J., 1990. Summary of selected characteristics of large reservoirs in the US and Puerto Rico, 1988: USGS open-file report 90-163.

Schnoebelen, D.J., Kalkhoff, S.J., Becher, K.D., Thurman, E.M., 2003. Water-quality assessment of the Eastern Iowa Basins—selected pesticides and pesticide degradates in streams, 1996–98: USGS water-resources investigations report 03–4075.

Scribner, E.A., Thurman, E.M., Goolsby, D.A., Meyer, M.T., Mills, M.S., Pomes, M.L., 1993. Reconnaissance data water of the midwestern US, 1989–1990: USGS open-file report 93-457.

Scribner, E.A., Goolsby, D.A., Thurman, E.M., Meyer, M.T. Pomes, M.L., 1994. Concentrations of selected herbicides, two triazine metabolites, and nutrients in storm runoff from nine stream basins in the midwestern US, 1990–1992: USGS open-file report 94-396.

Scribner, E.A., Goolsby, D.A., Thurman, E.M. Meyer, M.T., Battaglin, W.A., 1996. Concentrations of selected herbicides, herbicide metabolites, and nutrients in outflow from selected midwestern reservoirs: April 1992–September 1993: USGS open-file report 96-393.

Scribner, E.A., Goolsby, D.A., Thurman, E.M., Battaglin, W.A., 1998. A reconnaissance for selected herbicides, metabolites, and nutrients in streams of nine mdwestern states: 1994–1995. USGS open-file report 98-181.

Scribner, E.A., Battaglin, W.A., Goolsby, D.A., Thurman, E.M., 2000. Changes in Herbicide concentrations in midwestern streams in relation to changes in use. Sci. Total Environ. 248, 255–263.

Scribner, E.A., Battaglin, W.A., Dietze, J.E., Thurman, E.M. 2003. Reconnaissance data for glyphosate, other selected herbicides, their degradation products, and antibiotics in 51 streams in nine midwestern states, 2002: USGS open-file report 03-217.

Scribner, E.A., Dietze, J.E., Thurman, E.M., 2004. Acetamide herbicides and their degradation products in ground water and surface water of the United States, 1993-2003: USGS data series 88.

Scribner, E.A., Thurman, E.M. Goolsby, D.A., Meyer, M.T., Battaglin, W.A., Kolpin, D.W., 2005. Summary of significant results from studies of triazine herbicides and their degradation products in surface water, ground water, and precipitation in the midwestern United States during the 1990s: USGS scientific investigations report 2005-5094.

Scribner, E.A., Meyer, M.T., Kalkhoff, S.J., 2006. Occurrence of isoxaflutole, acetamide, and triazine herbicides and their degradation products in 10 Iowa Rivers draining to the Mississippi and Missouri Rivers, 2004: USGS scientific investigations report 2006-5169.

Scribner, E.A., Battaglin, W.A., Gilliom, R.J., Meyer, M.T., 2007. Concentrations of glyphosate, its degradation product, aminomethylphosphonic acid, and glufosinate in ground- and surface water, rainfall, and soil samples collected in the United States, 2001–2006: USGS scientific investigations report 2007-5122.

Shaner, D., Brunk, G., Nissen, S., Westra, P., 2006. Soil dissipation and biological activity of metolachlor and S-metolachlor in five soils. Pest Manag. Sci. 62, 613–617.

Stamer, J.K., Zelt, R.B., 1992. Distribution of atrazine and similar nitrogen containing herbicides, lower Kansas River basin. USGS Yearbook, 1991, pp. 76–79.

Stamer, J.K., Battaglin, W.A., Goolsby, D.A., 1998a. Herbicides in midwestern reservoir outflows, 1992–1993: USGS fact sheet 134–98.

Stamer, J.K., Goolsby, D.A., Thurman, E.M., 1998b. Herbicides and associated degradation products in rainfall across the midwestern and northeastern US, 1990–1991: USGS fact sheet 181–97.

Thurman, E.M., Cromwell, A.E., 2000. Atmospheric transport, deposition, and fate of triazine herbicides and their metabolites in pristine areas at Isle Royale National Park. Environ. Sci. Technol. 34, 3079–3085.

Thurman, E.M., Fallon, J.D., 1996. The deethylatrazine/atrazine ratio as an indicator of the onset of the spring flush of herbicides into surface water of the midwestern US. Int. J. Environ. Anal. Chem. 65, 203–214.

Thurman, E.M., Mills, M.S., 1998. Solid-Phase Extraction. John Wiley & Sons, New York.

Thurman, E.M., Scribner, E.A., 2008. A decade of measuring, monitoring, and studying the fate and transport of triazine herbicides and the degradation products in groundwater, surface water, reservoirs, and precipitation. In: LeBaron, H.M., McFarland, J.E., Burnside, O.C. (Eds.), The Triazine Herbicides. Elsevier, San Diego, CA, pp. 451–475.

Thurman, E.M., Meyer, M.T., Pomes, M.L., Perry, C.A., Schwab, A.P., 1990. Formation and transport of deethylatrazine and deisopropylatrazine in surface water. Anal. Chem. 62, 2043–2048.

Thurman, E.M., Goolsby, D.A., Meyer, M.T., Kolpin, D.W., 1991. Herbicides in surface waters of the midwestern US: the effect of spring flush. Environ. Sci. Technol. 25, 1794–1796.

Thurman, E.M., Goolsby, D.A., Meyer, M.T., Mills, M.S., Pomes, M.L., Kolpin, D.W., 1992. A reconnaissance study of herbicides and their metabolites in surface water of the midwestern US using immunoassay and gas chromatography/mass spectrometry. Environ. Sci. Technol. 26, 2440–2447.

Thurman, E.M., Meyer, M.T., Mills, M.S., Zimmerman, L.R., Perry, C.A., Goolsby, D.A., 1994. Formation and transport of deethylatrazine and deisopropylatrazine in surface water. Environ. Sci. Technol. 28, 2267–2277.

Thurman, E.M., Imma, Ferrer, Damia, Barcelo, 2001. Choosing between atmospheric pressure chemical ionization and electrospray ionization interfaces for the HPLC/MS analysis of pesticides. Anal. Chem. 73, 5441–5449.

Thurman, E.M., Ferrer, I., Furlong, E.T., 2003 TOF-MS and quadrupole ion-trap MS/MS for the discovery of herbicide degradates in groundwater. In: Ferrar, I., Thurman, E.M. (Eds.). Liquid Chromatography/Mass Spectrometry, MS/MS and Time-of-Flight MS: Analysis of Emerging Contaminants, American Chemical Society Symposium 850. Washington D.C. pp. 128–144.

Thurman, E.M., Goolsby, D.A., Aga, D.S., Pomes, M.L., Meyer, M.T., 1996. Occurrence of alachlor and its sulfonated metabolite in rivers and reservoirs of the Midwestern United States: the importance of sulfonation in the transport of choroacetanilide herbicides. Environ. Sci. Technol. 30, 569–574.

United Nations Population Fund, 2001. The State of World Population 2001. United Nations Population Fund Chapter 2, http://www.unfpa.org/swp/2001/english/ch02.htm

US Census Bureau, 2008. Population clocks. http://www.census.gov/

USDA, 1991–2004. Agricultural Chemical Usage: 1990–2004 Field Crops Summary. National Agricultural Statistics Service (NASS), US Department of Agriculture, Washington, D.C.

USEPA (2008) Atrazine background. http://www.epa.gov/opp00001/factsheets/atrazinebackground.htm (accessed 19.08.08.).

USGS (2006). National stream quality accounting network (NASQAN). http://www.water.usgs.gov/nasqan

UW Weed Science (1999). Herbicide mode of action reference. University of Wisconsin. http://www.128.104.239.6/uw_weeds/extension/articles/herbmoa.htm

Weed Science (2008). International surveys of herbicide resistant weeds. http://www.weedscience.org/ln.asp

Chapter 14

Monitoring Pharmaceutical Residues in Sewage Effluents

Zulin Zhang
The Macaulay Institute, Craigiebuckler, Aberdeen AB15 8QH, UK

Darren P. Grover, and John L. Zhou
Department of Biology and Environmental Science, University of Sussex, Falmer, Brighton BN1 9QG, UK

INTRODUCTION

Traditionally the impact of chemical pollution has focused almost exclusively on the conventional priority pollutants (e.g., PAHs, PCBs, and organochlorine pesticides). However, the growing use of pharmaceuticals worldwide, classified as the so-called emerging pollutants, has become a new environmental problem, which has raised great concern among scientists in the last few years. Although the first reports on pharmaceuticals in wastewater effluents and surface waters were published in the United States in the 1970s (Garrison et al., 1976; Kim et al., 2007), pharmaceuticals as environmental contaminants did not receive a great deal of attention until the link was established between a synthetic birth-control pharmaceutical (ethynylestradiol) and its impacts on fish (Desbrow et al., 1998; Jobling et al., 1998; Kim et al., 2007).

Over 3,000 chemical substances are used in human and veterinary medicine (Ternes et al., 2004). Such pharmaceuticals include antiphlogistics/anti-inflammatory drugs, contraceptives, β-blockers, lipid regulators, tranquilizers, antiepileptics, and antibiotics (Ternes et al., 2004; Petrovic et al., 2005). Some typical pharmaceuticals classified by groups according to therapeutic effect and physico-chemical properties are listed in Table 1.

Handbook of Water Purity and Quality
Copyright © 2009 by Elsevier Inc. All rights of reproduction in any form reserved.

TABLE 1 Typical Pharmaceuticals and Their Physicochemical Properties

Compound	Therapeutic Class	Log K_{ow}	pK_a	MW	Half-Life (h)	Molecular Formula
Ketoprofen		3.12	4.45	254	2–2.5	$C_{16}H_{14}O_3$
Naproxen		3.18	4.15	230	12–15	$C_{14}H_{14}O_3$
Ibuprofen	Analgestic/anti-inflammatories	3.97	4.91	206	1.5–2	$C_{13}H_{18}O_2$
Indomethacine		4.27	4.5	358	4.5	$C_{19}H_{16}ClNO_2$
Diclofenac		4.51	4.14	296	1–2	$C_{14}H_{10}Cl_2NO_2$
Meclofenamic acid		5.12	4.2	241	1–4.5	$C_{15}H_{15}NO_2$
Acetaminophen		0.46	9.38	151	1–4	$C_8H_9NO_2$
Propyphenazone		1.94	n/a	230	1–2	$C_{14}H_{18}N_2O$
Clofibric acid		n/a	n/a	214	18–22	$C_{10}H_{11}O_3Cl$
Gemfibrozil	Lipid regulators/cholesterol-lowering statin drugs	4.77	n/a	250	1.5	$C_{15}H_{22}O_3$
Bezafibrate		4.25	n/a	362	1–2	$C_{19}H_{20}ClNO_4$
Pravastatin		3.1	n/a	446	77	$C_{23}H_{36}O_7$
Mevastatin		3.95	n/a	391		$C_{25}H_{38}O_5$
Carbamazepine		2.47	7	236	25–65	$C_{15}H_{12}NO$
Fluoxetine	Psychiatric drugs	3.82	8.7	309	24–144	$C_{17}H_{18}F3NO$
Paroxetine		3.95	n/a	329	3–65	$C_{19}H_{20}FNO_3$
Lansoprazole	Antiulcer agent	2.58	8.73	369	1–1.5	$C_{16}H_{14}F_3N_3O_2S$
Loratadine	Histamine H_1 and H_2 receptor antagonists	5.20	n/a	383	8–12	$C_{22}H_{23}ClN_2O_2$
Famotidine		−0.64	n/a	337	2.5–4	$C_8H_{15}N_7O_2S_3$
Ranitidine		0.27	n/a	314	2–3	$C_{13}H_{22}N_4O_3S$

		3.06	8.8	734	1.5	$C_{37}H_{67}NO_{13}$
Erythromycin		3.06	8.8	734	1.5	$C_{37}H_{67}NO_{13}$
Azythromycin		4.02	8.74	749	68	$C_{38}H_{72}N_2O_{12}$
Sulfamethoxazole	Antibiotics	0.89	6.0	253	10	$C_{10}H_{11}N_3O_3S$
Trimethoprim		0.91	7.12	290	8–10	$C_{14}H_{18}N_4O_3$
Ofloxacin		n/a	n/a	361	9	$C_{18}H_{20}FN_3O_4$
Atenolol		0.16	9.6	266	6–7	$C_{14}H_{22}N_2O_3$
Sotalol		0.24	n/a	272	12	$C_{12}H_{20}N_2O_3S$
Metoprolol	β-Blockers	1.88	9.68	267	3–7	$C_{15}H_{25}NO_3$
Propranolol		1.2–3.48	9.5	260	4–5	$C_{16}H_{21}NO_2$
Meberverine	Gastrointestinal	n/a	n/a	429	2.5	$C_{25}H_{35}O_5$
Thioridazine	Antidepressant	n/a	n/a	371	7–20	$C_{21}H_{26}N_2S_2$
Tamoxifen	Anticancer	n/a	n/a	372	120–168	$C_{26}H_{29}NO$
Monensin	Growth promoters	2.75–3.89	6.65	692	1056	$C_{36}H_{61}NaO_{11}$

Log K_{ow} is the logarithm of the octanol–water partition coefficient, pK_a is the negative logarithm of the ionization constant, MW is the molecular weight, and n/a is not available.

Although their toxicity to aquatic and terrestrial organisms is relatively unknown, a number of reported investigations have shown that pharmaceutical compounds pose a real threat to the environment (Oaks et al., 2004; Fent et al., 2006; Lacey et al., 2008). For example, diclofenac, which is frequently detected in aquatic matrices, has been found to have adverse effects in both fish and birds (rainbow trout and vultures). Diclofenac accumulates, with a concentration factor of up to 2,732, in the liver of rainbow trout and causes histopathological alterations in both the kidneys and the gills (Schwaiger et al., 2004). In vulture populations, this drug has been shown to cause renal failure and has resulted in a population decline in Pakistan (Oaks et al., 2004). This highlights the potential danger to both terrestrial and aquatic lives. Moreover, it underlines the latent risk to humans.

Wastewater treatment plants (WWTPs) are major contributors of pharmaceuticals in the environment. Because of their high consumption, pharmaceuticals along with their metabolites are continuously introduced to sewage systems, mainly through excreta, disposal of unused or expired drugs, or directly from pharmaceutical discharges. Recently, research has shown that the elimination of some pharmaceutical compounds during wastewater treatment processes is rather low; for example, removal rates for carbamazepine in a German WWTP was 7% whereas the average removal rate for the 14 compounds investigated was 65% (Ternes, 1998). Of the β-blockers, Paxeus (2004) reported elimination rates of <10% for atenolol and Castiglioni et al. (2006) reported average elimination rates of atenolol to be 10% in the winter and 55% in the summer in an Italian WWTP. Vieno et al. (2007) observed that the elimination rates were <65% for β-blockers and the lowest elimination rates of 29% for metoprolol. Compounds not removed in WWTP effluent are eventually released to receiving water bodies such as rivers and as a result, they will contribute to contamination in surface, ground, and drinking waters. For this reason, pharmaceuticals may have the same exposure potential as persistent pollutants because even their high transformation and removal rates can be compensated by their continuous input into the environment. Nowadays, it is a well-established fact that WWTP effluents are the major source for the introduction of pharmaceuticals into the environment. Most often, these compounds occur at micrograms per liter or sub-micrograms per liter concentrations (Buchberger, 2007; Gros et al., 2006, Roberts and Thomas, 2006). Numerous papers reported the levels of pharmaceuticals in wastewaters. Table 2 provides an overview of the concentrations of several main classes of pharmaceuticals in WWTP effluents reported in the literature.

Many believe that of all the emerging contaminants, antibiotics are the biggest concern because of the potential for antibiotic resistance (Erickson, 2002). The increasing use of these drugs in livestock, poultry production, and fish farming during the last five decades has caused a genetic selection of more harmful bacteria, which is a matter of great concern. However, other pharmaceutical compounds, especially polar ones, such as acidic anti-inflammatory drugs and lipid regulators, also deserve particular attention. As described earlier, elimination of

TABLE 2 Occurrence of Pharmaceutical Residues in the WWTP Effluents

Compound	Concentration (µg/L) Median (Maximum)							
Reference	Andreozzi et al. (2003)	Ternes (1998)	Ashton et al. (2004)	Choi et al. (2008)	Spongberg and Witter (2008)	Lishman et al. (2006)	Nakada et al. (2006)	Metcalfe et al. (2003), Miao et al. (2004)
Antiphlogistics/anti-inflammatory drugs								
Ibuprofen	0.05 (7.11)[1]	0.37 (3.4)[2]	3.09 (27.3)[3]	–[4]	–[5]	0.353 (0.773)[6]	0.672 (1.130)[7]	4.0 (24.6)[8]
Naproxen	1.12 (5.22)	0.30 (0.52)	–	–	–	0.351 (1.189)	0.010 (0.023)	12.5 (33.9)
Ketoprofen	n.d (1.62)	0.2 (0.38)	–	–	–	0.114 (0.210)	0.208 (0.369)	n.d.
Diclofenac	0.68 (5.45)	0.81 (2.1)	0.42 (2.35)	–	0.031 (0.177)	0.140 (0.748)	–	n.d.
β-Blockers								
Propanolol	0.01 (0.09)	0.17 (0.29)	0.08 (0.28)	–	–	–	–	–
Metoprolol	0.08 (0.39)	0.73 (2.2)	–	–	–	–	–	–
Acebutolol	0.06 (0.13)	–	–	–	–	–	–	–
Oxprenolol	0.02 (0.05)	–	–	–	–	–	–	–

(*Continued*)

TABLE 2 (Continued)

Compound	Concentration (µg/L) Median (Maximum)							
Lipid regulators								
Gemofibrozil	0.84 (4.76)	0.40 (1.5)	–	–	0.063 (0.084)	0.255 (0.436)	–	1.3 (1.3)
Fenofibrate	0.14 (0.16)	n.d. (0.03)	–	–	–	n.d.	–	–
Bezafibrate	n.d. (1.07)	2.2 (4.6)	–	–	–	–	–	–
Clofibric acid	n.d. (0.68)	0.36 (1.6)	–	–	n.d.	n.d.	–	n.d.
Antiepileptic drug								
Carbamazepine	0.87 (1.20)	2.1 (6.3)	–	0.18 (0.20)	0.076 (0.111)	–	0.005 (0.270)	0.7 (2.3)
Antibiotics								
Trimetroprim	0.04 (0.13)	–	0.07 (1.29)	0.10 (0.17)	–	–	–	–
Sulfamethoxazole	0.05 (0.09)	–	<0.05 (0.13)	0.19 (0.49)	0.273 (0.472)	–	–	0.24 (0.87)
Erythromycin	–	–	<0.01 (1.84)	–	–	–	–	0.08 (0.84)

[1]*Seven WWTPs in France, Greece, Italy, and Sweden.*
[2]*Forty-nine WWTPs in Germany.*
[3]*Five WWTPs in the United Kingdom.*
[4]*Four WWTPs in Korea.*
[5]*One WWTP in the United States, over a 6-month period.*
[6]*Twelve WWTPs in Canada.*
[7]*Sixteen WWTPs in Japan.*
[8]*Fourteen WWTPs in Canada (eight WWTPs for antibiotics).*

pharmaceuticals in WWTPs was found to be rather low and consequently sewage effluents are one of the main sources of these compounds and their recalcitrant metabolites. Because of their physicochemical properties (high water solubility and often poor degradability), they are able to penetrate through all natural filtration steps and enter groundwater as well as drinking water, which will cause the potential adverse effects on aquatic and terrestrial organisms (Reddersen et al., 2002; Ternes, 2002; Petrovic et al., 2005).

Because of the recent awareness of the potentially dangerous consequences of the presence of pharmaceuticals in the sewage effluent, the analytical methodology for the determination in complex matrices is still evolving and the number of methods described in the literature has grown considerably. In many cases, the common procedures involve sampling, sample treatment (e.g., pre-concentration, cleanup step) by solid-phase extraction (SPE), or related techniques, followed by analysis using chromatography in combination with mass spectrometry (MS) as the detector. When residue analysis of pharmaceuticals became an important issue in the 1990s, gas chromatography (GC) was the preferred chromatographic technique together with various derivatization procedures for the analytes. Nowadays, GC–MS may still be the preferred technique for certain classes of pharmaceuticals (Togola and Budzinski, 2008), although high performance liquid chromatography (HPLC) hyphenated with atmospheric pressure ionization MS has established itself as a better choice for simultaneous determination of pharmaceuticals of widely differing structures (Buchberger, 2007). In this chapter, the analysis of pharmaceuticals in sewage effluents will be discussed, including the sampling, sample treatment and instrumental measurement and analytical quality control.

SAMPLING

To monitor the pharmaceuticals in sewage effluent, the first step is to collect samples from the selected WWTPs. The sample that is taken should be a representative one, the composition of which is as close as possible to the whole mass of whatever (e.g., sewage effluent) is being analyzed. Obtaining good samples is a crucial first step in the chemical analysis process. Prior to sampling, a sampling strategy should be drawn concerning the WWTPs under investigation, the number of samples to be taken, where to conduct replicate sampling, size of samples, and storage and transport of samples. Preparation should also ensure that all the *in situ* measurement equipment is calibrated, and necessary sampling tools and containers are cleaned appropriately before sampling.

There are two sampling methods, spot and passive sampling, which are complimentary to each other and the advantages and disadvantages of these methods, together with typical methodologies, are highlighted in Table 3. Currently, the most widely used technique for performing monitoring of organic contaminants is spot sampling followed by laboratory-based extraction and analysis. In general, spot sampling uses a glass sampler or stainless steel sampler such as

TABLE 3 A Compilation of Spot and Composite Sampling Methods

Technique	Methods and Mechanisms	Advantages	Disadvantages
Spot sampling	Manual collection by glass or stainless steel sampler	Easy to perform Requires limited expertise Usually inexpensive Allows multicontaminant analysis	Yields only an instantaneous measurement of pollutant levels Uncertainty over temporal variations Labour required increases with the number of samples collected
	Autosampling into glass or stainless steel sampler	Requires limited expertise Reduces labour requirement compared to manual collection Allows multicontaminant analysis	Yields only an instantaneous measurement of pollutant levels Uncertainty over temporal variations Can be expensive
Composite sampling	Autosampling into glass or stainless steel sampler	Allows continuous monitoring and amount of pollutant over a defined period of time Allows multicontaminant analysis	Can be expensive Indicates only the total amount (or a calculated average) of a pollutant over a defined period of time Can require large-volume sampling Requires a greater level of expertise Deployment duration dependent on size of collection vessel
	Passive sampling devices (e.g., POCIS)	Allows continuous monitoring and amount of pollutant over a defined period of time Deployment time is easily variable Reduced interference Can stimulate the behavior of aquatic organisms	Can be expensive Indicates only the total amount (or a calculated average) of a pollutant over a defined period of time Compounds that can be analyzed is limited by the sorbent Requires a greater level of expertise

bucket (Gulkowska et al., 2007). The sample volume is typically 200 mL–1 L (Gros et al., 2006; Gomez et al., 2007; Nakada et al., 2006; Kim et al., 2007; Togola and Budzinski, 2008). General biocides such as sodium azide (final concentration 0.02 M) are added to each sample on-site to inactivate bacteria and prevent sample degradation during storage and processing. The samples are stored in a refrigerator below 4°C until filtration and extraction. The samples will be filtered through a 0.7 μm glass fiber filter in order to remove particles that may interfere during the extraction procedure. Prior to use, all glasswares are thoroughly soaked with detergents (e.g., Decon-90) and cleaned with deionized water, before further treatment (e.g., rinsed with distilled solvents such as dichloromethane and methanol, or baked in a furnace). The procedure should be adjusted for the compounds to be analyzed, based on their physicochemical properties (e.g., solubility, polarity).

Spot sampling is a well-established technique that is easy to perform and inexpensive, and requires limited expertise. However, it yields only an instantaneous measurement of pollutant levels and suffers from the uncertainty of short- and long-term concentration variations, which occur in the aquatic environment. An increase in sampling frequency or the use of flow- and time-weighted automatic samplers may reduce such uncertainty; however, the associated increase in costs can be prohibitive. There has been rapid development in the use of passive sampling devices such as polar organic chemical integrative sampler (POCIS) (Alvarez et al., 2004), Chemcatcher® (Mills et al., 2007), and silicon rod (Paschke et al., 2007) that allow continuous monitoring of aqueous pollutants, acting rather like an organism (e.g., mussels) but without the disadvantages of using organisms (passive sampler could mimic the bioconcentration of pollutants in aquatic organisms but not suffer from adverse effects as organisms). Of the various passive sampling devices, the most widely used one is POCIS, which comprises a solid receiving phase (sorbent) sandwiched between two microporous polyethersulfone (PES) membranes (Figure 1; Alvarez et al., 2004; http://www.est-lab.com/pocis.php). POCIS samples from water and thereby enables the chemical concentration to be estimated as follows (Alvarez et al., 2004; Vrana et al., 2005):

$$M_s = C_w R_s t \qquad (1)$$

where M_s is the mass of analytes in the receiving phase at time t, C_w represents time-weighted average concentration in water during the deployment period, and R_s is the sampling rate of the system, which may be interpreted as the volume of water cleared of analyte per unit of exposure time by the device (Vrana et al., 2006; Zhang et al., 2008). Recently, the application of POCIS for pharmaceutical residue measurements in sewage effluent has been successfully reported (Macleod et al., 2007; Zhang et al., 2008).

The POCIS is versatile and by changing the sequestrating medium, specific chemicals or chemical classes can be targeted. It is common to have POCIS of various configuration deployed together to maximize the data obtained. There

(a)

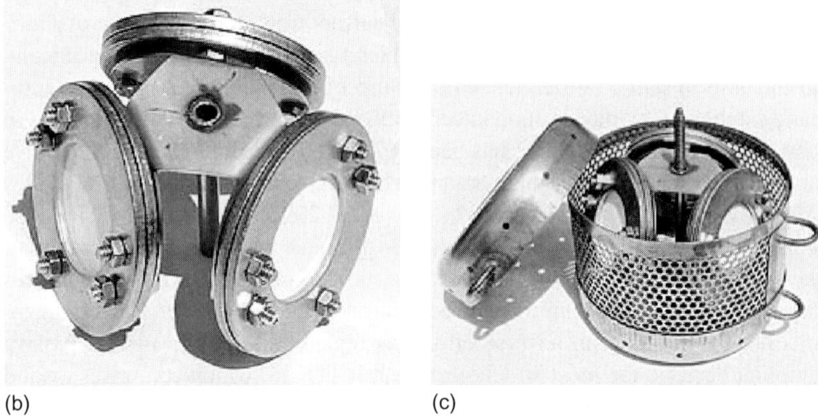

(b)   (c)

FIGURE 1 (a) A single POCIS, (b) three POCIS assembled together, and (c) the final setup for POCIS before deployment in the field. (From http://www.est-lab.com/pocis.php, with permission)

are two configurations of POCIS that are typically used. One is a generic system, which is useful for general hydrophilic organic contaminant purposes, and the other is for pharmaceutical sampling. The generic configuration contains the triphasic sorbent admixture of Isolute ENV$^+$ polystyrene divinylbenzene (Argonaut Technologies, Redwood City, CA) and Ambersorb 1500 carbon (Rohm and Haas, Philadelphia, PA) dispersed on S-X3 Biobeads (200–400 mesh, Bio-Rad, Hercules, CA). This mixture exhibits excellent trapping and recovery of many pesticides, natural and synthetic hormones, and other wastewater-related contaminants (Alvarez et al., 2004, 2005). The pharmaceutical configuration uses the Oasis-HLB sorbent (Waters, Milford, MA) for sequestering the chemicals of interest. This configuration is necessary as many pharmaceuticals, with multiple functional groups, have a tendency to strongly bind to the carbonaceous component of the sorbent admixture. The membrane acts as a semipermeable barrier, allowing chemicals of interest to pass through to the sorbent, while excluding particulate matter, biogenic material, and other large potentially interfering substances. The PES membrane (Pall Gelman Sciences, Ann Arbor, MI) contains water-filled pores, 0.1 μm in diameter, to facilitate transport of the hydrophilic chemicals. The POCIS was designed to mimic respiratory exposure of aquatic organisms to dissolved chemicals without the inherent problems of metabolism, depuration of chemicals, avoidance of contaminated areas, and mortalities of test organisms. Also, dietary uptake of polar organic compounds likely represents only a small fraction of residues accumulated in the tissues of aquatic organisms (Huckins et al., 1997). Thus, the POCIS provides a worst-case exposure scenario for aquatic organisms, enables concentration of sufficient amounts of bioavailable hydrophilic organic chemicals for some biomarker tests, and permits determination of the biologically relevant time-weighted average concentrations in water.

The POCIS devices can be deployed at the sampling site for a duration ranging from 1 week to 2 months (Alvarez et al., 2005; Zhang et al., 2008). Quality control (QC) is achieved using both fabrication and field blanks ($n = 3$) for each analytical technique. Fabrication blanks account for any background contribution due to interferences from POCIS components and for contamination incurred during laboratory storage, processing, and analytical procedures. Field blank POCIS are used as QC samples for transport, deployment, and retrieval procedures. (Note that these POCIS blanks are sealed again in the same shipping cans and stored frozen during the exposure period.) The field blank POCIS are treated identically as the deployed devices, with the exception that they are not exposed to waters at the monitoring sites.

The procedures for the recovery of sequestered chemical residues from the deployed POCIS are as follows: briefly, the POCIS is disassembled, and the sorbent is transferred into glass gravity-flow chromatography columns or glass beaker (Alvarez et al., 2005; Zhang et al., 2008). Chemical residues are recovered from the sorbent by organic solvent elution/extraction. Methanol is widely used to recover the pharmaceuticals. The extracts are reduced in volume by rotary evaporation and under a gentle stream of nitrogen, and then ready for further instrumental analysis.

Zhang et al. (2008) has described the use of POCIS for the analysis of emerging contaminants including pharmaceuticals, and compared the predicted compound concentrations in sewage effluent and river water with those measured by spot sampling. As shown in Figure 2, for the pharmaceuticals propranolol, sulfamethoxazole, carbamazepine, indomethacine, and diclofenac, their mean aqueous concentrations measured by spot sampling varied between 3.0 and 45.6 ng/L, <LOD and 17.6 ng/L, 16.6 and 539 ng/L, 0.4 and 7.2 ng/L, and 2.4 and 65.2 ng/L, respectively; whereas their concentrations predicted by POCIS were between 2.8 and 40.5 ng/L, <LOD and 18.2 ng/L, 26.3 and 427 ng/L, 0.5 and 11.9 ng/L, and 4.4 and 165 ng/L, respectively. It is apparent that for most samples, the predicted pharmaceutical concentrations by POCIS are similar to those by spot sampling.

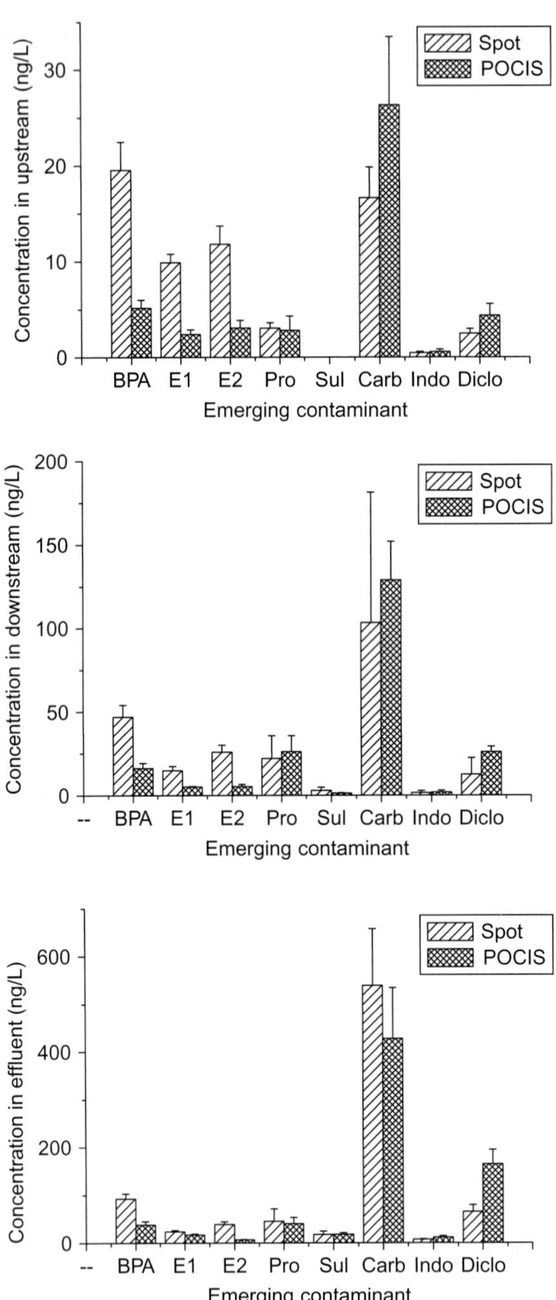

FIGURE 2 Comparison of the mean contaminant concentrations between spot sampling with those predicted by POCIS in sewage effluent from Sheffield Park WWTP, and its upstream and downstream in the River Ouse, West Sussex, UK (BPA, bisphenol A; E1, estrone; E2, 17β-estradiol; Pro, propranolol; Sul, sulfamethoxazole; Carb, carbamazepine; Indo, indomethacine; Diclo, diclofenac).

Also, POCIS was validated and deployed for monitoring pharmaceuticals and personal care products in wastewater and surface water, in which the time-weighted average concentrations derived from the POCIS were in good agreement with those from spot/grab sampling for pharmaceuticals (Macleod et al., 2007), confirming again the potential application of passive samplers for monitoring of sewage effluent.

SAMPLE PREPARATION

The sample preparation is an important step in analysis, particularly because the concentration levels of pharmaceuticals found in sewage effluent water samples are generally too low and the matrix is too complex to allow a direct injection into a chromatographic system. Therefore, efficient pre-concentration steps are necessary which should also result in some sample cleanup. Several techniques have been developed and optimized, with SPE being the most frequent. Also solid-phase microextraction (SPME), liquid-phase microextraction (LPME), and lyophilization have been applied (Fatta et al., 2007). In a review of 32 pharmaceutical studies, Fatta et al. (2007) found that most of them (28 studies) used SPE for extraction from wastewater and water samples. This extraction procedure can be based on multiple equilibria between the liquid phase and the sorbent in SPE cartridges.

Pharmaceuticals of adequate hydrophobicity can easily be pre-concentrated using any reverse-phase material such as alkyl-modified silica or polymer-based materials (Buchberger, 2007). Deprotonation of acidic compounds and protonation of basic compounds should be suppressed to ensure sufficient hydrophobicity of the analytes. Therefore, acidic pharmaceuticals should be pre-concentrated under acidic conditions, whereas basic analytes should be pre-concentrated at an alkaline pH. Alternatively, mixed-mode SPE materials can be used, which exhibit both reversed-phase and cation-exchange properties due to the presence of sulfonic acid groups on the hydrophobic surface of the particles. Using acidified sample solutions, acidic and neutral analytes could be extracted by hydrophobic interactions, whereas protonated basic analytes would interact via ion exchange mechanisms. Such an approach has been used among others by Stolker et al. (2004) for SPE of a set of 13 pharmaceuticals of different classes. Mixed-mode materials with reversed-phase and anion-exchange properties have been used under slightly basic conditions for antibiotics containing carboxylic acid functionality (Benito-Pena et al., 2006).

From the practical point of view, it might be desirable to extract pharmaceuticals from wastewater samples without any pH adjustment (Buchberger, 2007). Furthermore, various (neutral) pharmaceuticals may exhibit significant hydrophilic properties, which make it difficult to enrich them on conventional alkyl-modified silica materials. SPE procedures for the extraction of polar compounds from aqueous samples are still a big challenge in analytical chemistry.

A recent review has summarized new SPE materials that can improve the recoveries for polar analytes (Fontanals et al., 2005). These SPE cartridge materials are mainly polymeric sorbents that improve the retention of polar compounds either by novel functional groups in the polymeric structure (resulting in a hydrophilic–hydrophobic balance material) or by considerably increased surface area. Some of these new materials have turned out to be well-suited for multiclass analysis of pharmaceuticals in water samples. A number of different SPE stationary phases (Table 4) have been evaluated for the extraction of the selected pharmaceutical compounds (Hilton and Thomas, 2003; Weigel et al., 2004; Zhang and Zhou, 2007). Nowadays, one of the most widely used sorbents is a copolymer of divinylbenzene and vinylpyrrolidone, which has been commercialized under the trade name Oasis-HLB by Waters. Weigel et al. (2004) demonstrated that this sorbent can simultaneously extract acidic, neutral, and basic pharmaceuticals at neutral pH. Multiresidue methods for different classes of pharmaceuticals using Oasis-HLB at neutral pH have also been reported recently by Barceló and coworkers (Gomez et al., 2006; Gros et al., 2006). Trenholm et al. (2006) developed a comprehensive method for the analysis of 58 potential endocrine-disrupting compounds and pharmaceuticals using a single SPE step based on Oasis-HLB. Various other studies can be found which describe the successful use of Oasis-HLB for pharmaceuticals in wastewater (Petrovic et al., 2006; Nikolai et al., 2006). A typical multiresidue analysis would include filtration of the wastewater, conditioning of the Oasis-HLB material (between 60 and 500 mg packed into a suitable cartridge) by several milliliters of methanol and water, application of up to 2 L of sample at a flow rate of approximately 10 mL/min, rinsing the cartridge with several milliliters of water to remove dissolved interference, drying the SPE material by applying a vacuum to remove excess of water, elution with approximately 10 mL of methanol (it may be necessary to repeat the elution step), evaporation of the extracts under a gentle stream of nitrogen, and reconstitution in 0.1–0.3 mL of methanol or a mixture of methanol and water containing internal standards, which are then ready for instrumental analysis. The whole analytical procedure is illustrated in Figure 3 (Zhang and Zhou, 2007; Zhang et al., 2008). All the glasswares used for the extraction were baked at 450°C for 4 h to eliminate any organic contaminants. All the solvents used were purchased from Rathburn, which were of distilled-in-glass grade.

Hilton and Thomas (2003) have shown that Strata-X is useful for extracting selected pharmaceuticals, after comparing seven types of SPE cartridges. So in a few cases Strata-X (a polydivinylbenzene resin containing piperidone groups manufactured by Phenomenex) has been employed for generic SPE procedures (Hilton and Thomas, 2003; Roberts and Bersuder, 2006). Nebot et al. (2007) also used Strata-X successfully to determine the concentrations of a range of human pharmaceuticals in sewage effluent. Strata-X may have similar properties to Oasis-HLB, but at present there is not enough data in the literature to allow this comparison to be made.

TABLE 4 A Summary of Different Types of SPE Cartridges Being Used for Pharmaceutical Extraction

Cartridge	Description	Application	Manufacturer
DSC-C_{18}	Polymerically bonded, octadecyl		Supelco
DSC-Si	Unbonded acid washed silica sorbent		Supelco
DSC-SCX	Aliphatic sulfonic acid, Na^+ counterion		Supelco
DSC-SAX	Quaternary amine, Cl^- counterion		Supelco
Strata X-CW	Polymeric weak cation	Suitable for a range of pharmaceuticals by Hilton and Thomas (2003), Roberts and Bersuder (2006), and Nebot et al. (2007)	Phenomenex
Strata SDB-L	Styrene–divinylbenzene polymeric		Phenomenex
Chromabond-easy	Bifunctionally modified polystyrene–divinylbenzene adsorbent resin		Macherey-Nagel
Chromabond-C_{18} Hydra	Octadecyl-modified silica		Macherey-Nagel
Chromabond-drug	Modified silica	Applied to pharmaceuticals of adequate hydrophobicity (Buchberger, 2007); Required pH adjustment (Stolker et al., 2004)	Macherey-Nagel
Isolute C_{18}	Octadecyl		International Sorbent Technology
Isolute C_{18}/ENV^+	C_{18} hydroxylated polystyrene–divinylbenzene		International Sorbent Technology
Isolute C_8	Octadecyl functionalized silica		International Sorbent Technology

(Continued)

TABLE 4 (Continued)

Cartridge	Description	Application	Manufacturer
Oasis-HLB	Poly(divinylbenzene-*co*-*N*-vinylpyrrolidone)	Suitable for a range of pharmaceuticals by Weigel et al. (2004), Gomez et al. (2006), Gros et al. (2006), and Zhang and Zhou (2007)	Waters
Oasis MCX	Poly(divinylbenzene-*co*-*N*-vinylpyrrolidone, -SO$_3$H)		Waters
Varian bond elut C$_{18}$	Irregularly based acid-washed silica	Good for a range of pharmaceuticals by Hilton and Thomas (2003), subsequent to pH adjustment	Varian

SPE of pharmaceuticals is often done off-line, and is useful for on-site sampling (performing the pre-concentration step in the field, followed by the elution step in the lab). It is a technique which is also well-suited for online procedures and automation in the laboratory. The SPE cartridge can be installed in the injection valve instead of the injection loop and the pre-concentrated analytes directly eluted onto the analytical column. An example of this approach is the work of Pozo et al. (2006) who determined 16 antibiotics in surface and groundwater samples. In such a setup, the SPE cartridge is generally reused for a series of samples. Contrary to this configuration, fully automated SPE procedures with single-use cartridges can be realized by commercially available instrumentation such as the Symbiosis™ Environ manufactured by Spark Ltd. This robotic system includes an automated cartridge exchange module that transfers the cartridge after the pre-concentration step into the flow of mobile phase of HPLC (Rodriguez-Mozaz et al., 2007). This approach that has primarily been used with the analysis of pharmaceutical residues could be applied in sewage effluent monitoring in small volume (200 mL) combined with pre-filtering device (to remove the particles). There are some significant advantages with the approach (Fatta et al., 2007): (1) direct injection of untreated sewage effluent samples; (2) automatic sample cleanup and/or analyte enrichment; (3) elimination of conventional manual sample pre-treatment steps; (4) faster procedures; (5) methods are less prone to errors, resulting in better reproducibility; (6) reduction of health risks; and (7) samples can be run unattended (e.g., overnight or over the weekend).

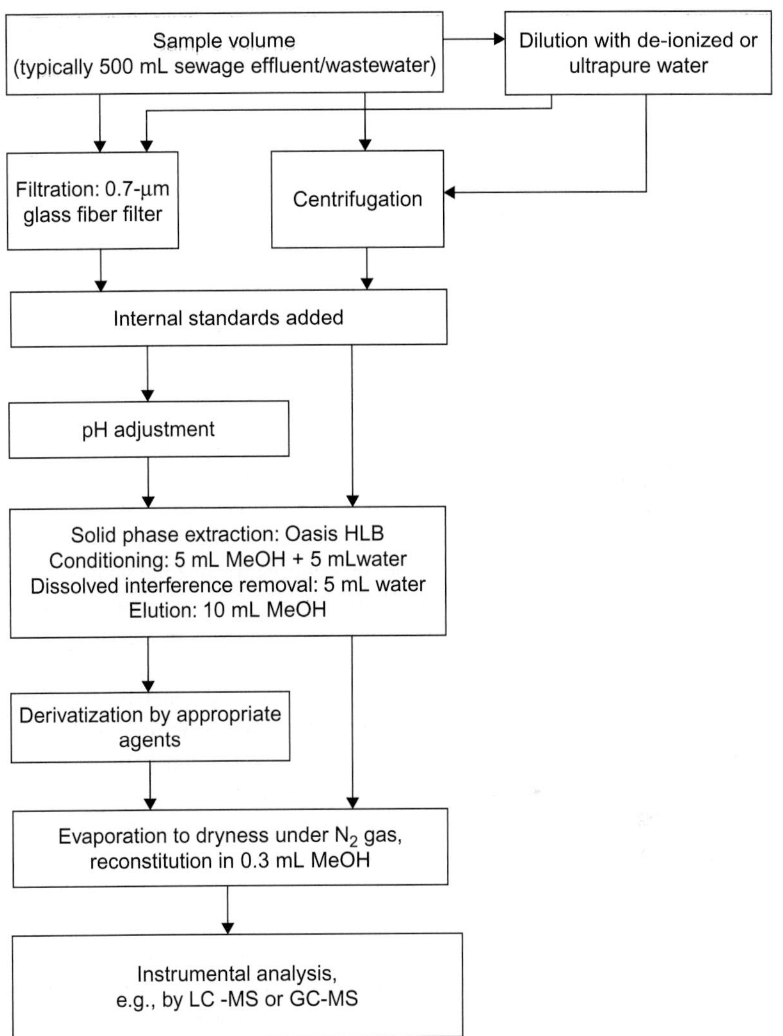

FIGURE 3　The procedures involved in the extraction and analysis of pharmaceuticals in sewage effluent/wastewater samples.

INSTRUMENTAL ANALYSIS

The majority of pharmaceuticals lacks sufficient volatility and as such are not directly compatible with GC analysis (Buchberger, 2007). Various groups of pharmaceuticals can be derivatized to make them suited for GC analysis. Although such procedures may be time-consuming and can introduce errors due to side reactions during the derivatization; however, they are still widely in use and well-established for routine work. The major advantage of GC hyphenated with MS is the fact that the usual ionization modes such as electron impact (EI) or chemical ionization (CI) are generally less affected by the matrix of the sample than ionization modes used by HPLC-MS, for example. Typical derivatization reagents for acidic pharmaceuticals include pentafluorobenzylbromide (Reddersen and Heberer, 2003), methyl chloromethanoate (Weigel et al., 2004), methanol/BF$_3$ (Verenitch et al., 2006), or tetrabutylammonium salts (for derivatization during injection) (Lin et al., 2005). Phenazone-type drugs have been derivatized by silylation using N-*tert*-butyldimethylsiyl-N-methyltrifluoroacetamide (MTBSTFA) (Zuhlke et al., 2004). Silylation procedures are also commonly used for synthetic estrogens (Quintana et al., 2004; Fernandez et al., 2007), although careful selection of the reagent and the reaction conditions is necessary to avoid side reactions (Shareef et al., 2006). Derivatization reactions are useful for sorptive extraction combined with thermal desorption GC.

Generally, the use of GC–MS seems to be a well-established approach for residue analysis of pharmaceutical. Currently, there is a trend toward tandem-MS techniques as the MS component of choice for this type of analysis. The advantages of such instruments will be discussed in more detail in the context of HPLC-MS.

Despite the indisputable merits of GC procedures for residue analysis of certain classes of pharmaceuticals, HPLC shows much more universal applicability (Buchberger, 2007). In some cases, when just a few analytes of a certain class are to be analyzed, even a simple UV absorbance detection may be feasible. This has been demonstrated for residues of oxytetracycline in water, which can be detected at 360 nm (Himmelsbach and Buchberger, 2005). Table 5 highlights the advantages and disadvantages of liquid chromatography (LC) and gas chromatography (GC) hyphenated with MS and tandem-MS. Fluorescence detection may also have some benefits as shown for the determination of some other compounds such as anthracycline cytostatics and fluoroquinolones (Mahnik et al., 2006, Golet et al., 2001, 2002). Nevertheless, MS detection involving atmospheric pressure ionization such as electrospray ionization (ESI) is nowadays state of the art.

Although single-quadrupole instruments have been successfully used when HPLC-MS procedures started to be developed for pharmaceutical residue analysis (Ahrer et al., 2001), more sophisticated mass analyzers are nowadays commonly employed that allow an unequivocal confirmation of the identities of the analytes. Triple quadrupole (QqQ) MS instruments have become widely used with HPLC for environmental analysis. When using a QqQ instrument, false-positive results can be avoided if the ions of at least two ion–ion transitions are used in combination with at least one ion intensity ratio. Several studies have dealt with HPLC-QqQ/MS for multiclass analysis of pharmaceuticals (Castiglioni et al., 2005; Rodriguez-Mozaz et al., 2004; Miao et al., 2004; Ternes et al., 2005; Gomez et al., 2006; Gros et al., 2006; Zhang and Zhou, 2007). Precursor and product ions used for quantification and confirmation purposes have been compiled for a wide range of pharmaceutical compounds (Petrovic et al., 2005). The HPLC-QqQ/MS could monitor the specific compound by detecting a parent to fragment transition, which is known as multiple reaction monitoring (MRM). It makes us more certain that the peak is the analyte we seek and the background

TABLE 5 A Comparison of Different Analytical Techniques Involving Coupling of Mass Spectrometry with Chromatography

Analytical Technique	Advantages	Disadvantages
GC-MS	Less susceptible to matrix interferences than LC techniques Widely available	Derivatization required as few pharmaceuticals are volatile May lack selectivity compared with MS/MS techniques
GC-MS/MS	Less susceptible to matrix interferences than LC techniques Improved selectivity Department of Biology and Environmental Science, University of Sussex, Falmer, Brighton, BN1 9QG, UK, single-stage MS detection	Derivatization required, as few pharmaceuticals are volatile Can be expensive to set up Development of MS/MS techniques requires greater expertise and time
LC-MS	Suitable for the analysis of a large range of pharmaceuticals, without the need for derivatization	May lack selectivity as compared to MS/MS techniques May be more susceptible to matrix interferences than GC techniques
LC-MS/MS	Suitable for the analysis of a large range of pharmaceuticals, without the need for derivatization Improved selectivity Department of Biology and Environmental Science, University of Sussex, Falmer, Brighton, BN1 9QG, UK, single-stage MS detection	May be more susceptible to matrix interferences than GC techniques Can be expensive to set up Development of MS/MS techniques requires greater expertise and time

interference is minimal. An example of a typical multipharmaceutical residue analysis by HPLC-QqQ/MS is shown in Figure 4 (Zhang and Zhou, 2007). The LC separation system was equipped with a Waters Symmetry C_{18} column (4.8 mm $\times$ 75 mm, particle size 3.5 μm). The mobile phase was made of eluent A (0.1% formic acid in ultrapure water), eluent B (acetonitrile), and eluent C (methanol). The flow rate of the mobile phase was 0.2 mL/min. The gradient elution was operated with 10% of eluent B, followed by a 25-min gradient to 80% of eluent B and a 3-min gradient to 100% of eluent B, and then changed to 100% of eluent C within 8 min and held there for 10 min. The system was reequilibrated for 10 min between runs. Typically the injection volume was 10–20 μL.

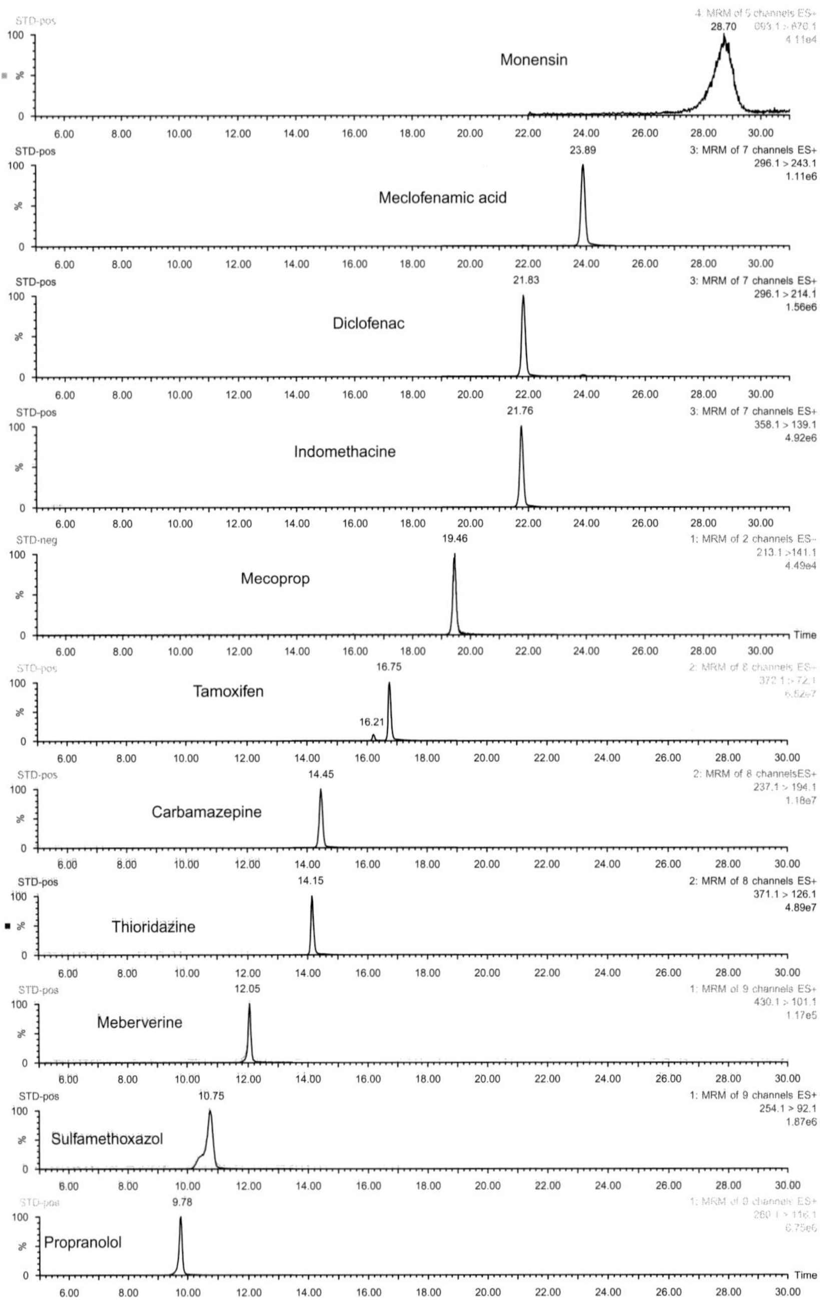

FIGURE 4 Chromatograms for the extracted and overlapped MRM of selected pharmaceutical compounds.

A general and well-known problem of HPLC-ESI/MS is ionization suppression due to matrix components eluting at the same time as an analyte. On the one hand, ionization suppression effects can reduce the sensitivity of the method considerably; on the other hand, special care must be taken to achieve reliable quantitation. Ideally, isotopically labeled analytes should be used as internal standards but are available only in few cases for pharmaceuticals. Otherwise, standard addition methods must be applied to obtain correct quantitative data; however, such methods increase the length of the analytical procedure considerably. In all cases, it makes sense to optimize the efficiency of the sample cleanup protocols in order to minimize interferences. New generation MS instruments, in some cases, allow a dilution of the sample extract before injection into the HPLC system thereby minimizing any matrix effects. Furthermore, other MS detectors such as time-of-flight (TOF) are more selective than QqQ, hence more suited for samples of highly complex matrix such as wastewater or biological extracts. However, the sensitivity of TOF currently is not as good as QqQ and requires further development.

ANALYTICAL QUALITY CONTROL

To ensure data quality, all the analytical process should be subject to strict QC procedures to determine systematic and random errors. QC measures in relation to sewage effluent water analysis include the collection of blank samples derived from laboratory-grade or organic-free water to determine if sampling procedures, sampling equipment, field conditions, sample shipment and storage (field blank), or laboratory procedures (laboratory blank) introduce target analytes into environmental samples. The spiked water samples are used to check the precision and recoveries. The blank and spiked water samples are typically made by deionized water taken from a Milli-Q system. Typically, several blank and spiked samples are produced with each set of real samples (10 samples for each set). In addition, the random errors involved in sampling are assessed by carrying out replicate sampling of water at the same site and the analysis of sample extracts. Internal standards (usually the target compounds labeled by stable isotopes such as ^{13}C or ^{2}H) are used to compensate for losses involved in the sample extraction and workup to further characterize the method performance. Prior to use, all glasswares are rinsed with dichloromethane ($\times$2) and methanol ($\times$2), or baked at 450°C for 4h. All the solvents used are of distilled-in-glass grade. All of these processes are carried out to minimize the cross-contamination and loss of analytes through adsorption onto the surface of sampling vessels and the extraction apparatus. As sewage effluent has a very different matrix from other environmental waters, more appropriate blanks should be used in future for wastewater studies. As an example, synthetic sewage effluent prepared by dissolving appropriate amounts of organic substances (e.g., organic matter filtered from wastewater) in deionized water at a concentration identical to that in real sewage effluent with high organic content should be a more appropriate matrix for making blank samples and spiked recovery samples.

To assess systematic errors, most would use the so-called recovery experiments by spiking known amounts of each target compound in sewage effluent, followed by extraction and analysis. This gives a good indication of how reliable the measurement values are. In addition, certified reference material (CRM) for pharmaceuticals in wastewater should be prepared in the future, which can identify the closeness between a measured value (from individual laboratory) and a certified value (from the supplier). As CRM is vigorously tested under varying environmental conditions by the supplier and independently verified by laboratories worldwide, it becomes a calibration tool for the international community on pharmaceutical research. Ideally, the relative difference between measured and certified values should be as small as possible (e.g., $\pm 10\%$). Such material will ensure that everyone follows the right procedures and generates data of the highest quality. In addition, this practice will ensure monitoring data obtained from any sewage effluent samples anywhere can be compared against each other, so as to identify hotspots of pharmaceutical pollution, emergence of new pharmaceuticals, and temporal variation of pharmaceutical concentrations in the sewage effluent from different WWTPs.

OTHER TECHNIQUES

Immunoassays show attractive features for organic trace analysis because of the fact that they require little sample pre-treatment, exhibit high sensitivity, and are inexpensive in comparison to the instrumental analysis described earlier. A considerable number of immunoassays have been developed and used for residue analysis of pesticides in water samples, but immunoassays for pharmaceuticals in the aquatic environment are still quite rare. Although test kits for pharmaceuticals are commercially available, these kits are in most cases optimized for samples such as blood, urine, or food. The applicability to environmental samples has not been investigated in the majority of cases. Deng et al. (2003) developed a highly sensitive and specific indirect competitive enzyme-linked immunosorbent assay (ELISA) for the determination of diclofenac in environmental water samples. When the assay was applied to analysis of diclofenac in tap and surface water as well as wastewater in Austria and Germany, they showed that ELISA-derived diclofenac concentrations in wastewater samples were about 25% higher than those using GC-MS. It suggested that the technique will be applicable to sewage effluent matrix, although the sensitivity and selectivity should be further improved.

As a result of the need to reduce matrix interferences, particularly in the case of samples from wastewater, molecularly imprinted polymers or molecularly imprinted solid-phase extraction (MISPE) have received some attention as a highly selective preparatory technique in the analysis of a limited range of pharmaceutical compounds. Sun et al. (2008) and Jun et al. (2008) successfully applied the technique to diclofenac, with results comparable to other techniques but with reduced error and detection limits. Gros et al. (2008) reported results with similar recoveries but improved selectivity (and hence detection limits) of a range of β-blockers, compared with Oasis-HLB cartridges. To date, the widest application of MISPE to wastewater analysis of pharmaceuticals has been by Zorita et al. (2008) who satisfactorily analyzed a range of acidic pharmaceuticals in various matrices by LC-MS/MS, following MISPE.

SUMMARY AND CONCLUSIONS

Analysis of emerging pollutants such as pharmaceutically active compounds in the aquatic environment including sewage effluent was reviewed in this chapter. As pharmaceutical compounds are usually present at trace levels (e.g., ng/L–µg/L) in a complex matrix (e.g., wastewater), it is a common practice to develop extraction and analytical methods that can concentrate the target compounds while minimizing matrix interference. The analytical procedure involves many interrelated steps including sample pre-treatment (e.g., filtration), pre-concentration (e.g., SPE), and analysis by advanced techniques (e.g., HPLC-tandem-MS). Residues of pharmaceuticals have most probably been present in our aquatic environment since their application, but only recently advances in analytical chemistry and instrument performance have allowed analysis of such compounds at low (ng/L) concentration. The development of advanced mass spectrometric detectors for chromatography has made a significant contribution to these achievements. The limits of detection of analytical methods may be improved even further during the next few years. Residues of pharmaceuticals in aquatic systems are not yet included in regular monitoring programs. The high costs of instrumental analysis may be prohibitive to more extended studies. A focus on a limited set of pharmaceuticals that are representative in regard to toxic effects may be advantageous (but a final selection of such a set has not yet been done). The importance of reliable and inexpensive biosensors may increase in the future, provided that they meet the criteria of analytical QC in the same way as traditional techniques do.

In addition, general QC procedures must be followed, including appropriate replicate sampling, sample preservation at 4°C, application of isotopically labeled internal standards, suitable blanks, satisfactory recovery of the target compounds, and eventually use of CRM (either in-house or commercial ones). Although pharmaceutical residues in the environment is a major concern and has been widely studied in sewage effluents, the residues of pharmaceuticals in aquatic systems are not yet included in the regular monitoring programs of regulatory bodies. Further research on emerging pollutants such as pharmaceuticals is needed especially on their behavior in the WWTP process, where our understanding of pharmaceutical removal behavior is not well-developed but is improving. The low cost and robustness of passive sampling such as POCIS is recommended for routine monitoring of pharmaceuticals and other similar pollutants by governmental agencies.

REFERENCES

Ahrer, W., Scherwenk, E., Buchberger, W., 2001. Determination of drug residues in water by the combination of liquid chromatography or capillary electrophoresis with electrospray mass spectrometry. J. Chromatogr. A 910, 69–78.

Alvarez, D.A., Petty, J.D., Huckins, J.N., Jones-Lepp, T.L., 2004. Development of a passive, in situ, integrative sampler for hydrophilic organic contaminants in aquatic environments. Environ. Toxicol. Chem. 23, 1640–1648.

Alvarez, D.A., Stackelberg, P.E., Petty, J.D., Huckins, J.N., Furlong, E.T., Zaugg, S.D., et al., 2005. Comparison of a novel passive sampler to standard water-column sampling for organic contaminants associated with wastewater effluents entering a New Jersey stream. Chemosphere 61, 610–622.

Andreozzi, R., Raffaele, M., Nicklas, P., 2003. Pharmaceuticals in STP effluents and their solar photodegradation in aquatic environment. Chemosphere 50, 1319–1330.

Ashton, D., Hilton, M., Thomas, K.V., 2004. Investigating the environmental transport of human pharmaceuticals to streams in the United Kingdom. Sci. Total Environ. 333, 167–184.

Benito-Pena, E., Partal-Rodera, A.I., Leon-Gonzalez, M.E., Moreno-Bondi, M.C., 2006. Evaluation of mixed mode solid phase extraction cartridges for the preconcentration of beta-lactam antibiotics in wastewater using liquid chromatography with UV-DAD detection. Anal. Chim. Acta 556, 415–422.

Buchberger, W.W., 2007. Novel analytical procedures for screening of drug residues in water, waste water, sediment and sludge. Anal. Chim. Acta 593, 129–139.

Castiglioni, S., Bagnati, R., Calamari, D., Fanelli, R., Zuccato, E., 2005. A multiresidue analytical method using solid-phase extraction and high-pressure liquid chromatography tandem mass spectrometry to measure pharmaceuticals of different therapeutic classes in urban wastewaters. J. Chromatogr. A 1092, 206–215.

Castiglioni, S., Bagnati, R., Fanelli, R., Pomati, F., Calamari, D., Zuccato, E., 2006. Removal of pharmaceuticals in sewage treatment plants in Italy. Environ. Sci. Technol. 40, 357–363.

Choi, K., Kim, Y., Park, J., Park, C., Kim, M., Kim, H., et al., 2008. Seasonal variations of several pharmaceutical resideues in surface water and sewage treatment plants of Han River, Korea. Sci. Total Environ. 405, 120–128.

Deng, A.P., Himmelsbach, M., Zhu, Q.Z., Frey, S., Sengl, M., Buchberger, W., et al., 2003. Residue analysis of the pharmaceutical diclofenac in different water types using ELISA and GC-MS. Environ. Sci. Technol. 37, 3422–3429.

Desbrow, C., Routledge, E.J., Brighty, G.C., Sumpter, J.P., Waldock, M., 1998. Identification of estrogenic chemicals in STW effluent. 1. Chemical fractionation and in vitro biological screening.. Environ. Sci. Technol. 32, 1549–1558.

Erickson, B.E., 2002. Analyzing the ignored environmental contaminants. Environ. Sci. Technol. 36, 140A–145A.

Fatta, D., Nikolaou, A., Achilleos, A., Meric, S., 2007. Analytical methods for tracing pharmaceutical residues in water and wastewater. Trends Anal. Chem. 26, 515–533.

Fent, K., Weston, A.A., Caminada, D., 2006. Ecotoxicology of human pharmaceuticals. Aquat. Toxicol. 76, 122–159.

Fernandez, M.P., Ikonomou, M.G., Buchanan, I., 2007. An assessment of estrogenic organic contaminants in Canadian wastewaters. Sci. Total Environ. 373, 250–269.

Fontanals, N., Marce, R.M., Borrull, F., 2005. New hydrophilic materials for solid-phase extraction. Trends Anal. Chem. 24, 394–406.

Garrison, A.W., Pope, J.D., Allen, F.R., 1976. In: Keith C.H. (Ed.), Identification and analysis of organic pollutants in water, Ann Arbor Sci. Publ, Ann Arbor, MI, p. 517.

Golet, E.M., Alder, A.C., Hartmann, A., Ternes, T.A., Giger, W., 2001. Trace determination of fluoroquinolone antibacterial agents urban wastewater by solid-phase extraction and liquid chromatography with fluorescence detection. Anal. Chem. 73, 3632–3638.

Golet, E.M., Strehler, A., Alder, A.C., Giger, W., 2002. Determination of fluoroquinolone antibacterial agents in sewage sludge and sludge-treated soil using accelerated solvent extraction followed by solid-phase extraction. Anal. Chem. 74, 5455–5462.

Gomez, M.J., Petrovic, M., Fernandez-Alba, A.R., Barcelo, D., 2006. Determination of pharmaceuticals of various therapeutic classes by solid-phase extraction and liquid chromatography-tandem mass spectrometry analysis in hospital effluent wastewaters. J. Chromatogr. A 1114, 224–233.

Gomez, M.J., Martinez Bueno, M.J., Lacorte, S., Fernandez-Alba, A.R., Aguera, A., 2007. Pilot survey monitoring pharmaceuticals and related compounds in a sewage treatment plant located on the Mediterranean coast. Chemosphere 66, 993–1002.

Gros, M., Petrovic, M., Barcelo, D., 2006. Development of a multi-residue analytical methodology based on liquid chromatography-tandem mass spectrometry (LC-MS/MS) for screeningand trace level determination of pharmaceuticals in surface and wastewaters. Talanta 70, 678–690.

Gros, M., Pizzolato, T., Petrovic, M., Lopez de Alda, M., Barcelo, D., 2008. Trace level determination of β-blockers in waste waters by highly selective molecularly imprinted polymers extraction followed by liquid chromatography–quadrupole-linear ion trap mass spectrometry. J. Chromatogr. A 1189, 374–384.

Gulkowska, A., He, Y.H., So, M.K., Yeung, L.W.Y., Leung, H.W., Giesy, J.P., et al., 2007. The occurrence of selected antibiotics in Hong Kong coastal waters. Mar. Pollut. Bull. 54, 1287–1306.

Hilton, M.J., Thomas, K.V., 2003. Determination of selected human pharmaceutical compounds in effluent and surface water samples by high-performance liquid chromatography-electrospray tandem mass spectrometry. J. Chromatogr. A 1015, 129–141.

Himmelsbach, M., Buchberger, W., 2005. Residue analysis of oxytetracycline in water and sediment samples by high-performance liquid chromatography and immunochemical techniques. Microchim. Acta 151, 67–72.

Huckins, J.N., Petty, J.D., Thomas, J., 1997. Bioaccumulation: How Chemicals Move from the Water into Fish and Other Aquatic Organisms. American Petroleum Institute (API); API publication number 4656, Washington, D.C.

http://www.est-lab.com/pocis.php. (2009) Product: POCIS and its Depolyment. Environmental Sampling Technologies Inc., LorimorStudios. XHTML, CSS & 508.

Jobling, S., Noylan, M., Tyler, C.R., Brighty, G., Sumpter, J.P., 1998. Widespread sexual disruption in wild fish. Environ. Sci. Technol. 327, 2498–2506.

Jun, Z., Schussler, W., Sengl, M., Niessner, R., Knopp, D., 2008. Selective trace analysis of diclofenac in surface and wastewater samples using solid-phase extraction with a new molecularly imprinted polymer. Anal. Chim. Acta 620, 73–81.

Kim, S.D., Cho, J., Kim, I.S., Vanderford, B.J., Snyder, S.A., 2007. Occurrence and removal of pharmaceuticals and endocrine disruptors in South Korean surface, drinking, and waste waters. Water Res. 41, 1012–1021.

Lacey, C., McMahon, G., Bones, J., Barron, L., Morrissey, A., Tobin, J.M., 2008. An LC-MS method for the determination of pharmaceutical compounds in wastewater treatment plant influent and effluent samples. Talanta 75, 1089–1097.

Lin, W.C., Chen, H.C., Ding, W.H., 2005. Determination of pharmaceutical residues in waters by solid-phase extraction and large-volume on-line derivatisation with gas chromatography-mass spectrometry. J. Chromatogr. A 1065, 279–285.

Lishman, L., Smyth, S., Sarafin, K., Kleywegt, S., Toito, J., Peart, T., et al., 2006. Occurrence and reductions of pharmaceuticals and personal care products and estrogens by municipal wastewater treatment plants in Ontario, Canada. Sci. Total Environ. 367, 544–558.

Macleod, S.L., McClure, E.L., Wong, C.S., 2007. Laboratory calibration and field deployment of the polar organic chemical integrative sampler for pharmaceuticals and personal care products in wastewater and surface water. Enviorn. Toxicol. Chem. 26, 2517–2529.

Mahnik, S.N., Rizovski, B., Fuerhacker, M., Mader, R.M., 2006. Development of an analytical method for the determination of anthracyclines in hospital effluents. Chemosphere 65, 1419–1425.

Metcalfe, C.D., Koenig, B.G., Bennie, D.T., Servos, M., Ternes, T.A., Hirsch, R., 2003. Occurrence of neutral and acidic drugs in the effluents of Canadian sewage treatment plants. Environ. Toxicol. Chem. 22(12), 2872–2880.

Miao, X.S., Bishay, F., Chen, M., Metcalfe, C.D., 2004. Occurrence of antimicrobials in the final effluents of wastewater treatment plants in Canada. Environ. Sci. Technol. 38, 3533–3541.

Mills, G.A., Vrana, B., Allan, I., Alvarez, D.A., Huckins, J.N., Greenwood, R., 2007. Trends in monitoring pharmaceuticals and personal-care products in the aquatic environment by use of passive sampling devices. Anal. Bioanal. Chem. 387, 1153–1157.

Nakada, N., Tanishima, T., Shinohara, H., Kiri, K., Takada, H., 2006. Pharmaceutical chemicals and endocrine disrupters in municipal wastewater in Tokyo and their removal during activated sludge treatment. Water Res. 40, 3297–3303.

Nebot, C., Gibb, S.W., Boyd, K.G., 2007. Quantification of human pharmaceuticals in water samples by high performance liquid chromatography–tandem mass spectrometry. Anal. Chim. Acta 598, 87–94.

Nikolai, L.N., McClure, E.L., MacLeod, S.L., Wong, C.S., 2006. Stereisomer quantification of the blocker β-blockers drugs atenolol, metoprolol, and propranolol in wastewaters by chiral high-performance liquid chromatography-tandem mass spectrometry. J. Chromatogr. A 1131, 103–109.

Oaks, J.L., Gilbert, M., Virani, M.Z., Watson, R.T., Meteyer, C.U., Ridesut, B.A., et al., 2004. Diclofenac residues as the cause of vulture population decline in Pakistan. Nature 427, 630–633.

Paschke, A., Brummer, J., Schuurmann, G., 2007. Silicone rod extraction of pharmaceuticals from water. Anal. Bioanal. Chem. 387, 1417–1421.

Paxeus, N., 2004. Removal of selected non-steroidal anti-inflammatory drugs (NSAIDs), gemfibrozil, carbamazepine, β-blockers, trimethoprim and triclosan in conventional wastewater treatment plants in five EU countries and their discharge to the aquatic environment. Water Sci. Technol. 50, 253–260.

Petrovic, M., Hernando, M.D., Diaz-Cruz, M.S., Barcelo, D., 2005. Liquid chromatography-tandem mass spectrometry for the analysis of pharmaceutical residues in environmental samples: a review. J. Chromatogr. A 1067, 1–14.

Petrovic, M., Gros, M., Barcelo, D., 2006. Multi-residue analysis of pharmaceuticals in wastewater by ultra-performance liquid chromatography-quadrupole-time-of-flight mass spectrometry. J. Chromatogr. A 1124, 68–81.

Pozo, O.J., Guerrero, C., Sancho, J.V., Ibanez, M., Pitarch, E., Hogendoorn, E., et al., 2006. Efficient approach for the reliable quantification and confirmation of antibiotics in water using on-line solid-phase extraction liquid chromatography/tandem mass spectrometry. J. Chromatogr. A 1103, 83–93.

Quintana, J.B., Carpinteiro, J., Rodriguez, I., Lorenzo, R.A., Carro, A.M., Cela, R., 2004. Determination of natural and synthetic estrogens in water by gas chromatography with mass spectrometric detection. J. Chromatogr. A 1024, 177–185.

Reddersen, K., Heberer, T., 2003. Multi-compound methods for the detection of pharmaceutical residues in various waters applying solid phase extraction (SPE) and gas chromatography with mass spectrometric (GC-MS) detection. J. Sep. Sci. 26, 1443–1450.

Reddersen, K., Heberer, T., Dunnbier, U., 2002. Identification and significance of phenazone drugs and their metabolites in ground- and drinking water. Chemosphere 49, 539–544.

Roberts, P.H., Bersuder, P., 2006. Analysis of OSPAR priority pharmaceuticals using high-performance liquid chromatography-electrospray ionisation tandem mass spectrometry. J. Chromatogr. A 1134, 143–150.

Roberts, P.H., Thomas, K.V., 2006. The occurrence of selected pharmaceuticals in wastewater effluent and surface waters of the lower Tyne catchment. Sci. Total Environ. 356, 143–153.

Rodriguez-Mozaz, S., Lopez de Alda, M.J., Barcelo, D., 2004. Picogram per liter level determination of estrogens in natural waters and waterworks by a fully automated on-line solid-phase extraction-liquid chromatography-electrospray tandem mass spectrometry method. Anal. Chem. 76, 6998–7006.

Rodriguez-Mozaz, S., Lopez de Alda, M.J., Barcelo, D., 2007. Advantages and limitations of on-line solid phase extraction coupled to liquid chromatography-mass spectrometry technologies versus biosensors for monitoring of emerging contaminants in water. J. Chromatogr. A 1152, 97–115.

Schwaiger, J., Ferling, H., Mallow, U., Wintermayr, H., Negele, R.D., 2004. Toxic effects of the non-steroidal anti-inflammatory drug diclofenac Part 1: histopathological alterations and bioaccumulation in rainbow trout. Aquat. Toxicol. 68, 141–150.

Shareef, A., Angove, M.J., Wells, J.D., 2006. Optimization of silylation using N-methyl-N-(trimethylsilyl)-trifluoroacetamide,N,O-bis-(trimethyl.silyl)-trifluoroacetamide and N-(tert-butyldimethylsilyl)-N-methyltrifluoroacetamide for the determination of the estrogens estrone and 17 alpha-ethinylestradiol by gas chromatography-mass spectrometry. J. Chromatogr. A 1108, 121–128.

Spongberg, A., Witter, J., 2008. Pharmaceutical compounds in the wastewater process stream in Northwest Ohio. Sci. Total Environ. 397, 148–157.

Stolker, A.A.M., Niesing, W., Hogendoorn, E.A., Versteegh, J.F.M., Fuchs, R., Brinkman, U.A.T., 2004. Liquid chromatography with triple-quadrupole or quadrupole-time of flight mass spectrometry for screening and confirmation of residues of pharmaceuticals in water. Anal. Bioanal. Chem. 378, 955–963.

Sun, Z., Schussler, W., Sengl, M., Niessner, R., Knopp, D., 2008. Selective trace analysis of diclofenac in surface and wastewater samples using solid-phase extraction with a new molecularly imprinted polymer. Anal. Chim. Acta 620, 73–81.

Ternes, T.A., 1998. Occurrence of drugs in German sewage treatment plants and rivers. Water Res. 32, 3245–3260.

Ternes, T.A., 2002. Removal of pharmaceuticals during drinking water treatment. Environ. Sci. Technol. 36, 3855–3863.

Ternes, T.A., Joss, A., Siegrist, H., 2004. Scrutinizing pharmaceuticals and personal care products in wastewater treatment. Environ. Sci. Technol. 38, 392A–399A.

Ternes, T.A., Bonerz, M., Herrmann, N., Loffler, D., Keller, E., Bago Lacida, B., et al., 2005. Determination of pharmaceuticals, iodinated contrast media and musk fragrances in sludge by LC/tandem MS and GC/MS. J. Chromatogr. A 1067, 213–223.

Togola, A., Budzinski, H., 2008. Multi-residue analysis of pharmaceutical compounds in aqueous samples. J. Chromatogr. A 1177, 150–158.

Trenholm, R.A., Vanderford, B.J., Holady, J.C., Rexing, D.J., Snyder, S.A., 2006. Broad range analysis of endocrine disruptors and pharmaceuticals using gas chromatography and liquid chromatography tandem mass spectrometry. Chemosphere 65, 1990–1998.

Verenitch, S.S., Lowe, C.J., Mazumder, A., 2006. Determination of acidic drugs and caffeine in municipal wastewaters and receiving waters by gas chromatography-ion trap tandem mass spectrometry. J. Chromatogr. A 1116, 193–203.

Vieno, N., Tuhkanen, T., Kronberg, L., 2007. Elimination of pharmaceuticals in sewage treatment plants in Finland. Water Res. 41, 1001–1012.

Vrana, B., Mills, G.A., Allan, I.J., Dominiak, E., Svensson, K., Knutsson, J., et al., 2005. Passive sampling techniques for monitoring pollutants in water. Trends Analyt. Chem. 24, 845–868.

Vrana, B., Mills, G.A., Dominiak, E., Greenwood, R., 2006. Calibration of the Chemcatcher passive sampler for the monitoring of priority organic pollutants in water. Environ. Pollut. 142, 333–343.

Weigel, S., Kallenborn, R., Huhnerfuss, H., 2004. Simultaneous solid-phase extraction of acidic, neutral and basic pharmaceuticals from aqueous samples at ambient (neutral) pH and their determination by gas chromatography-mass spectrometry. J. Chromatogr. A 1023, 183–195.

Zuhlke, S., Dunnbier, U., Heberer, T., 2004. Detection and identification of phenazone-type drugs and their microbial metabolites in ground and drinking water applying solid-phase extraction and gas chromatography with mass spectrometric detection. J. Chromatogr. A 1050, 201–209.

Zhang, Z.L., Zhou, J.L., 2007. Simultaneous determination of various pharmaceuticals in water by solid-phase extraction-liquid chromatography-tandem mass spectrometry. J. Chromatogr. A 1154, 205–213.

Zhang, Z.L., Hibberd, A., Zhou, J.L., 2008. Analysis of emerging contaminants in sewage effluent and river water: comparison between spot and passive sampling. Anal. Chim. Acta 607, 37–44.

Zorita, S., Boyd, B., Jonsson, S., yilmaz, E., Svensson, C., Mathiasson, C., et al., 2008. Selective determination of acidic pharmaceuticals in wastewater using molecularly imprinted solid phase extraction. Anal. Chim. Acta 626, 147–154.

Monitoring for Terrorist-Related Contamination

Dan Kroll

Hach Homeland Security Technologies, 5600 Lindbergh Drive, Loveland, CO 80539, USA

INTRODUCTION

The direction of the last 100 years of analytical science, as it pertains to drinking water, took a dramatic shift on 11 September, 2001, and in the subsequent anthrax attacks of 18 September, 2001. Prior to these dates, analytical emphasis was on the detection and removal of naturally occurring or accidental

Handbook of Water Purity and Quality
Copyright © 2009 by Elsevier Inc. All rights of reproduction in any form reserved.

contaminants that found their way into drinking water supplies. After the terrorist attacks, a new fear dawned in the water supply industry. What if someone were to intentionally introduce a contaminant into the drinking water? The vast array of potential contaminants that could be used by a terrorist, the innumerable sites at which an attack could occur, and the potential consequences of not rapidly detecting such an event demanded a sea change from the old monitoring paradigm of collecting occasional grab samples and monitoring for a small suite of potential contaminants. The challenges entailed in this endeavor and some of the technologies that are becoming available to protect our water supplies from deliberate attack are the subjects of this chapter.

WHAT IS THE TERRORIST THREAT TO WATER?

According to Ronald Dick, FBI Deputy Assistant Director for Counter Terrorism, "In reality, targeting the water supply may prove difficult. In order to be successful, a terrorist would have to have large amounts of agent."(Dick, 2001). Then EPA director Christie Todd Whitman in October 2001 stated, "People are worried that a small amount of some chemical or biological agent—a few drops for instance—could result in significant threats to the health of large numbers of people. I want to assure people—that scenario can't happen. It would take large amounts to threaten the safety of a city water system. We believe that it would be very difficult for anyone to introduce the quantities needed to contaminate an entire system" (Whitman 2001). Are these statements true? Do we, in fact, have nothing to worry about as far as terrorists attacking our water supplies?

Legendary FBI director J. Edgar Hoover thought otherwise. In 1941 he wrote, "Among public utilities, water supply facilities offer a particularly vulnerable point of attack to the foreign agent, due to the strategic position they occupy in keeping the wheels of industry turning and in preserving the health and morale of the American populace. Obviously, it is essential that our water supplies be afforded the utmost protection" (Hoover, 1941). Who is correct in assessing the potential for terrorist activities directed at our water supplies, Mr. Hoover or Ms Whitman or Mr. Dick? In a sense it turns out that they are all correct.

The answer depends entirely on which portion of the water supply system you are referring to. The systems that supply our water have various components, each of which has a distinct level of susceptibility to contamination. Mr. Dick and Ms. Whitman were referring to the poisoning of large reservoirs or rivers that we utilize as raw water sources, when they made their statements about the water supply being safe. Even a small reservoir has enormous volume and it would require large quantities of most potential threat agents to contaminate such a large volume of water to toxic levels. In effect, they were relying upon the old adage that, "The solution to pollution is dilution." In most cases, as far as raw water supplies are concerned, this reliance is justified. While a contaminant attack on raw water supplies is possible, the threat of mass casualties resulting from such an attack is minimal.

Another layer of protection is afforded by the design of modern water supply systems found in the United States. Treatment plants are designed to remove accidental or naturally occurring pollutants, and they also serve a similar function for most intentional contaminants. The plants themselves are also potential targets, but their compact size and the ability to equip them with physical security measures to limit access helps to ameliorate this threat to some extent.

It was not long after 9/11 when government officials and industry experts realized that the vulnerability to contamination was not in the source water or the treatment plants, but rather, in the distribution system. By October 2003, a GAO report to the Senate stated that the distribution system was the area most vulnerable to attack (GAO-04-29, 2003). Assuming that an attack with chemical, biological, or radiological (CBR) agents would most likely take place somewhere in the distribution system—several misconceptions about this type of attack still persist. Many experts have contended that such attacks require the assistance of several technicians, are expensive to carry out, and require complicated and expensive pumping equipment to inject contaminants into a pressurized system. More recent studies by the Army Corps of Engineers and Hach Homeland Security Technologies (HST), among others, show that CBR attacks could in fact be carried out for 50 cents or less per lethal dose, that a single individual can obtain or produce effective contaminants in quantity, and that contaminants can be introduced into the distribution system with the aid of inexpensive and easy to obtain pumping equipment via a method called backflow attack (Kroll, 2003; Army Corps of Engineers Calculations; Waterborne CBR Agent Building Protection, 2003; Allman, 2003).

WHAT IS A BACKFLOW ATTACK?

A backflow attack occurs when a pump or siphon is used to overcome the pressure gradient that is present in a distribution system's pipes. This is usually around $80 \, lbs/in^2$ or less and can be easily overcome by using pumps available for rent or purchase at most home improvement stores, or by using a siphon that injects using the Bernoulli effect. After the pressure has been overcome or the siphoning begins, a contaminant is introduced into the flowing system and the normal movement of water in the system acts to disseminate the contaminant throughout the network, affecting areas surrounding the introduction point. The introduction point can be anywhere in the system, such as a fire hydrant, commercial building, or residence. Studies conducted by the U.S. Air Force and Colorado State University have shown this to be a very effective means of contaminating a system (Allman, 2003). A few gallons of highly toxic material was enough, if injected at a strategic location, to contaminate an entire system supplying a population of 150,000 people in a matter of a few hours. A terrorist could launch such an attack and be on a plane out of the country before the first casualties occur.

Currently, monitoring of drinking water supplies in the distribution system is limited. The most common practice is to take widely scattered grab samples once-a-month and monitor for disinfectant residual and bacterial contamination. Previous to the terrorist threat, more comprehensive monitoring regimens were not undertaken as it was not a priority.

The ability to detect an event in the distribution system and then its identification would be of incomparable value in responding to an incident in a timely and proper manner. Such ability would also serve the purpose of mapping a system for cleanup, and afterward, it could be used as a forensic tool to identify the source of an event. Prior to 9/11, there were no devices capable of detecting such an event and alerting the system's managers so that the effects of an attack or accidental event could be detected or contained. The general scientific consensus was that no practical, available, or cost-effective real-time technology then existed to detect and mitigate intentional attacks or accidental incursions in drinking water distribution systems. The development of such an early warning system was listed by a panel of experts and industry leaders as a top priority in enhancing water security (Office of Science and Technology Policy, 2003).

One of the crucial but often overlooked elements of an effective early warning system is the response to an indicated event. How one responds to an indicated event is often dependent on the confidence one has in the data, implying that an event has occurred. The ability to detect extremely rare transient events such as an attack on water supplies absolutely requires the utilization of continuous online monitoring techniques. One of the problems with detecting these rare events is that the false alarm rate or unknown alarm rate will exceed the number of true events. This inevitably leads to alarms being ignored unless there is a rapid and reliable method for validating and verifying alarms generated by the online systems.

This verification is most commonly performed by further testing using an alternative field or laboratory method. This effectively divides the categories of technology that will be discussed into two groups. The groups are methodologies suitable to online deployment and technologies suitable for verification. The remainder of the discussion will explore a number of such technologies.

ONLINE MONITORING

What Should an Early Warning System Look Like?

An early warning system would be a device capable of monitoring the safety of the water supply in question, analyzing, interpreting, and communicating that data to the appropriate people so that decisions could be made to protect the public's health. To meet these goals an early warning system should have certain characteristics. According to the EPA, these desired characteristics should include the following (USEPA, 2005):

- A rapid response.
- Be capable of detecting a sufficiently wide range of potential contaminants.

- Exhibit a significant degree of automation including automatic sample archiving.
- Allow acquisition, maintenance, and upgrades at an affordable cost.
- Require low skill and limited training to operate.
- Identify the source of the contaminant and allow accurate prediction of the location and concentration downstream of the detection point.
- Demonstrate sufficient sensitivity to detect contaminants at the levels of interest.
- Experience minimal false positives/false negatives.
- Exhibit robustness and ruggedness in continually operating in a water environment.
- Allow remote operation and adjustment.
- Function continuously.
- Allow for third-party testing, evaluation, and verification.

There are many technical hurdles in designing a system that meets all of these goals.

What Should a Monitoring System Detect?

One of the principal problems when designing such a monitoring system for water is the vast number of chemical agents that could be utilized by a terrorist to compromise a water supply system. This diversity tends to preclude monitoring on an individual chemical basis. Chemical warfare agents such as VX, Sarin, Soman, etc.; commercially available herbicides, pesticides, and rodenticides; street drugs such as LSD and heroin; heavy metals; radionuclides; cyanide, and a host of other industrial chemicals could be exploited as weapons. There are also various biological agents and biotoxins that could be problematic. It is still a matter of conjecture as to which of the possible agents would be the most likely to be deployed in a terrorist assault. The possible number of chemical and biological substances that could be used in an attack is immense (Kroll, 2004). There are various lists publicly available in the current literature, such as the list compiled by the Center for Disease Control (CDC) (CDC Emergency Preparedness and Response Web site) and the military Tri-Services list (U.S. Army Center for Health Promotion and Preventive Medicine, 1999), specifying likely agents. Some lists that have been compiled are unavailable, because of security reasons, such as the list compiled by the EPA. Many of these lists are similar in composition, but no two lists are identical and in several cases, they are contradictory.

To be truly effective, a monitoring device needs to be able to detect any and all of the possible agents that could be encountered. A dedicated device capable of detecting botulinum toxin for instance is interesting but not very practical, as it could be thwarted by the terrorist use of another agent. Another example is the GC type of system that may be very effective for detecting volatile organics but would offer no protection against the introduction of a heavy metal such as mercury.

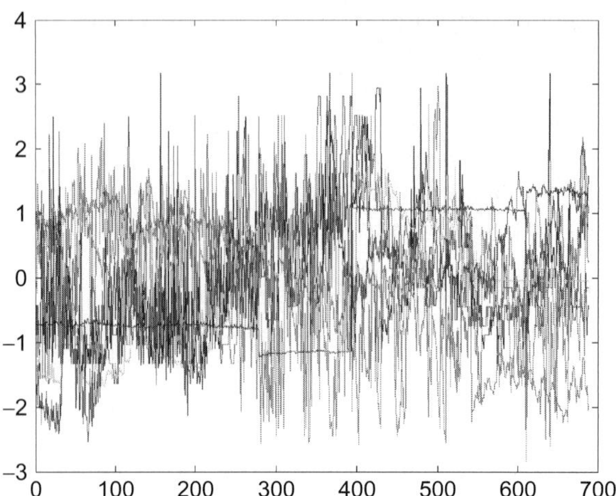

FIGURE 1 Real world water quality is highly variable. Graph courtesy of Hach HST (see Plate 3 of Color Plate section).

This need to detect diverse contaminants requires a realignment of thinking from the traditional development of a sensor for a specific compound or agent. Toxicity testing is one route being investigated. Sensor arrays on a chip or the use of analytical instrumentation capable of detecting this variety is also being developed but present many challenges. Another approach is to use chemometrics to detect and characterize changes in basic water quality parameters and correlate them with threat agent introduction.

Water Analysis Presents Many Problems

A common misconception concerning analysis in water is that the system is stable with little variation over time or from location to location. This is probably due to most analysts that are not specifically involved in water research being exposed to laboratory grade deionized water (DI) as the norm when running experiments. In the real world, even after treatment, there is great diversity in the water found in distribution systems. For a parameter as simple as pH that we would expect to be around 7 ± 1 pH units, the diversity is much greater and can run from 3 to near 11 pH units. In a given system there is great variation over time in basic conditions such as pH, turbidity, conductivity, and so on. Figure 1 is representative of the diversity that can be found in the real world in these types of parameters over time.

On top of the great diversity of water quality that may be present in the distribution system, the water distribution system environment is also very harsh. The water conditions may be corrosive or scaling in nature. This can lead to degradation of anything placed in the system or the formation of a coating of various types of materials (Figure 2). There is also something present in most pipe systems known as biofilm. This is a thin layer of bacteria and their associated

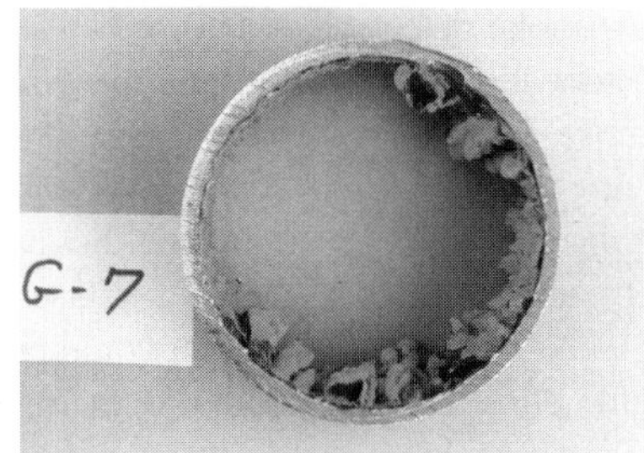

FIGURE 2 The interior of many distribution system pipes can present a very hostile environment for sensor deployment. Photos courtesy—US Army Corps of Engineers (see Plate 4 of Color Plate section).

glycocalyx (slimey film) that coats the inside of pipes and anything else present in the system. This layer of growth can coat sensors and render them unable to function properly. It can also clog small tubes and pipes used to draw off samples resulting in erroneous readings. Any detection system designed to function over long periods of time in the distribution system must be capable of handling these harsh conditions. There is also the problem of aging infrastructure. Many of the water pipes in our major cities are over 100 years old and are occluded with rust, crumbling concrete, and other debris. Some of the pipes are actually still the original wooden pipes installed when the systems were first built. This raises concerns for installation and sampling for a distribution system monitoring platform as well as the functioning of many of the new and emerging technologies involving microfluidic microprocessors and other nanotechnologies.

Another criterion for monitoring water supplies is speed of response. Most distribution systems are designed to flow at a range of 1 to 3 ft/s. This puts extraordinary pressure on detection systems to recognize, identify, and confirm contamination as rapidly as is possible. With the above-mentioned flow rates and the fact that pipe diameters can range from a few inches to many feet, the volume of water that could pass a monitoring point could be quite large if the response is not as rapid as possible.

The poor condition of the distribution system has implications other than just presenting a harsh environment for sensors. Because of the aging infrastructure plaguing most municipal water supply systems, drinking water and wastewater infrastructure investment costs over the next 20 years may range from $492 billion to $820 billion, according to a Congressional Budget Office (CBO) report (Water Infrastructure Network News). This huge expenditure for needed upgrades leave little funding room for things such as security. This makes it very important that any sensor system be very cost-effective. This goal of cost-effectiveness can be addressed in two different ways. One is to design a very inexpensive system that could be deployed for a low cost per customer and the other is to develop a system that is capable of providing data that could be useful in decreasing the day-to-day operating costs of the system and improving general water quality, thus making it a recoverable cost.

A smoke detector can be used as an analogy. The relative low cost of smoke detectors allows their wide spread deployment to protect against an unlikely event. If smoke detectors were to cost several thousand dollars each, few locations would be equipped with them. A water system protection device is similar. Unless it was very inexpensive, few municipalities would fund a system designed to protect solely against terrorist events, because of the low likelihood of their occurrence in a given location. The market could bear a higher cost for a dual use device that would help streamline general operations and help provide increased water quality, hence providing real value even if a terrorist event never occurs.

With these goals and constraints in mind, many unique approaches have been developed recently to address the problem of detecting intentional contamination in our drinking water supplies.

Toxicity Monitoring

Toxicity is the ability of a substance to cause a living organism to undergo adverse effects upon exposure. These effects can include negative impacts on survival, growth, behavior, and reproduction among others. Toxicity tests are an attempt to measure toxicity in a sample by analyzing the results that the exposure produces on standard test organisms (Kroll, 2007). Toxicity can be acute, subchronic, or chronic: http://www.medterms.com/script/main/art.asp?articlekey=34093

- Acute toxicity involves harmful effects in an organism through a single or short-term exposure.
- Subchronic toxicity is the ability of a toxic substance to cause effects for more than 1 year but less than the lifetime of the exposed organism.
- Chronic toxicity is the ability of a substance or mixture of substances to cause harmful effects over an extended period, usually upon repeated or continuous exposure, sometimes lasting for the entire life of the exposed organism.

Toxicity testing in the realm of security monitoring holds much promise because of its ability to detect a wide variety of potential threats. This ability has led to the development of a number of online toxicity monitoring devices as well as field verification kits that utilize a number of diverse organisms and methods to detect problems in the water supply. As a general rule, the closer an organism is to humans on the evolutionary tree the closer the organism's response will be to humans for a given compound. That is why much medical research is done using primate models. Working with higher organisms can be complicated and that is why trade-offs are made and lower organisms are used for most toxicity testing.

Fish-Based Systems

Fish monitoring is a system that dates back to antiquity. Through experience, our ancestors generally understood that it was a bad idea to drink out of a lake that had a layer of dead fish floating on the surface, and it was not long before people began to use this type of early warning phenomena to detect problems with their water supplies. For centuries, many areas in Asia have used a simple avoidance system where fish are kept in tanks and are normally fed in the tank closest to the incoming water supply. If a problem occurs with the incoming water, the fish move to other tanks progressively farther away from the food and influent in an attempt to avoid exposure to the toxin. A simple glance to see which tank the majority of the fish are occupying is all that is needed to determine if there is a problem. Homemade systems using this technique have been installed in a number of U.S. cities. These systems are simple and work well but are effective only in detecting gross changes in water quality. Recognition of subtle changes requires the coupling of the rudimentary fish technologies with modern technology.

The use of fish to monitor water quality really took off about 20 years ago after a major spill of toxic chemicals polluted the water supplies for a number of countries in Europe. There are a number of fish biomonitors that utilize advanced data collection systems such as CCD cameras and algorithmic-based interpretation of the data to determine if there is a problem.

One such system is the ToxProtect64 manufactured by bbe Moldaenke. The ToxProtect64 monitors the swimming activity of up to 20 fish by measuring the frequency of interruption of an array of light barriers. The result is given in interrupts per minute per fish. Immobile fish at the bottom and in the upper region of the aquarium are also registered. In the event of the values falling below a given threshold for a certain period of time, the alarm verification process is activated. Because of naturally occurring random variations in fish behavior, each alarm criterion is fulfilled from time-to-time by accident. Hence, in order to prevent false alarms, a verification system is required. This is achieved by increasing the illumination inside the aquarium during the verification period. Normally, this leads to a dramatic increase in fish activity. Under toxic conditions, however, this is less likely to occur. Hence, by monitoring the increase in activity during modified illumination, it is possible to accept or reject the initial alarm. The fish species used can be selected by the user, with recommendations given in the specifications; however, all fish must be active and 4–6 cm in length (bbe Maldaenke, 2007).

The research and development for another advanced fish-based system was completed by the U.S. Army Center for Environmental and Health Research (USACHER). This system is more advanced than systems that measure simple fish movement. The key to this system is the cough reflex experienced by bluegill fish when they are under stress. The system simultaneously monitors the normal respiratory actions and movements of the fish held in the tanks of the systems. The parameters that the system monitors include ventilation rate, strength of ventilation, gill purge (cough) rate, and body movement. When at least six out of the total of eight fish exhibit unusual behavior, the computer sounds the alarm, notifies appropriate personnel, and takes a sample for further forensic investigation (Aquatic Biomonitoring).

The USACHER systems are commercially available from the Intelligent Automation Corporation as the IAC 1090 Intelligent Aquatic BioMonitoring System. The systems have been deployed on the source water of a number of

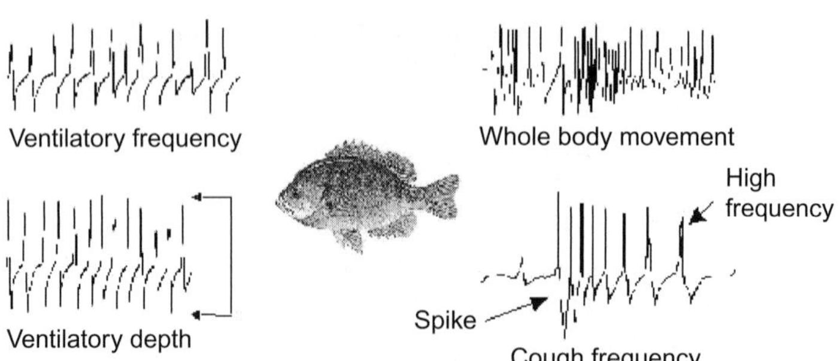

FIGURE 3 The USACHER system utilizes variations in electrical field strength to determine if fish breathing is labored.

large cities including San Francisco, Washington, D.C., and New York. New York City has been testing its system since 2002 and is seeking to expand it. The New York City Department of Environmental Protection reported at least one instance in which the system caught a toxin before it made it into the water supply: The fish noticed a diesel spill 2 h earlier than any of the agency's other detection devices (Wholsen and Marcus, 2006) (Figure 3).

These are just a couple of examples of the many fish-based monitors that are available. Most of these monitors were originally designed to be deployed in monitoring source water; however, many manufacturers have adapted the systems with sample preconditioning systems that allow their deployment in the distribution system. The drawbacks of sample preconditioning will be discussed in the section Problems with Toxicity Testing.

Bivalve-Based Systems

There are several systems available that are based on the response of mollusk bivalve organisms to the presence of toxins. These include the MosselMonitor® from Delta Consult based in the Netherlands and Sybio form PROTE Technologies for the Environment based in Poland. These systems use various electronic methods to monitor the behavior of the bivalve species used. These organisms tend to close their shells for protection when unfavorable water conditions exist. This opening and closing of the shells is used to correlate with the toxicity present in the water (USEPA, 2005) (Figure 4).

Daphnia and Other Invertebrate-Based Systems

Daphnia, or water fleas, are small crustaceans between 0.2 and 5 mm in length. They are native to various aquatic environments. They are highly sensitive to various toxins and have been used for a number of years in standard laboratory methods for determining toxicity.

One of the online instruments that utilizes the water flea is the Daphnia Toximeter produced by bbe Moldaenke. The system measures the mobility and agility of the daphnia that are continuously exposed to the water sample to be

FIGURE 4 Bivalve behavior can be used to detect toxins in water. Courtesy—USDA.

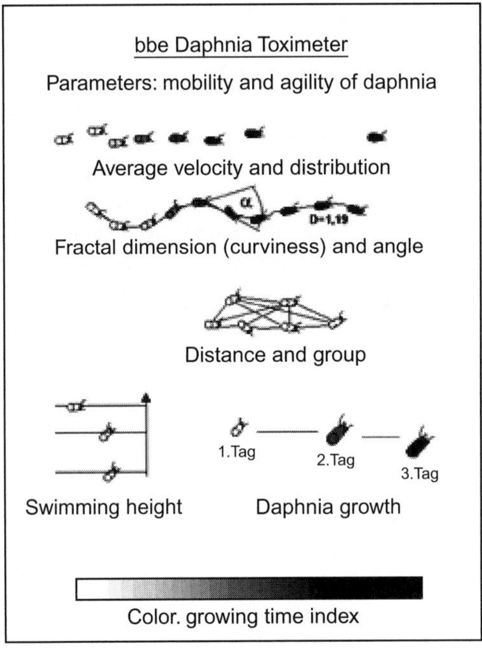

FIGURE 5 The Daphnia toximeter simultaneously monitors a number of components of daphnia movement. Courtesy—bbe Moldaenke.

evaluated at a rate of 0.5–2.0 L/h. The daphnia are observed via a CCD camera and the signals are processed in an integrated PC. A change in the movement of the Daphnia is calculated on a number of different parameters. If a statistically significant change is detected, an alarm is generated (bbe Moldaenke; Figure 5).

There are a number of toxicity methods that have not yet been put online but have value as confirmatory field or lab methods. Strategic diagnostics offers various test kits in their MicroBioTest line. These tests make use of various organisms including *Tetrahymena thermophila*, *Brachionus calyciflorus*, *B. plicatilis*, *Thamnocephalus platyurus* and several *Daphnia* species. Toxicity in many of these tests is judged by the ingestion or failure to ingest red microspheres that are clearly visible in the organism's digestive tract. This is a much easier end point to judge than traditional assays that look for lethality or changes in behavior.

Aqua Survey Inc. of Flemington, New Jersey, under the brand name IQ Toxicity Test, provides another simplification of classic invertebrate toxicity testing. This method is based on fluorescent tagging of a sugar molecule that is placed in the daphnia's food. If the daphnia are healthy and actively metabolizing, they will ingest the sugar and cleave the molecule. This cleaving of the molecule causes the organisms to become fluorescent under UV light. It is basically a variation of; the classic MUG test for *E. coli* bacteria. If the daphnia glow, there is no toxicity if they do not, there is toxicity. Daphnia testing is extremely sensitive and may actually be overly sensitive to common drinking water constituents in some cases.

Algae-Based Methods

An online method for monitoring toxicity in source water utilizing indigenous algae has been developed by scientists at Oak Ridge National Laboratory (ORNL).

AquaSentinel is an automated and field-deployable real-time technology for the detection of source water environmental toxins that is based on the fluorescence induction properties of algae that grow naturally in water. The algal biosensors can be used as a sentinel alarm system based on toxin-induced fluorescence readout as the characteristic signature for identification and verification of environmental pollutants in source drinking waters. The system is reagentless and uses a self-contained optoelectronic detection. It utilizes an original algorithm for performing the analysis of the readings from the biosensors. The approach is based on differential offset between the fluorescence signatures of healthy algae and that of the poisoned algae. The technique yields a set of time-dependent numbers that uniquely map the transformation of the normal or healthy fluorescence induction curve to that of the poisoned curve. The set of numbers generates a characteristic signature that can be used to group and identify the specific pollutant that caused the alteration of the fluorescence (Greenbaum and Rodriguez, 2006). By using native algae as the test organism, this system is limited to source waters but could be adapted to treated waters if a source of algae were supplied (Figure 6).

Bacterial Luminescence-Based Methods

There are a number of bacteria-based methods. Many of them are based on the ability of photobacteria to produce fluorescence. Two of the bacteria that are utilized are *Vibrio fischeri* and *Photobacterium leiognathi*. Luminous bacteria

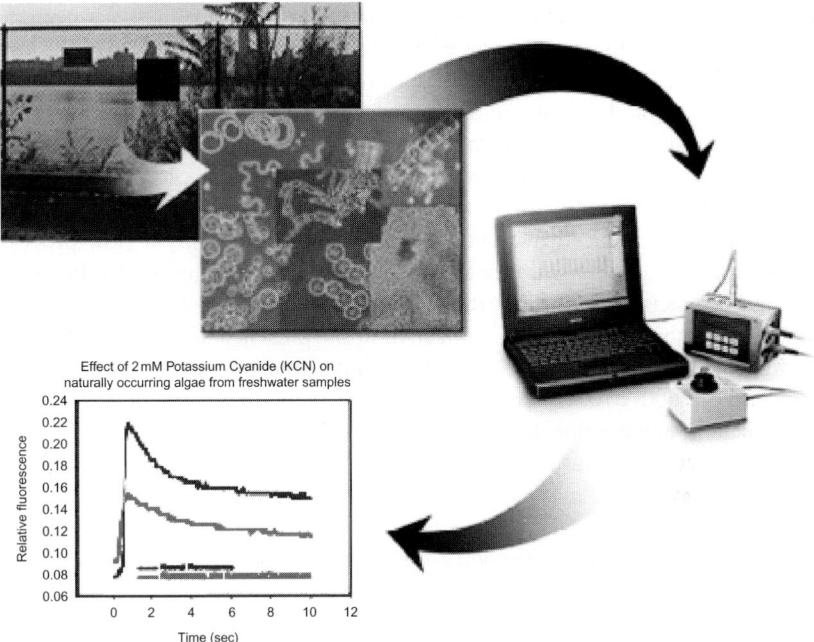

FIGURE 6 The Oak Ridge system utilizes the fluorescent response form naturally occurring algae. Courtesy—Elias Greenbaum, Oak Ridge National Lab.

produce and emit light as a by-product of cellular respiration. Factors that influence the respiration cycle, cell membrane integrity, and function etc. in the bacteria promptly alter the level of luminescence produced. By comparing the level of luminescence produced in a suspected toxic sample to that produced in a known nontoxic control, a measure of toxicity can be achieved. If a toxin is present, the bioluminescence of the bacteria is decreased. There are many field and laboratory methods based on this technology including the MicroTox and DeltaTox models from Strategic Diagnostics and the Toxscreen system from CheckLight. These technologies utilize a luminometer to detect the changes in bioluminescence. These methods would be readily adaptable to online monitoring, and active programs are under way to achieve an effective online instrument.

Bacterial Respiration-Based Methods

Another method for measuring toxicity is the inhibition of bacterial respiration. PolyTox™ sold by InterLab Supply uses standard dissolved oxygen electrodes to measure respiration of a specially formulated bacterial culture that is sold under the PolyTox™ brand. Oxygen consumption is a good measurement of overall bacterial health; however, this method does present some problems. Drawbacks are that the test requires the use of a dissolved oxygen electrode, a touchy piece of equipment. The samples must be aerated for 30 min before starting the test. This aeration could result in the loss of volatile toxins from the sample. Finally, only one test can be run at a time; multiple tests require long periods of time, coupled with extensive sample manipulation.

Another method based on bacterial respiration is the ToxTrak™ Rapid Toxicity Testing System produced by Hach Co. Rather than directly measuring oxygen consumption, the ToxTrak™ system utilizes a colorimetric system based on the rate of reduction of resazurin dye. As the bacteria actively metabolize, the dye is reduced from blue to pink. This is normally a slow process, but the Hach system makes use of a patented accelerator solution that increases the rate of reaction and allows the test to be completed in as little as 45 min. Inhibition of this rate change is indicative of the presence of toxicity. The color change can be measured with any spectrophotometer or colorimeter capable of measuring at 600–610 nm. The color change is actually capable of being measured visually and a color disk comparator method is available. This system is very inexpensive and has been shown to react to a wide variety of toxins. While virtually any bacteria may be used in the test, one drawback is that cultures must be grown and maintained in advance of testing. Also, the duration of the test makes these difficult to modify for online use but they are very effective and inexpensive for field use.

Inhibition of Photosynthetic Enzymes

Bell labs of Canada have developed a system based on the inhibition of chlorophyll fluorescence by photosynthetic systems. The combined use of photosynthetic enzyme complexes (PECs), isolated from higher plants, and whole photosynthetic organisms (algae) allows a wider range of toxic inhibitors to be detected in just 10–15 min since photosynthetic light reactions are sensitive to various pollutants including metal ions, PAH, herbicides, cyanide, etc. They have developed both a field kit known as the LuminoTox and an online method called the Robot LuminoTox that take automated measurements every 30 min. (Bellemare et al., 2005)

Inhibition of Chemiluminescence

A final method of monitoring toxicity is based on chemiluminescence. This method is the one used in the Severn Trent Services Eclox™ kit. The reaction of luminal and an oxidant in the presence of horseradish peroxidase, which results in the chemical production of light or chemiluminescence, can be used to detect the presence of toxins. Any free radical scavengers or antioxidants such as those contained in feces or urine will interfere with the reaction, thus reducing the light emission. Substances such as phenols, amines, heavy metals, or compounds that attack or coat the enzyme will also reduce the light output. The light output is plotted over time and produces characteristic curves. Results are compared to DI. Samples containing pollution will give lower light levels (Figure 7).

The Eclox™ method is extremely robust and has been hardened for field use by the British military. This method has been shown to be effective against

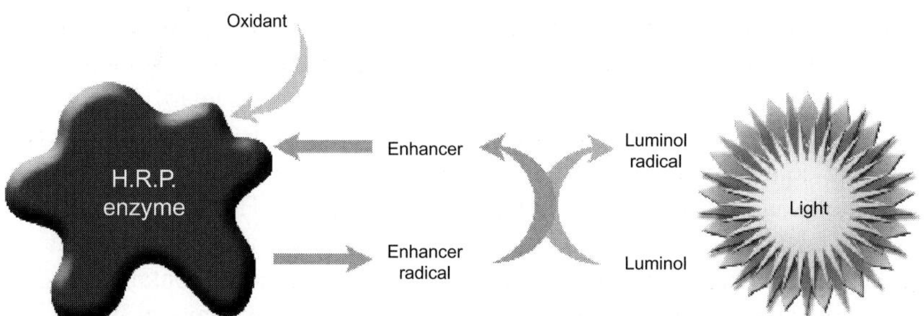

FIGURE 7 The detection of toxicity by the Severn Trent Services Eclox™ kit is based on inhibition of the enzyme horseradish peroxidase's ability to cause luminol to produce light. Figure Courtesy—Severn Trent Services.

a wide variety of substances that could be potential risks. No online iteration currently exists.

Problems with Toxicity Testing

One problem with toxicity testing is that no matter which organism or method is chosen, there will be significant differences between the suite of toxicants that will elicit a response and the degree of toxicity exhibited in the test organisms and the modeled organism (i.e., humans). Also, while toxicity tests are fairly adept at detecting chemical toxins, they are for the most part ineffective against biological agents such as bacteria and viruses. The largest draw back to all invertebrate testing is culture maintenance. For emergency testing, a usable culture needs to be maintained at all times. In some cases, organisms in a specific stage of development or state of hunger are needed. This is hard to maintain for an emergency program and is probably better suited to an ongoing program of laboratory testing, where the organisms are more easily maintained in the proper state.

All toxicity-testing methods require knowledge of a baseline. Although this is fairly straightforward for online methods, it may be problematic for field methods and requires significant time expenditure to build up a database. Also, some toxicity methods may be too sensitive when testing in the distribution system. Water treatment chemicals or simply common constituents of drinking water such as trace metals that are not toxic to humans may adversely affect them. Treatment chemicals such as chlorine, chloramines, fluoride, and others can affect the response of the test organisms or kill them outright.

For a toxicity monitoring system to function appropriately in a treated water matrix, constituents such as chlorine must first be removed from the water before the organisms are exposed to it. Chlorine can be easily removed by treatment with dechlorinating agents such as thiosulfate. The problem with this is that the process of removing the chlorine can alter the toxicity of the water. For example, if a system contaminated with a mercury compound is dechlorinated with sodium thiosulfate, binding to the thiosulfate (for which it has a very great affinity) will also sequester the mercury. This binding to the thiosulfate can render the mercury unavailable and hence mask the toxicity of the water.

As a whole, the use of toxicity measurements to safeguard source water has significant merit. These systems have the ability to detect a wide variety of potential contaminants. They offer a good first-line of defense to prevent contaminated source water from entering treatment plants. The drawbacks have to do with cost, false alarms from normal background occurrences not related to contamination, their unproven ability to alarm on all possible contaminants, and maintenance and culturing problems. In many cases, they may not be the ideal solution because of these drawbacks. A lower cost less hassle solution may be required in some deployment scenarios so that more sites can be monitored.

Another drawback to toxicity monitoring in the distribution system is the variable environment that can come into play in the network. Some organisms such as fish can be quite sensitive to changes in the general surroundings. Increases in vibrations and noise levels at times of peak traffic could become problematic and lead to false alarms. Shielding organisms from this type of upset can be costly or can severely limit your options for deployment. For use in distribution systems where widespread deployment is required, toxicity monitoring is usually not a viable option. However, toxicity monitoring is applicable to monitoring source water where monitoring points are fewer and more easily controlled.

Sensor Arrays and Lab-on-a-Chip Technologies

Another method of solving the problems of multiple target analytes is to use miniaturization and large arrays of various types of sensors to detect diverse targets.

Microelectromechanical systems (MEMS) also known as lab-on-a-chip technologies are an innovation that is rapidly growing and finding use in the medical technology field. These are basically microscale devices that attempt to miniaturize and streamline traditional analytical and biochemical methods, and mass-produce them, utilizing many of the same fabrication techniques that facilitate production in the computer chip and microelectronics industry. These mass production techniques allow the manufacture of very low cost devices when compared to traditional instrumentation. As the technology becomes more robust, it has been branching out from its traditional role in medical diagnostics and finding use in other analytical applications. Water is no exception and much research is being done to amend these types of devices to water analysis.

One of the projects currently under development is the MicroBio Chem Lab being produced by the Sandia National Laboratory (Kroll, 2004). This device makes use of microfluidics and microchemical techniques to sample and analyze water for various components including the presence of harmful bacteria and viruses. These types of devices are in effect miniaturized discrete analyzers that test for specific substances. They can use various detection techniques such as miniaturized gas chromatographs, immunoassay techniques, microcantilevers,

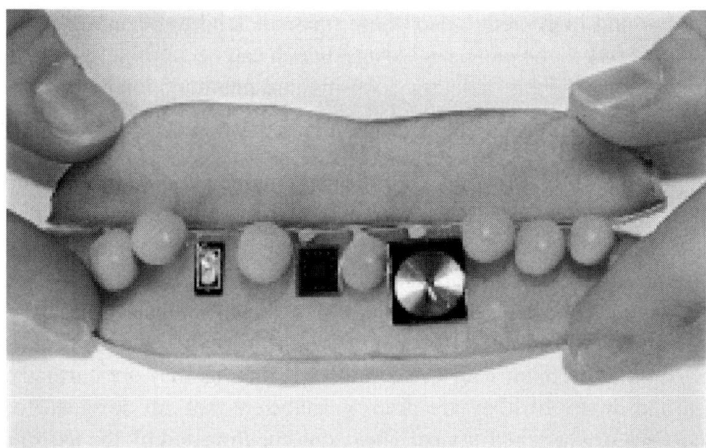

FIGURE 8 The three principal components of Sandia's microchem lab for gas-phase detection and analysis are small enough to fit easily inside a snow-pea pod. The left-most component is the surface acoustic wave sensor array, the lab's detection mechanism. The center one is a preconcentrator that absorbs or adsorbs chemical vapors. The right one that looks like a tiny CD is a miniature gas chromatograph column. Together they can collect, concentrate, and analyze a chemical sample weighing less than a single bacterium. Other Sandia microchem labs analyze liquids. Photo courtesy—Randy Montoya.

surface acoustic wave devices, and proteomics or gel electrophoresis systems. Recently, projects involving Sandia and an Australian company called Tenix have been initiated to mature this technology from an independent hand held device to an online configuration to do monitoring of water supplies (Figure 8).

It is yet to be seen as to whether these efforts to bring the system into the realm of online distribution monitoring will be successful. If it were, the low cost of the instrumentation could allow monitoring at a huge number of sites in a cost-effective manner. There are, however, some problems with these types of systems that must be overcome before they can be successful in their new role as water analyzers.

These devices rely upon microfluidic techniques to draw samples and perform analyses. The sample size is extremely, and this can lead to problems in regard to testing for biological agents. Unlike chemical agents that tend to be fairly evenly dispersed in a water sample, bioagents are discrete particles that do not have an equal probability of being found in every sample. Also, in many cases, for bioagents a single infective particle can represent a lethal dose. These factors have led the EPA and other regulatory bodies to require large samples when testing for microorganisms to ensure their absence in a water supply. For example *E. coli* testing requires a 100 mL sample and testing for *Cryptosporidium* may require a sample of up to 1,000 L. This requires extensive sample pretreatment before a micro-type device scan be utilized for this testing. Extensive filter systems and centrifuge techniques have been utilized but can be cumbersome and inefficient. Also, these methods tend to accumulate superfluous debris as well as the target organisms, which can be a problem. A number of newer techniques such as dielectric focusing and manipulation of particles with sound waves are being investigated but have not as yet reached the commercial market place.

Beyond the problems caused by sample concentration, the distribution system, by its very nature, is not a friendly environment for such techniques. The aging distribution infrastructure is plagued with the problem of particulate matter. Rust particles and debris of various sorts is a common component of the water in these systems. These particles could easily block the microchannels in these devices. Attempts to prefilter the sample could alter the characteristics of the sample. Another problem with these devices is that, as they are currently being designed and deployed, they are discrete analyzers that are designed to detect specific toxins or classes of toxins. They could be thwarted by the use of a toxic substance that the instrumentation was not designed to detect.

Bulk Parameter Monitoring

Bulk parameter monitoring is the method of monitoring common water quality parameters and then looking for anomalies that may be indicative of a water contamination event. Immediately after 9/11 the concept of deploying common sensors to act in just such a manner was investigated for water security monitoring. A number of government (EPA, 2006), academic (Byer and Carlson, 2005), and private industry studies (Kroll, 2002) evaluated various sensors to see if they would respond to the contaminants most likely to be used by a terrorist in an attack.

Various instrument manufacturers have developed multiple parameter water quality monitors for both source water and distribution system water. These systems encompass a diverse selection of different sensors and can be tailored to meet monitoring needs.

The current state of bulk parameter online monitoring with existing instrumentation is that significant actual events should be detectable. The problem is what to do with all of this data. Enormous amounts of streaming data need to be processed. Another problem is the minute-to-minute variability that is present in a system. How are we to determine if alterations in water quality parameters are significant against a background of dynamic changes? In the words of an anonymous sixth grader, "I have heard that you can tell what time it is by looking at the sun. I myself have never been able to make out the numbers." Unless a full-time team of statisticians is to be employed to make sense of this information, there is a need for intelligent algorithms to streamline the process. Intelligent algorithms should be capable of detecting the subtle changes in bulk parameter readings that are indicative of an incursion into the system. They should also be capable of discriminating the unique pattern of responses that are elicited by different classes of agent. These differences may

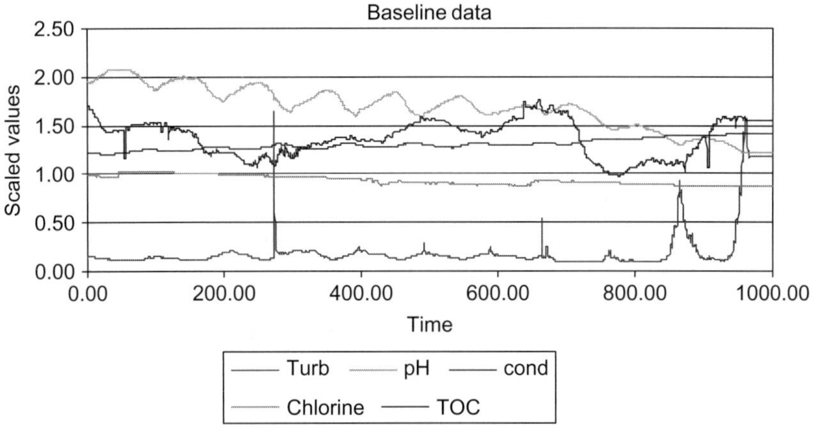

FIGURE 9 Actual real-world baseline date. The variability in bulk water parameters that is common in the distribution system requires that any algorithm should contain a workable baseline estimator. Graph courtesy—Hach HST (see Plate 5 of Color Plate section).

be enough to identify the class of an event and possibly fingerprint the most likely members of that class.

An assortment of sophisticated algorithms for interpreting online data and recognizing threats is being developed by a number of private and public entities including Sandia National Labs and the EPA in their Threat Ensemble Vulnerability Assessment (TEVA) program. One such commercially available system designed by HST makes use of five common bulk parameters that are monitored simultaneously in real time. The parameters that are monitored are pH, conductivity, total organic carbon, turbidity, and residual chlorine. When measured in real time, these parameters can show a lot of variability in a given system. That is why a baseline estimator that is sensitive to small perturbations and yet is resilient enough to not be constantly alarming because of normal fluctuations is required when developing such a system. Many classical methods of baseline determination result in poor sensitivity or high false alarm rates. The proprietary baseline estimator used in this system seems to address these problems (Figure 9).

In the system as it is designed, the signals from all of the instruments are processed from a five-paramater measure into a single-scalar trigger signal in an event monitor computer system that contains the algorithm. The signal then goes through the crucial proprietary baseline estimator. A deviation of the signal from the estimated baseline is then derived. Then a gain matrix is applied that weights the various parameters based on experimental data for a wide variety of probable threat agents. The magnitude of the deviation signal is then compared to a preset threshold level. If the signal exceeds the threshold, the trigger is activated. Figure 10 shows the same data from Figure 9 processed through the algorithm.

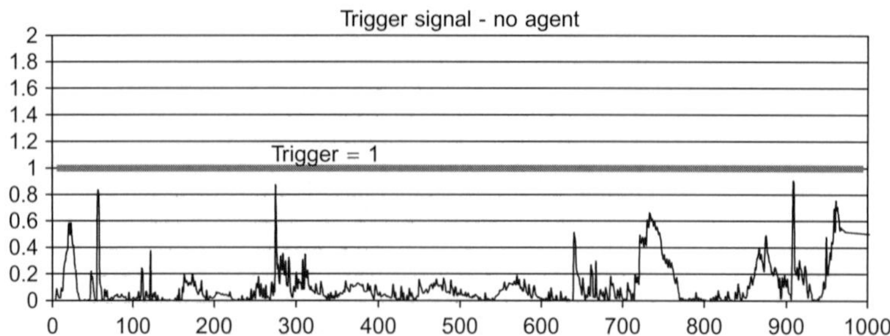

FIGURE 10 The noisy data from Figure 9 become easy to interpret when processed through an algorithm. In this case, no significant events above a threshold of one are occurring; therefore, no trigger is initiated (see Plate 6 of Color Plate section).

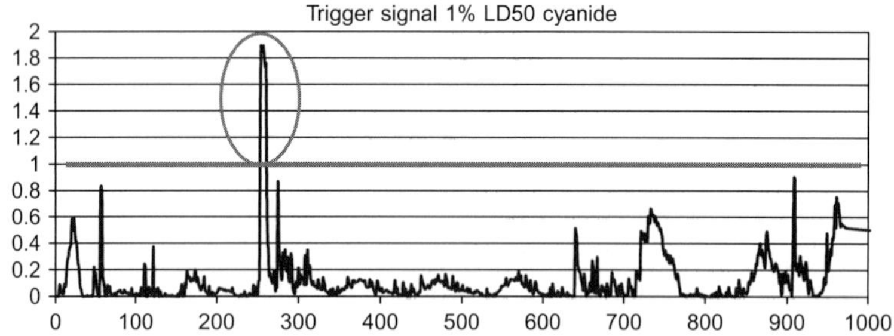

FIGURE 11 The algorithm system's ability to differentiate and trigger on low levels of contaminants against a noisy background is shown (see Plate 7 of Color Plate section).

Even with noisy data, the system does not trigger at a threshold level set at 1. Therefore, during normal operation, with no agent present, the process deviation should not be large enough to produce a trigger signal >1. However, when the data for a cyanide incursion at 1% of the LD-50 or approximately 2.8 mg/L is superimposed on the system, the trigger level of 1 is easily exceeded (Figure 11). Other contaminants exhibit similar results.

The unknown alarm rate, when the system is tracking real world data, is also quite low. The system is equipped with a learning algorithm, so that as unknown alarm events occur over time, the system has the ability to store the signature that is generated during the event. The operator can then go into the program and identify that function and associate it with a known cause such as the turning on of a pump or the switching of water sources, etc. The next time that event occurs, it will be recognized and identified appropriately. Over time as the system learns, the probability of an unknown alarm that has not been previously encountered and identified will continue to decrease and will eventually approach zero. The probability of an unknown alarm because of a given event depends on the frequency of the occurrence of such an event and the time that the algorithm has had to learn that event. Events that occur frequently will be quickly learned whereas rare or singular events will take longer to be learned and stored. This results in a fairly rapid drop off in the number of unknown alarms as common events are quickly learned.

The deviation vector that is derived from the trigger algorithm contains significantly more data than what is needed to simply trigger the system. The deviation vector's magnitude relates to concentration and trigger signal, whereas the deviation vector direction relates to the agent characteristics. Seeing that this is the case, laboratory agent data can be used to build a threat agent library of deviation vectors. A deviation vector from the water monitor can be compared to agent vectors in the threat agent library to see if there is a match within a tolerance. This system can be used to classify what agent is present. Each vector results in a vector angle in *n*-space. The fact that the direction of the vector is unique for a given agent allows the use of an algorithm to classify the cause of a trigger being set off. When the event trigger is set off, the library search begins. The agent library is given priority and is searched first. If a match is made, the agent is identified. If no match is found, the library of learned responses called *the plant library* is then searched, and the event is identified if it matches one of the vectors in the plant library. If no match is found, the data are saved and the operator can enter an ID when one is determined. The agent library is provided with the system, and the plant library is learned on-site.

This type of system has unique advantages that should be noted. First, as soon as the system is turned on, it will be actively working and will have the ability to trigger and classify immediately if the signature of a known threat agent is encountered. Second, if a completely unknown agent is introduced at levels that exceed the threshold signal level, the system will trigger and classify it as an unknown agent that warrants further investigation. This attribute of the system is unique and the only available method that will trigger an alarm on contaminants not previously known to the system or its developers.

There has also been conjecture that perhaps simply monitoring pH and chlorine levels is adequate for a security system. Laboratory experimentation has generated data that indicate a significant number of compounds that are threat agents that would be expected to respond with a change in chlorine levels, in fact, do not. This may be due to complete unreactivity or extremely slow kinetics of the reaction between chlorine and the compound. This effect may be exacerbated in systems that use monochloramine rather than free chlorine as a residual disinfectant. This adds importance to the collection of experimental data and the addition of supplemental functions above and beyond chlorine and pH to a monitoring system. To enhance these measurements, other parameters such as total organic carbon are needed.

A number of similar multiparameter measurement platforms without the addition of intelligent algorithms have been evaluated for such applications (EPA, 2006). These systems appear to be a good choice for detecting water quality excursions that could be linked to water security events. There are a number of advantages in using such systems. The chief advantage is that these instruments are not new. They are common everyday parameters that the average industry worker is quite familiar, thus adding a degree of comfort in operations not afforded by other new technology. As existing technologies, these instruments have been proven to be robust and dependable in prior field deployments. They represent measurements that would be of interest and use to water utility personnel above and beyond their role as water security devices. This in turn could allow improvements in system operation that may result in cost savings and definitely will result in a higher quality product being delivered to the consumer.

One of the largest advantages to this type is the multiparameter array's ability to detect such a wide variety of potential threat agents from metals to organics to bioagents. The ability to trigger on unique unknown events is also a major advantage. Some of the disadvantages are that there are some events that occur during normal operation that may trigger an unknown alarm. This, however, can be an advantage if the information is used to streamline operational procedures and lower costs while improving quality. Nonetheless, this learning phase does generate "unknown" alarms associated with normal system maintenance and requires an input of time and effort to investigate and classify these alarms, so that they can be placed in the plant library. Another disadvantage of such systems is that while they detect biological events, they are not as sensitive to such events as some other methods. The majority of the detection capability comes form the growth media that may or may not be cointroduced along with the bioagents. They do not perform cell counts nor do they carry out individual bacterial identification. One bioevent tends to look pretty much like another. Such systems are not likely to detect low levels of bacteria in the system and, for the EPA requirement of <1 coliforms per 100 mL sample, they would not respond. However, one of the likely forms a bioagent attack could take would include growth media both from an ease of use point and as a means to degrade chlorine levels until the bacteria could survive. In these cases, the instruments would respond.

Another problem has to do with deployment. Many of these instrument packages tend to be somewhat large and require a suitable site for deployment. Many also generate a waste stream that needs to be dealt with. These size and waste constraints can limit where these types of systems can be deployed. There are, however, options for other means of measuring these parameters than those of traditional wet chemistry and optics. These include electrochemical and microscale devices that can be inserted directly into pipes. Microchemical-based devices tend to suffer from the problems of robustness detailed in the section Sensor Arrays and Lab-on-a-Chip Technologies. These and other electrochemical methods tend to offer less sensitivity than more traditional means of measuring bulk parameters. They may be more constrained as to what water conditions they require for proper functioning (e.g., electrochemical chlorine measurement may only be effective in a limited pH range); however, they may be the only option for some deployment scenarios.

OTHER ONLINE TECHNOLOGIES

Although the three technologies discussed earlier are the best currently available options for dealing with the wide diversity of potential threat agents that may be encountered in a deliberate contamination event, other new and emerging online technologies may be better at detecting specific threats or classes of threats.

UV Absorption and Fluorescence

The tendency of various compounds to absorb light in the UV spectrum and for some materials to fluoresce when exposed to this light can be utilized as a detection mechanism. Such methods have long been used in air to detect various biological contaminants such as anthrax.

One instrument the Spectro::Lyser™, produced by Messtechnik GmbH of Austria, uses UV absorption at various wavelengths to detect organic contaminants. They use algorithms to interpret the incoming absorption spectra to determine when a contaminant is present. The system can detect organics that have an absorption signal in the UV range; it allows up to eight different alarm parameters to be set. The identification of a single substance or group of substances is limited to those that are detectable in the UV spectrum and implemented in the setup procedure. While it can detect many substances such as phenols, benzene, toluene, xylene, some pesticides, some nerve gases, oils, and others, it would be ineffective in detecting short-chain aliphatic compounds or inorganic compounds such as metals (S::CAN).

Particle Counting and Characterization with Optical Methods

There are a number of optical methods based on particle counting and/or flow cytometry that are being utilized to characterize potential biological pathogens in the water supply. Many of these methods require the utilization of fluorescent labels to detect the various pathogens. This requires an added step and the complex problem of developing suitable labels.

One method that does not rely upon labels is the BioSentry™ device produced by JMAR Technologies Inc. The BioSentry™ device uses laser-produced, multiangle light scattering (MALS) technology to generate unique microorganism bio-optical signatures. Similar to a laser turbidimeter, the device uses lasers to interrogate a water sample and analyze how a particle in the water refracts the light.

The difference lies in the "multiangle" part in the MALS technology. Rather than just reading at an angle of 90°, a number of different angles are monitored at once. This allows the generation of a three-dimensional (3D) pattern that represents the structure and size of the particle in the laser's path. This allows a pattern to be formed that is representative of the particle in question in the same way that a shadow of an object gives some indication as to its size and structure. This pattern can then be compared to a library of patterns for different types of organisms. The pattern can then be classified using JMAR's pathogen detection library (JMAR).

This appears to be an effective method for monitoring water for biological contamination. However, it would be ineffective against chemical contaminants. Another potential drawback is sample size. Because of the very small path capable of being monitored by the laser, only a small sample can be analyzed. It would be quite possible to miss bacterial or protozoan contaminants that were present at very low concentrations. This instrumentation would be able to sense large contamination events, but, chances are, low-level contamination events may be missed.

Gas Chromatography

Various manufacturers have modified gas chromatography methods to be online tools that work in a batch mode. Gas chromatography is a chromatographic technique that can be used to separate organic compounds that are volatile. A gas chromatograph consists of a flowing mobile phase, an injection port, a separation column containing the stationary phase, a detector, and a data recording system. The organic compounds are separated because of differences in their partitioning behavior between the mobile gas phase and the stationary phase in the column.

One such instrument is the INFICON Scentograph CMS200 portable water and air analysis/monitoring. The Scentograph analyzes VOCs in water using a modified EPA purge and trap protocol. This is made possible by the SituProbe, which performs the purge in the water. No pumps, valves, or cells are exposed to the water matrix, eliminating the need for sample pretreatment or filtration. Since the sample matrix does not affect the system's performance, even difficult water samples can be analyzed (Inficon).

The largest drawback to this technique is the limited scope of compounds that are detected. Only volatile organics are amenable to being analyzed by this method. Also, some of the instrumentation can be touchy and the cost per deployed unit can be quite high. This technology may actually be of more use as a confirmatory technology, when the presence of a volatile organic contaminant is suspected.

Technologies Currently More Suited for Field Confirmatory Analysis

There are a number of technologies that are being developed or are already commercially available that have not been adapted for online use. Much of this work has been funded by the Department of Defense (DOD) as rapid checks for tactical water supplies in battlefield situations. This water is generally prepared via special treatment processes such as reverse osmosis (RO) and as such tends to be relatively consistent and static in its characteristics. Some of these technologies may have more difficulty when exposed to the variability of water supplies treated to municipal standards. While some technologies that can be used either online or in the field have already been described, the following technologies at this point are field use only, although some of them may be adapted for online monitoring in the future.

Immunoassays

Bacteria and biotoxins are not always easily detectable through conventional means. Immunoassays make use of the interaction between antibodies and antigens to detect the presence of specific organisms or compounds. These tests come in various formats and are capable of detecting a wide variety of bacteria, viruses, biotoxins, and specific chemicals. One of the most common formats for these tests is that of a lateral flow assay.

We are all familiar with the lateral flow format as it is the same one that is used in commercially available home pregnancy tests. After a sample is applied to the sample reservoir end of the strip, the liquid begins to migrate down the length of the strip through wicking and capillary action. In the course of moving down the strip, it comes into contact with regions of the test strip that have specific antibodies for the antigen expressed by the target being tested for impregnation on them. These antibodies are tagged with colored or fluorescent labels. They reach a certain area in the strip that binds using other or the same antibodies. The result is a colored or fluorescent line if the target antigen is present. Usually, a control line is included to verify that everything has run properly. These tests are quite simple to use and can usually be read with the naked eye (Figure 12).

A number of manufacturers produce these sorts of test strip assays for the analytes of interest. One of the problems with this sort of assay is that they are very specific for the antigen being tested. Therefore, you need to know what you are looking for and run the appropriate test. Another problem is specificity related. Cross-reactivity with other microbes or antigens can result in false positives. A final problem with these tests is that their sensitivity is not always all that could be desired. Some compounds such as botulinum toxin can be detected at fairly low levels, whereas others such as ricin require more of the compound to be present. Also, bacterial detection levels are usually fairly high. This means that it is advisable to use a sample preconcentration device or preculture before performing the test. There are several preconcentration methods currently under study, but they are presently commercially unavailable. Preculture methods can be time-consuming and may not be the best choice for emergency response situations.

Polymerase Chain Reaction Technology

Polymerase chain reaction (PCR) is a technique for detecting living organisms by extracting and multiplying the DNA specific to that organism. The technique allows a small amount of the DNA molecule to be copied over and over, thus amplifying it many times in an exponential manner. With more DNA available, analysis is made much easier. The DNA can be detected via various methods including fluorescent gene probes, fluorescent melting curves, or electrophoresis. PCR is commonly used in medical and biological research labs for various tasks. PCR is not a simple procedure and normally relies on advanced laboratory techniques.

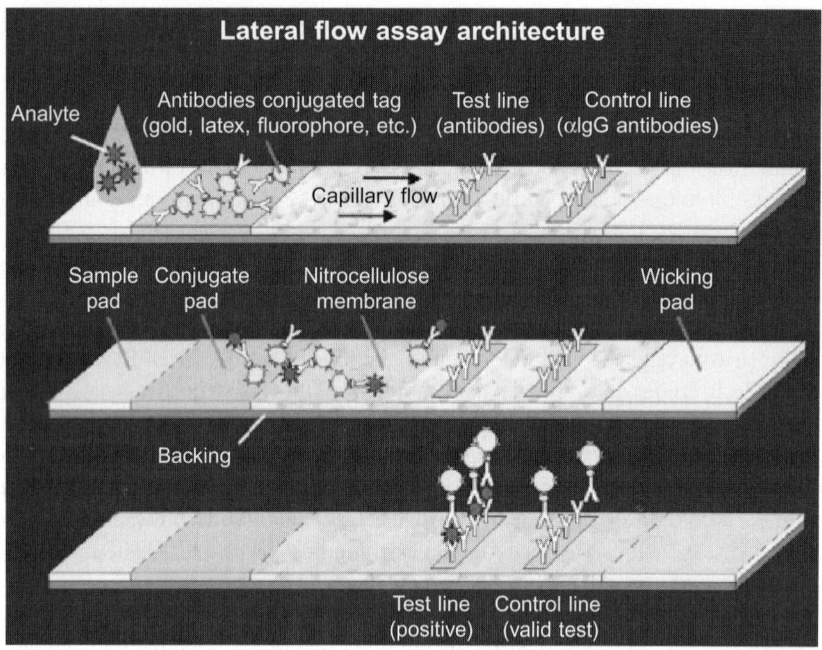

FIGURE 12   The quantitative lateral flow assay (QLFA) is a "test strip" for identifying biological organisms in a sample, which flows from the sample pad to the wicking pad. On the conjugate pad, specific antibodies (Y shapes) tagged with chemical markers (ovals) bind to the target antigen (sunbursts) in the sample and flow toward the wicking pad. At the test line, other immobilized antibodies bind the antigens to produce a positive test result, revealed as fluorescing color. A control line antibody confirms that the test ran successfully; that is, the sample flowed through the length of the test strip. Courtesy—NASA. http://www.spaceresearch.nasa.gov/general_info/homeplanet.html

In recent years, manufactures have developed various automated techniques that remove much of the expertise required to perform this technique and has allowed it to move into the field and be operated by relatively unskilled technicians. One of these methods is the Ruggedized Advanced Pathogen Identification Device or RAPID manufactured by Idaho Technologies of Salt Lake City, Utah. It uses freeze-dried reagents and can screen for up to eight different targets at the same time. These targets can be chosen from a list of potential viral or bacterial pathogens. This is a very sensitive technique and can detect the presence of even a small number of organisms in a sample, for example, as few as five *Bacillus anthracis* cells.

Another, even smaller, version is produced by Idaho Technologies and is known as the RAZOR. It can detect up to 12 analytes at a time and also uses freeze-dried reagents. These instruments are rugged and easy to use. They are also capable of detecting very low levels of the analytes in question. The main problem has to do with price. The RAPID unit cost around $55,000 with reagents for each test running $50.

Adenosine Triphosphate Detection

Adenosine triphosphate (ATP) is the component found in cells that is responsible for energy transfer. Therefore, all living cells contain ATP, and it should be possible to look for changes in ATP levels in water as an indication of biological contamination. ATP measurement has long been used in the clean room industry as an indicator of proper cleaning and sterilization of work surfaces.

Several manufactures have adapted this test for use in water samples. These systems are usually based on the ATP's role in providing the energy source for bioluminescence. This method is a good candidate for detecting gross changes in the level of ATP present in a sample. The problem lies in the fact that the water in our distribution system is not sterile and always contains some level of ATP. It is therefore imperative to be well aware of baseline as in toxicity testing. There is also the problem of distinguishing if a rise in ATP levels is due to a general slough-off of biofilm or a bacterial attack. Some differential methods for lyses of the cells of certain organisms are currently under study, which would make this method more specific for the designated bacterial types.

Rapid Tests for Cholinesterase-Inhibiting Substances (Nerve Agents and Pesticides)

There are a number of rapid tests for detecting substances that inhibit the activity of the enzyme cholinesterase. One type is based on a test strip that can be used to detect nerve agents and pesticides available from Severn Trent services. These test strips work on the basis of inhibition of the enzyme acetochlinesterase to orchestrate a color change in a dye. Many nerve agents and pesticides are capable of inhibiting this change and are thus detected by the strips.

The test is a qualitative test for the detection of pesticides, based on their inhibition of cholinesterase. One side of the ticket contains a disk that is saturated with cholinesterase, an enzyme present in most living organisms, except plants, and whose main function is to control muscle performance. If the enzyme is altered or dies, so does the organism. Pesticides can inhibit an organism's ability to produce cholinesterase, and therefore, kill the organism. If enough pesticide is present in the tested sample, it will inactivate the cholinesterase that is chemically bonded to the ticket and prevent a chemical reaction which, when pesticides are absent, turns the disk blue. A white color result indicates a positive result for the presence of pesticides or nerve agents. In the absence of pesticides or nerve agents, the cholinesterase hydrolyzes an ester to form a colored compound. Inhibiting compounds interfere with the reaction and stop hydrolyzation, preventing further color development. This test is very sensitive to nerve agents; pesticides, however, are usually designed in such a way as to decrease their ability to inhibit cholinesterase. They often have a moiety that needs to be oxidized to increase its sensitivity to the test. This oxidation occurs spontaneously in chlorine-treated water but requires an additional reagent addition in neutral or reducing waters such as surface or groundwaters with no chlorine.

ICX agentase produces another iteration of this type of test. This technology is based on the ability to effectively incorporate enzymes within polyurethane foams. Agentase has effectively extended this platform technology to multiple enzyme systems with a diverse range of applications. Numerous benefits such as improved stability, reusability, and environmental resistance are incurred when enzymes are effectively copolymerized within polyurethane polymers.

The agentase chemical agent detection (CAD) kit is a highly selective measurement system to detect chemical warfare agents. The CAD kit measures not only nerve agents but also uses other enzymes for blood agents and blister agents on surfaces and in liquid samples. The increased stability imparted to the enzymes used in these methods lends credence to the possibility of placing a system based on this technology online. Agentase has undertaken such a program.

A bench top prototype model has proved to rapidly respond to contamination in water at relevant concentrations at live agent tests conducted at government facilities. It has shown to be resistant toward chemical and environmental interference, and to operate for extended periods without user intervention. Further development of an online system is under way (Agentase).

Infrared Spectroscopy

The fact that only volatiles are detected by gas chromatography may be addressed by using it in conjunction with infrared (IR) spectroscopy. IR spectroscopy has traditionally been limited in water because the water in the samples interfered directly in the test method. This meant that aqueous samples that contained less than about 10% of products were not measurable by IR spectroscopy. Smith Detection's SensIR Technologies produces a portable IR spectrophotometer called the HazMat ID™. This system uses Fourier transform IR attenuated total reflection spectroscopy. It has been designed to detect and identify weapons of mass destruction, toxic industrial chemicals, narcotics, and explosives in nonaqueous samples in a HazMat role. To supplement this ability and expand it into aqueous samples, the manufacturers have developed a system called the ExtractIR™. This tool allows the extraction of nonvolatile organic compounds from water, so they can be analyzed via IR. Even with this extraction system, detection limits are fairly high in the order of 100 ppm in water (Smiths Detection).

Multiparameter Hand-held Devices

Many different versions of multiparameter monitors reduced to a lab-on-a-chip configuration are starting to become available. One such version is the WaterPoint 855 multiparameter water-quality analyzer produced by Sensicore of Ann Arbor, Michigan. This chip-based instrument is a hand-held device capable of measuring 14 separate parameters in one quick test. These parameters include

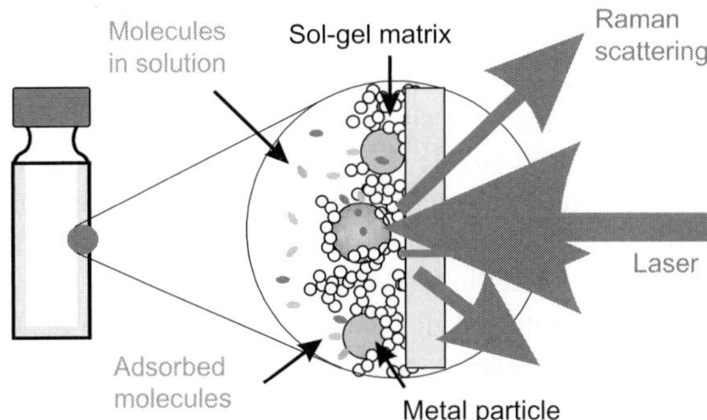

FIGURE 13 Metal molecules coating the inside of sample vials result in an increase in Raman scattering, allowing this method to be used for trace detection. Courtesy—Real Time Analyzers.

pH, ORP, conductivity, total dissolved solids, temperature, free chlorine, total chlorine, monochloramine, calcium, total hardness, carbon dioxide, total alkalinity, ammonium, and Langelier saturation index (Sensicore).

Research is currently under way to develop a communication module that could directly download field measurements to a central location where they can be compiled for decision making in a more real-time mode. This instrument offers the capability of quickly measuring several parameters at once. Drawbacks are that many of these parameters are not independent, such as conductivity and total dissolved solids, which are directly related. Also, many of these parameters have little or no bearing upon security monitoring. Performing superfluous tests adds little or no value to an instrument and only increases the per test cost. It may be simpler and more cost effective to simply perform the tests of interest in separate analysis in a more conventional way using simple electrochemical or colorimetric methods.

Surface-Enhanced Raman Spectroscopy

Surface-enhanced Raman spectroscopy (SERS) is an effect in which the Raman scattering is found to be greatly enhanced when it is close to a rough metal surface. The enhancement can be huge (10^{14} or so), and it enables Raman spectroscopy to be a sensitive technique. It is now known that SERS is observed for molecules found close to silver or gold nanoparticles because of surface plasmon resonance. Other metals may be used, but with a reduction in enhancement. The mechanism by which the enhancement of the Raman signal is provided is from a local electromagnetic field enhancement provided by an optically active nanoparticle (Figure 13).

Work is progressing toward development of a real-time analyzer employing these techniques. They have shown it to be effective in detecting warfare agents and cyanide compounds at the levels of interest. Work is progressing toward development of a real-time analyzer employing these techniques.

Ion Mobility Spectroscopy

Ion mobility spectroscopy (IMS) is a technique for identifying and measuring volatile compounds. An ambient air or vapor sample is drawn over a semipermeable membrane. Smaller volatile compounds pass through the membrane into the detection cell, where the sample is ionized by a weak plasma formed by a nickel radioactive source. The ionized sample molecules drift through the cell under the influence of an electric field. An electronic shutter grid allows periodic introduction of the ions into a drift tube where they separate, based on charge, mass, and shape. Smaller ions move faster than larger ions through the drift tube and arrive at the detector sooner. The amplified current from the detector is measured as a function of time and a spectrum is generated. A microprocessor evaluates the spectrum for the target compound, and determines the concentration based on the peak height. IMS is used in explosives detection equipment at airport security checkpoints. There are several portable IMS sensors for chemical detection, but all have been designed for use with air/vapor samples.

Some companies such as Smiths Detection have developed thermal desorption modules for their instruments that allow the testing of liquid or solid materials. For liquid samples, either temperature controlled ramping to evaporate the sample or a fiber solid-phase microextraction (SPME) probe is needed (USEPA, 2005).

The problem in IMS has always been resolution of the complex mobility spectra resulting from the inherent sensitivity of the method. The broad, tailing, and sometimes overlapping peaks of the spectra have necessitated the development of complex peak deconvolution and recognition algorithms or the introduction of preseparation methods, which lead to loss of sensitivity and increased analysis time. The challenge in IMS is to maintain the chemical sensitivity and response time while increasing the spectral resolution to aid in accurate identification of target analytes.

In common IMS instruments, spectra are generated by pulsing open the entrance to the drift region for 0.2 ms and then monitoring the ion current after a 20-ms drift time period. This brief entrance pulse represents only 1% of the 20-ms duty cycle. A longer "gate" pulse would allow more ions to be collected and increase the signal, but short entrance pulse duration is necessary to prevent unacceptable peak broadening and decreased resolution. The poor resolution typical of conventional IMS instruments is the result of this trade-off for increased sensitivity. Even so, acceptable signal-to-noise (S/N) is achieved by repeating and storing many scans and then summing the signals with a computer.

Signal-averaging IMS has limited ability to resolve adjacent peaks in a complex spectrum, which is critical to unambiguous sample identification and elimination of false positives. An alternative to the signal-averaging methodology is Fourier transform ion mobility spectrometry (FT-IMS). In this mode of operation, two ion gates are employed—an entrance gate and an external exit gate. The entrance gate admits ions into the drift region of the spectrometer in a manner similar to signal-averaging IMS with the distinction being the 50% duty cycle of the gates in FT-IMS. The downstream exit gate is pulsed synchronously with the entrance gate at increasing frequency to interact with the flowing ion stream and generate a frequency domain interferogram.

This interferogram that contains velocity information about all of the ions in the spectrum is then fast Fourier transformed to recover the complete time domain mobility spectrum. The increased S/N results from the 50% duty cycle of the gates. Also, ion–molecule reactions and labile clustering during ion transit through the drift tube results in peak broadening because of random variations in ion velocities. The phasing action of the gates in FT-IMS eliminates this noise signal and greatly improves spectral resolution (Tarver, 2000). Like IR methods, the sample needs to be in the vapor phase so heating or nebulization is required.

Surface Acoustic Wave Technology

Surface acoustic wave technology (SAW) technology has been used for decades in transceiver technology and cell phone technology. For chemical detection, SAW sensors can be configured in a microarray, with each element uniquely coated. Mass changes in a subset of elements because of interaction with a particular volatile chemical causes surface acoustic waves ($\sim 10\,\text{Å}$ in amplitude, 1–$100\,\mu\text{m}$ in wavelength), which are detected by piezoelectric materials.

The subset of elements that respond to a specific VOC can be recognized by software included in the sensor, allowing for a diverse list of detectable analytes.

Each sensor is coated with different polymers that provide a multipattern sensor response (fingerprint) to indicate the presence of contaminants in vapor samples. These sensors can detect and identify trace amounts of chemical warfare agents, including nerve and blister agents, and can be configured to detect phosgene and/or hydrogen cyanide (USEPA, 2005).

SUMMARY AND CONCLUSIONS

Monitoring is a critical component of any water security program. There is no other feasible way to address the severe vulnerability presented by the threat of an intentional contamination event especially in the distribution system. It is imperative that early detection of any such event be achieved to decrease the horrendous potential for mass casualties. Although preservation of human life is the number one priority, it is not the only imperative.

The ability to contain and isolate an incident is critical in limiting the number of casualties, but it is also exigent to limit cleanup of any incident. The anthrax cleanup for the Hart Office building after the contaminated mail incident cost the EPA over $27 million from its super fund site. It is possible that some agents could not be cleaned up and piping will need to be replaced.

This could be a very expensive proposition when it is considered that not only main pipes may need to be replaced but some household plumbing as well. Also, if the agent is widely disseminated in buildings because of aerosolization, many structures may need to be abandoned. Therefore, the need to rapidly detect and contain is critical in reducing casualties and in limiting cleanup costs.

As of today, no perfect monitoring system has been designed. The choice of what to deploy is not necessarily an either-or decision. The best choice may be a network configuration that deploys different types and cost ranges of sensors in different areas to give the optimum in coverage and capabilities. Although not every point will receive complete protection, a network approach has the best chance of detecting an event early in its onset and alerting the operators of the system so that they can make the crucial decisions that will be needed to limit the damage being done. If an attack is detected early, consumers can be warned not to use the water. Also, though the turning off of valves, it may be possible to isolate the contaminant plume to a small area before the entire system becomes unusable.

In an attempt to bring all of the various aspects of water security monitoring into a working system, the EPA has launched a pilot program to design and build such a system using off the shelf technology in a number of pilot communities. This program is called Water Sentinel. Direct monitoring of water quality is not the only approach that is being investigated to provide detection and warning of an attack. Syndromic surveillance is a concept that comes originally from the medical profession. In the medical case, the term refers to surveillance using health-related data that precede diagnosis and signal a sufficient probability of a case or an outbreak to warrant further public health response. Though historically syndromic surveillance has been utilized to target investigation of potential cases in a disease outbreak, its utility for detecting outbreaks associated with bioterrorism is increasingly being explored by public health officials. In the Homeland Security realm, as it pertains to water, syndromic surveillance is the concept of using advanced computational techniques and data mining algorithms to monitor a number of nonspecific indicators of a possible attack. These include such data as hospital admissions, 911 calls, pharmacy sales, and complaints to the utility. These data streams are directed to a centralized computing system that correlates all of the factors and extrapolates the probability of an attack using advanced algorithms. Once an attack has been indicated, appropriate response actions can be initiated to treat the potential victims.

Although much useful information could theoretically be extrapolated from such a monitoring program, there are severe drawbacks. Syndromic surveillance, by its very nature, is directed toward thwarting naturally occurring outbreaks of disease. The results of an intentional contamination event using water as a vector may spread quickly enough to make detection by such a mode redundant and unnecessary. Also, the reliance on such a mode of detection delays the reporting of the hypothetical event until actual exposures have occurred. This may be adequate in cases of a bacterial contaminants that may have a fairly long incubation period and can be treated with antibiotics. It is, however, woefully inadequate in the case of a chemical or biotoxin contamination event.

By the time such an attack is detected via syndromic surveillance, it is too late to do anything to decrease the number of casualties and the damage incurred. Ironically, under Federal Superfund statutes, if industries were to use such means to monitor for public safety, such an approach would be viewed as illegal, or the equivalent of "using the public as guinea pigs." The use of such technology as a stand-alone method becomes nothing more than a means to keep track of damage rather than to prevent it. Syndromic surveillance does have some merit when the stream of data being analyzed includes real-time water quality monitoring results. Recognizing this, the EPA is not relying solely on syndromic surveillance as some have advocated, but rather is using it as a supplement to water monitoring data.

The problem with detecting a water contamination emergency, whether it is terrorist-related or accidental, is how to respond effectively to limit the damage to life and property. The simplest answer is that if we suspect a problem, we will simply shut the water off. This is an unacceptable answer for many reasons. It is unlikely that the public would accept frequent disruptions in their water supplies for false alarms. This requires certainty before we take any action as drastic as stopping supplies.

Some water delivery pipes are in such a poor state of repair that the reduced pressure that would result from the water supply being disrupted could lead to major pipe failure. Other utilities, especially those in large metropolitan areas, cannot shut down because of the necessity of maintaining a pressure head for fire suppression, sanitation, and other needed functions. However, if no action is taken, there is a grave risk of an event causing mass casualties. The problem then is basically one of the controlled responses. We must carefully weigh and balance the problems of overresponding with that of under-responding to ensure that proper steps are taken during each phase of a potential emergency. The key to initiating proper action is confidence in our analytical results. That is why redundant systems using more than one technology, along with rapid confirmatory tests, are critical.

With the current state of technology, there is no need for us to operate our water systems in the unsecured mode of the past. Admittedly, the instrumentation available today is not perfect, but it will show us a clear enough picture to avoid many of the hazards that we would surely encounter if we left the old paradigm in place.

REFERENCES

Agentase Web site. http://www.agentase.com/products.php

Allman, T.P., 2003. Drinking Water Distribution System Modeling for Predicting the Impact and Detection of Intentional Contamination. Masters Thesis. Summer 2003. Department of Civil Engineering. Colorado State University. Fort Collins, CO.

Aquatic biomonitoring, USACHER. http://www.usacher.detrick.army.mil/Aquatic%20Biomonitor%20Product.html

Army Corps of Engineers, Calculations on threat agents and requirements and logistics for mounting a successful backflow attack.

bbe moldaenke Web site. http://www.bbe-moldaenke.de/toxprotect64.html (accessed 5.02.07).

bbe moldaenke. http://www.bbe-moldaenke.co/EN/Biomonitring?daphnia_Fish_Toximeters.html

Bellemare, F., Rouette, M.E., Lorrain, L., Perron, E., Boucher, N., LuminoTox: a rapid, portable and effective tool for toxicity screening in water. Abstract for SETAC 2005. http://www.abstracts.co.allenpress.com/pweb/setac2005/document/?ID = 55641

Byer, D., Carlson, K.H., 2005. Real-time detection of internal chemical contamination in the distribution system. J. Am. Water Works Assoc. 97 (1), 58–61.

CDC emergency preparedness and response Web site. http://www.bt.cdc.gov/agent/agentlistchem.asp and http://www.bt.cdc.gov/agent/agentlist.asp

Dick, R.L., Statement for the Record of Ronald L. Dick, Deputy Assistant Director, Counter Terrorism Division, and Director, National Infrastructure Protection Center, Federal Bureau of investigation Before the House Committee on Transportation and Infrastructure Subcommittee on Water Resources and Environment October 10, 2001.

EPA ETV verification report for eclox. http://www.hach.com/fmmimghach?/CODE%3AECLOXETV7699%7C1

EPA, 2006. Environmental test verification (ETV). http://www.epa.gov/etv/verifications/verification-index.html

GAO-04-29 Drinking Water Security: Experts' views on how future federal funding can best be spent to improve security; October 2003.

Greenbaum, E., Rodriguez, M., 2006. AquaSentinel: a real-time reagentless biosensor system for standoff detection and classification of toxins in source water. Abstract from the south east regional meeting of the Americal Chemical Society, November 1.

Hoover, J.E., 1941. Water supply facilities and national defense. J. Am. Water Works Assoc. 33 (11), 1861.

http://www.medterms.com/script/main/art.asp?articlekey=34093

Inficon Web site. http://www.inficonchemicalmonitoringsystems.com/en/Scentographcms200.html

JMAR Web site. http://www.jmar.com/2004/about.shtml

Kroll, D., 2002. Results of threshold beaker testing on chemical threat agents: is on-line water security monitoring feasible? Internal Hach HST Report. September 12, 2002.

Kroll, D., 2003. Confidential paper, "Mass Casualties on a Budget," 2003, Hach HST.

Kroll, D., 2004. Utilization of a new toxicity testing system as a drinking water surveillance tool. In: Lauer, W.C. (Ed.), Water Quality in the Distribution System. AWWA Press, Denver, CA.

Kroll, D., 2007. The Role of Rapid Toxicity Testing in Security Breach Evaluations. Waterworld, PennWell Publication, Tulsa, OK.

Office of Science and Technology Policy, The White House, "The National Strategy for the Physical Protection of Critical Infrastructures and Assets," February 2003.

S::CAN Web site. http://www.s-can.at/index.php?id=58

Sensicore Web site. http://www.sensicore.com/public/default.aspx

Smiths detection Web site. http://www.sensir.com?newsensir/Brochure?ExtractIR%20Product%20Note.pdf

Tarver, E., Stamps, J.F., Jennings, R.T., Siems, W.F., 2000. External Exit Gate Fourier Transform Ion Mobility Spectrometry. International Society for Ion Mobility Spectrometry. Sandia National Laboratories, Livermore, CA.

U.S. Army Center for Health Promotion and Preventive Medicine, 1999. Short-term chemical exposure guidelines for deployed military personnel.

USEPA Office of Water Office of Science Technology Health and Ecological Criteria Division, 2005. Technologies and techniques for early warning systems to monitor and evaluate drinking water quality: a state of the art review.

Water infrastructure network news. http://www.win-water.org/win_news/112702article.html

Waterborne CBR Agent Building Protection; Hock, V.F., Cooper, S., Van Blaricum, V., Kleinschmidt, J., Ginsberg, M.D., Lory, E., 2003. Proceedings of the National Association of Corrosion Engineers Exposition. Orlando, FL.

Whitman, C., 2001. EPA Press Release Thursday, October 18, 2001. Environmental News Whitman Allays Fears for Water Security; Possibility of Successful Contamination is Small.

Wholsen, M., 2006. Fish sentries monitor for terrorism, Discovery Channel News, September 18, 2006. http://www.dsc.discovery.com/news/2006/09/18/bluegill_ani.html?category=animals& guid=20060918150030&dcitc=w19-506-ak-0001

Groundwater Arsenic Removal Technologies Based on Sorbents: Field Applications and Sustainability

Sad Ahamed and Abul Hussam
Center for Clean Water and Sustainable Technologies, Department of Chemistry and Biochemistry, George Mason University, Fairfax, VA 22030, USA

Abul K.M. Munir
Manob Sakti Unnayan Kendro (MSUK), Kushtia, Bangladesh

Handbook of Water Purity and Quality
Copyright © 2009 by Elsevier Inc. All rights of reproduction in any form reserved.

INTRODUCTION

Existence of toxic levels of arsenic in groundwater and its severe health effects are prevalent in many countries around the world (Bhattacharya et al., 2002; Chakraborti et al., 2002, 2004; Mandal and Suzuki, 2002; Smedley and Kinniburgh, 2002; Mukherjee et al., 2006; Rahman et al., 2006). Drinking of arsenic-contaminated water for a long time causes illnesses such as hyperkeratosis on the palms or feet, fatigue, and cancer of the bladder, skin, or other organs. Figure 1 shows an example of an arsenicosis patient with hyperkeratosis on the palms. It is believed that 1 in every 10 people ingesting high levels of arsenic ($100\,\mu g/L$) could die of cancer triggered by arsenic poisoning. Groundwater is the primary source of drinking water in countries that have high levels of arsenic in water. It is estimated that more than half a billion people in the world may be drinking groundwater containing arsenic. To mitigate the sufferings of people and the public health crisis in the affected areas, supplying clean potable drinking water is the only solution. Assuming groundwater has no toxic organic compounds, the potable water should conform to the inorganic water-quality parameters as shown in Table 1.

The sources of arsenic-safe potable water may be classified into two major categories: (i) alternative sources such as surface water, dug wells, rainwater harvesting, and (ii) filter technologies that can be used to remove toxic arsenic species from contaminated water (Milton et al., 2006). The primary intention of this chapter is to discuss various technologies based on sorption of arsenic species from water as a rapid and inexpensive means for the purification of water.

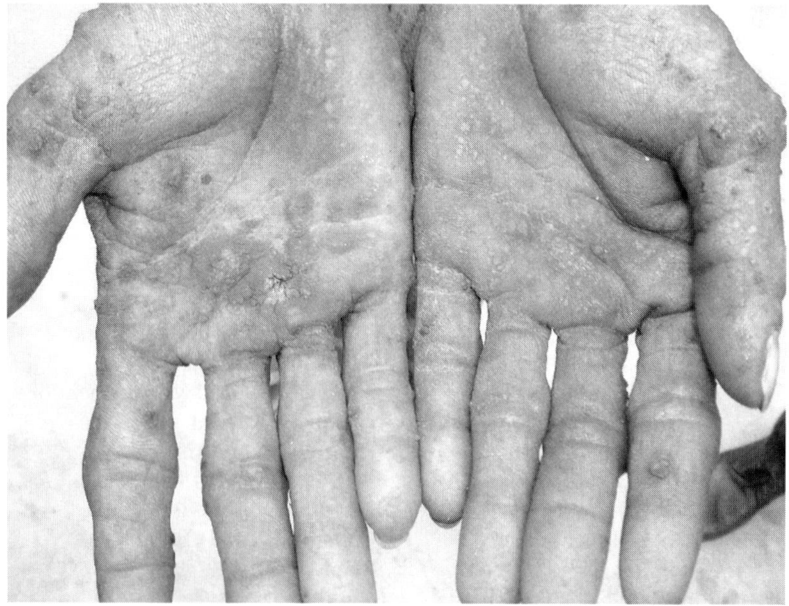

FIGURE 1   Arsenicosis patient with hyperkeratosis on the palm.

In particular, we are interested in iron-based technologies for arsenic removal because they have specific advantages over other sorbents and are environmentally benign. The majority of studies found in the literature were proof-of-concept work, mostly confined to the laboratory and had limited field test data or did not go through the environmental technology verification (ETV) tests by independent organizations. Here, we emphasize the technologies that were tested and used in the field, and some of them have gone through the ETV projects for arsenic mitigation (ETVAM).

The elimination of arsenic from drinking water, an urgent need for millions of people, was the subject of the inaugural Grainger Prize for sustainability, which was funded by the Grainger Foundation and administered by the National

TABLE 1 Water Quality Parameters for Inorganic Species Accepted by USEPA, World Health Organization (WHO), and Bangladesh Standards in Comparison to Typical Groundwater Quality in Bangladesh

Constituent	USEPA (MCL)	WHO Guideline	Bangladesh Standard[1]	Typical Groundwater Composition
Arsenic(total) (μg/L)	10	10	50	5–4,000
Arsenic(III) (μg/L)				5–2,000[2]
Iron(total) (mg/L)	0.3	0.3	0.3 (1.0)	0.2–20.7
pH	6.5–8.5	6.5–8.5	6.5–8.5	6.5–7.5
Sodium (mg/L)		200		<20.0
Calcium (mg/L)			75 (200)	120±16
Manganese (mg/L)	0.5	0.1–0.5	0.1 (0.5)	0.04–2.00
Aluminum (mg/L)	0.05–0.2	0.2	0.1(0.2)	0.015–0.15
Barium (mg/L)	2.0	0.7	1.0	<0.30
Chloride (mg/L)	250	250	200 (600)	3–12
Phosphate (mg/L)			6	<12.0
Sulfate (mg/L)			100	0.3–12.0
Silicate (mg/L)			–	10–26

1 mg/L = 1,000 μg/L.
[1] Bangladesh standard values are given as maximum desirable concentration with maximum permissible concentration in parentheses.
[2] In some wells As(III) concentrations could exceed 90% of the As(total).

Academy of Engineering (NAE). The goal of the challenge was to encourage invention of affordable, reliable, low-maintenance, electricity-free technologies for reducing arsenic in drinking water to an acceptable level for human consumption. In February 2006, NAE announced three winners of the Grainger Prize (http://www.nae.edu/nae/grainger.nsf). The first place was awarded to SONO filtration system, which is based on a composite iron matrix (CIM). The SONO system, which has been extensively tested and used in Bangladesh, meets or exceeds local government guidelines for arsenic removal. NAE recognized this innovative technology for its affordability, reliability, ease of maintenance, social acceptability, and environmental friendliness. The second place was awarded to a team for developing a community water treatment system based on activated alumina. The third place was awarded to Procter & Gamble for its PUR technology, which uses calcium hypochlorite (bleach) to kill a wide range of microbial pathogens and ferric sulfate to remove arsenic through flocculation–precipitation. These innovations generally cover the viable processes for arsenic removal in small-scale application and have potential for scaleup.

The aim of this chapter is to provide general descriptions of arsenic removal mechanisms and current literature review on the techniques in treating or removing arsenic from water. Tools involving surface-complexation reactions, sorption equilibrium and kinetics, field applications of arsenic removal technologies, and an evaluation of sustainability based on technical merits and technology verification criteria are presented.

PHYSICOCHEMICAL BASIS OF ARSENIC REMOVAL

This section is an introduction to theoretical tools necessary to study sorbent materials for arsenic removal. It covers, in brief, some basic reactions and mechanisms for arsenic removal and the dynamics pertaining to arsenic separation in small scale.

Speciation of Arsenic in Water

Arsenic in water exists in the form of arsenious acids (H_3AsO_3, $H_3AsO_3^-$, $H_3AsO_3^{2-}$), arsenic acids (H_3AsO_4, $H_3AsO_4^-$, $H_3AsO_4^{2-}$), arsenites, arsenates, methylarsenic acid, and dimethylarsinic acid (Bodek et al., 1998; Smedley et al., 2002). Inorganic forms of arsenic are the most common species in groundwater. Pentavalent, As(V), species predominate and are stable in oxygen-rich aerobic environments, whereas trivalent, As(III), arsenites predominate in reducing or anaerobic environments such as groundwater (Greenwood and Earnshaw, 1984). Arsenic species are also pH-sensitive to mobilization (pH 6.5–8.5) under both oxidizing and reducing conditions (Smedley and Kinniburgh, 2005; Wang and Mulligan, 2006). The E_h–pH dependence on speciation of arsenic is discussed elsewhere (Ringbom, 1963; Gupta and Chen, 1978; Ghosh and Yuan, 1987; Brookins, 1988; USEPA, 2000). Figure 2 shows pH-dependent speciation of inorganic arsenic species in the drinking water pH range. Assuming 50 μg/L

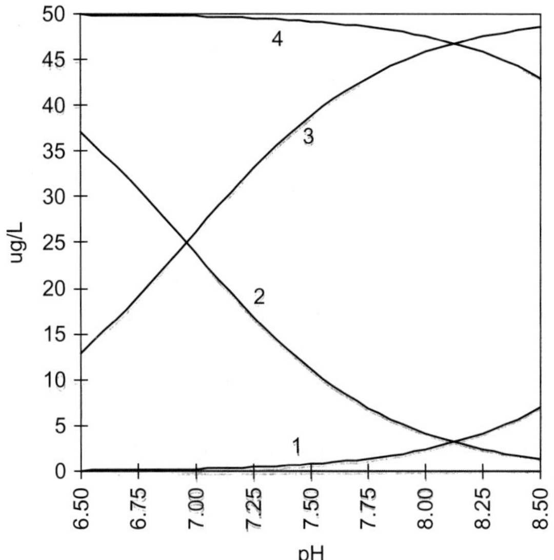

FIGURE 2   Inorganic arsenic species (1, $H_2AsO_3^-$; 2, $H_2AsO_4^-$; 3, $HAsO_4^{2-}$; 4, H_3AsO_3) near groundwater pH range.

each of As(III) and As(V), we calculate the distribution of most stable oxoanion arsenic species as a function of groundwater pH values. Clearly, three major species, H_3AsO_3, $H_2AsO_4^-$, $HAsO_4^{2-}$ are evident. Even at the 50% distribution level, the concentration of H_3AsO_3 remains relatively unchanged at groundwater pH values (6.9–8.0) as the predominant species, whereas, the distributions of $H_2AsO_4^-$ and $HAsO_4^{2-}$ are significantly dependent on the pH. Therefore, an effective filtration system must remove these species quantitatively, leaving total arsenic in the effluent water to $<10\,\mu g/L$.

Surface-Complexation Reactions

Surface complexation of arsenite and arsenate is the primary mode of arsenic removal with solid sorbents that contain iron and alumina. The primary active material in iron-based filters is made from hydrolysis of iron salts or from partial rusting of Fe(0) or Fe(0) amended materials, sometimes through proprietary processes. These sorbents can remove inorganic arsenic species quantitatively through possible reactions shown in Table 2. Infrared spectroscopy (IRS) (Manning et al., 1998) and extended x-ray absorption fine structure (EXAFS) (Waychunas et al., 1993) show that arsenate and arsenite form bidentate, binuclear surface complexes with =FeOH (or =FeOOH or hydrous ferric oxide [HFO]) as the predominant species tightly immobilized on the iron surface. The primary reactions are as follows: $=FeOH + H_2AsO_4^- \rightarrow\, = FeHAsO_4^- + H_2O$ $(K = 10^{24})$ and $=FeOH + HAsO_4^{2-} \rightarrow\, = FeAsO_4^{2-} + H_2O$ $(K = 10^{29})$.

TABLE 2 Chemical and Surface-Complexation Reactions in Iron-Based Filters

Description	Reactions
Oxidation of soluble iron Oxidation of ferrous to ferric through active oxygen species	$Fe(II) + O_2 \rightarrow O_2^- + Fe(III)OH_2^+$ $Fe(II) + O_2^- \rightarrow Fe(III) + H_2O_2$ $Fe(II) + CO_3^- \rightarrow Fe(III) + HCO_3^-$
Oxidation of As(III) (Equations are balanced for reactive species only.)	$As(III) + O_2^- \rightarrow As(IV) + H_2O_2$ $As(III) + CO_3^{\cdot-} \rightarrow As(IV) + HCO_3^-$ $As(III)OH^- \rightarrow As(IV)$ $As(IV) + O_2^- \rightarrow As(V) + O_2^-$
Formation of HFO in the presence of Fe(III) Fe(III) complexation and precipitation	$=FeOH + Fe(III) + 3H_2O \rightarrow Fe(OH)_3$ $(s, HFO) + =FeOH + 3H^+$ ($=FeOH$ is surface of hydrated iron)
Surface complexation of arsenates Surface complexation and precipitation of anionic species As(V) on HFO. log K values are shown in ( ). ψ is the surface potential.	$=FeOH + AsO_4^{3-} + 3H^+ \rightarrow = FeH_2AsO_4 + H_2O$ (29.31) $=FeOH + AsO_4^{3-} + 2H^+ - \exp(-F\psi/RT) \rightarrow$ $=FeHAsO_4^- + H_2O$ (23.51) $=FeOH + AsO_4^{3-} + H^+ - 2\exp(-F\psi/RT) \rightarrow$ $=FeAsO_4^{2-} + H_2O$ $=FeOH + AsO_4^{3-} - 3\exp(-F\psi/RT) \rightarrow$ $=FeOHAsO_4^{3-} + H_2O$ (10.58)
Precipitation of other metals Surface precipitation of arsenate with soluble metal ions if surface concentrations exceed solubility limits. Many metal ions are also quantitatively removed this way.	$=FeOHAsO_4^{3-} + Al(III) \rightarrow = FeOHAsO_4Al(s)$ $=FeOHAsO_4^{3-} + Fe(III) \rightarrow = FeOHAsO_4Fe(s)$ $=FeOH \cdot HAsO_4^{2-} + Ca(II) \rightarrow = FeOH \cdot HAsO_4Ca(s)$ $M(III) + HAsO_4^{2-} \rightarrow M_2 (HAsO_4)_3 (s), M = Fe, Al,$ $M(II) + HAsO_4^{2-} \rightarrow M(HAsO_4) (s)$ and other arsenates $M = Ba, Ca, Cd, Pb, Cu, Zn$, and other trace metals
Surface complexation of silicate species Reactions with iron surfaces and silicates can produce a porous solid matrix with extremely good mechanical stability for long-term use.	$=FeOH + Si(OH)_4 \rightarrow = FeSiO(OH)_3(s) + H_2O$ $=FeOH + Si_2O_2(OH)_5^- + H^+ \rightarrow = FeSi_2O_2$ $(OH)_5(s) + H_2O$ $=FeOH + Si_2O_2(OH)_5^- \rightarrow = FeSi_2O_3(OH)_4^-(s) + H_2O$

Source: Wilkie and Hering (1996), Dzombak and Morel (1990), Schecher and McAvoy (1998), MINTEQA2 Model System (2001), Stephen et al. (2001), and Davis et al. (2002). All surface species are indicated by = X.

These intrinsic equilibrium constants indicate very strong complexation and immobilzation of inorganic arsenic species.

A surface-complexation model (SCM) involves three steps: surface ionization of solid surface, formation of an electrical double layer on the solid surface, and formation of a complex between surface and the ionic species (anionic arsenates). In SCM, the double layer assumed to consist of an inner compact layer and an outer plane in which the diffuse layer starts. The inner layer is where the ions are located to form surface complexes. This model is appropriate for low ionic strength solutions such as groundwater or surface water. The surface potential term ψ can be obtained from the Poisson–Boltzmann equation (Dzombak and Morel, 1990). Computational models such as MINTEQA (MINTEQA2 Model System, 2001) allow a speciation calculation based on SCM when the surface area of the solid sorbent, the double-layer capacity, site types, and site densities are known. The basic principle of MINTEQA is shown in Figure 3a. An example application of MINTEQA, using surface-complexation reactions, in understanding the breakthrough capacity of HFO(s) as solid sorbent for the removal of arsenate is shown in Figure 3b. The figure also shows that a system open to air CO_2 does not change the arsenic removal capacity with HFO. These calculations are based on existing thermodynamic database in the program. Despite the limitations (assumed values for surface area, site density, and double-layer capacity) in using such models, one can gain some preliminary insight on sorbent capacity and possible interferences from ions such as phosphate, silicate, carbonate sulfate on the sorption capacity. The MINTEQA model also allows one to obtain natural attenuation of arsenic species from groundwater, which is critical in formulating synthetic water for filter testing, to identify possible mineral phases in the filter, and to model leaching of species from the mineral-saturated sorbents (Hussam et al., 2003).

Adsorption Isotherm Models for the Evaluation of Sorption Data

Equilibrium-Based Sorption Models

The equilibrium isotherm models describe the adsorption process in terms of mathematical equations. The Langmuir and Freundlich models are the most frequently used ones to fit the isotherm data as shown in Table 3. There exist other models that are unnecessarily complicated, empirical, and provide no total capacity and no predictive advantages over simple models. This is specifically true for sorption of ionic species that form surface complexes and precipitate out on the surface at high concentrations in a continuum.

Kinetic Models for the Evaluation of the Sorption Data

Chemical reactions happen over a range of time from microsecond to years, depending on the type of reaction. Kinetic studies of solutes in aqueous

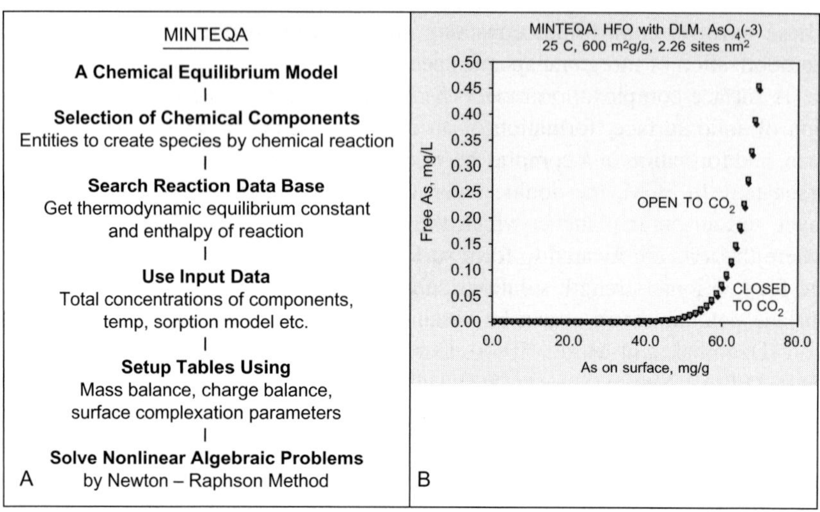

FIGURE 3   (a) Basic principle for MINTEQA-based speciation calculation. (b) Breakthrough of total arsenic (as arsenate) based on MINTEQA calculation. Input arsenic 0.075–7.5 mg/L, surface area 600 m^2/g, and site density 2.26 sites/nm^2. Figure shows that a 50 ug/L breakthrough occurs when the solid HFO sorbed 60 mg As(total)/g sorbent. It also shows that a system open to the atmospheric CO_2 does not alter sorbent capacity.

TABLE 3 Sorption Isotherm Models Used to Study Removal of Arsenic Species from Solution

Isotherm	Equations	Comments
Freundlich (1906)	$q_e = K_d C_e^{1/n}$ Linear form: $\log q_e = \log K_d + (1/n) \log C_e$	An empirical model. Does not predict sorption maximum. Single K_d indicates adsorption energy is independent of surface coverage. Multiple slopes should not be over-interpreted. The reciprocal of "n" is called heterogeneity factor, and its value ranges from 0 to 1. The more the surface is heterogeneous, the closer to zero value of $1/n$ is.
Langmuir (1918)	$q_e = K_l C_e/(1 + \alpha_L C_e)$ Linear form: $C_e/q_e = 1/K_l + (\alpha_L/K_l) C_e$	Monolayer coverage represents maximum coverage, adsorption is reversible, no local movement of adsorbed species, and adsorption energy is independent of surface coverage

Explanation of terms and symbols: q_e = amount of sorbate per unit mass of adsorbent (μg/g), K_d = equilibrium constant indicative of adsorption capacity (L/g), C_e = equilibrium solute concentration in solution (μg/L), $1/n$ = indicative of adsorption intensity in Freundlich equation, K_l = solute adsorptivity (L/g), α_L = a function of adsorption energy (L/μg), and K_l/α_L = monolayer adsorption capacity.

solution and solid adsorbent involve both chemical kinetics and transport kinetics, and in most cases, transport kinetics is the rate-limiting step. The transport phenomenon involves transport in the bulk solution phase, particle diffusion, and film diffusion. The literature is scanty on the later aspect with respect to arsenic sorption on solid, presumably because of our inability to uncouple these processes. Sorption of arsenic species on a solid support follows rate processes, as shown in Table 4. The first-order process rarely depicts the experimental measurements. The pseudo-second-order model appears to fit many experimental data relatively well (Ho and McKay, 1998). It is difficult to impart physical significance of constants in this model. The power function model appears to fit a very wide range of experimental data, but its physical significance is not interpretable unless $x = 0.5$. Rate constants are often normalized to surface area of the sorbent for a relative comparison. There is a tendency in literature

TABLE 4 Kinetic Models Used to Evaluate Sorption Dynamics for Arsenic Species (Only Those Used for Arsenic Species)

Model	Equation	Comments
Pseudo-first order (Lagergren, 1898)	$q_t = q_e[1 - \exp(-k_1 t)]$	A version of first-order process that fits data in dilute solution and monolayer sorption. Extensively used with other contaminants.
Pseudo-second order (Ho and McKay, 1998)	$1/q_t = (1/(k_2 q_e^2))/t + (1/q_e)$	A plot of $1/q_t$ vs. $1/t$ is a straight line. Parameters q_e and k_2 can be estimated from the slope $1/(k_2 q_e^2)$ and the intercept $(1/q_e)$. Appears to fit a large number of experiments.
Elovich (1957)	$q_t = (1/\beta) \ln(\alpha/\beta) + (1/\beta) \ln(t)$	Originally developed for heterogeneous sorption of gases on solids. It appears to work for aqueous species. The slope may depend on solution/solid ratio.
Parabolic diffusion (Weber and Morris, 1963)	$q_t/q_e = R_D t^{1/2} + $ constant	Diffusion is rate limiting on a uniform cylindrical particle surface.
Power function (Kuo and Lotse, 1974)	$q_t = kt^n$	Empirical equation turns into parabolic diffusion equation at $n = 1/2$.

Explanation of terms and symbols: q_e and q_t are the sorption capacity ($\mu g/g$) of the adsorbent at equilibrium and at time t (s), respectively; k_1 is the pseudo-first-order sorption rate constant (s^{-1}); k_2 is the pseudo-second-order sorption rate constant (g s/μg); t is the time (s); α is the initial sorption rate constant ($\mu g/g$ s); β is the desorption constant (g/μg); R_D is a diffusion parameter ($s^{-1/2}$); and k and x are constants.

to over-interpret kinetic models and extract mechanistic details. Such models should be used only for data fitting and interpolating some predictions.

Scaleup Approach: Fixed-Bed Column Design by the Kinetic Approach

This approach considers kinetics of surface diffusion to the inside of the adsorbent pore. The kinetic approach can be used to determine the scaleup size for a known breakthrough volume. If the volumetric flow is low and an instantaneous equilibrium is assumed, the following equation is a good approximation for the breakthrough curve (Noble and Terry, 2004)

$$\ln (C_0/C - 1)\, (k_1\, q_0\, M)/Q - (k_1\, C_0\, V)/Q$$

where C_0 = influent solute concentration, C = effluent solute concentration, k_1 = adsorption rate constant assuming Langmuir isotherm, q_0 = maximum concentration of solute in the solid adsorbent (g/g), M = mass of the adsorbent (g), Q = fluid flow rate, and V = volume of effluent. The left-hand side of the equation vs. V is a straight line from which k_1 and q_0 are obtained. These parameters can be used to calculate the mass of adsorbent for scaleup. A plot of the left-hand side of the equation vs. V is a straight line from which k_1 and q_0 can be obtained. These parameters can be used to calculate the mass of adsorbent for scaleup. Figure 4 shows efficiency of a typical iron matrix–based filter column in laboratory experiments. It shows that water containing 500 μg/L As(total) can be filtered to produce 50 μg/L As(total) for about 2 years using 10 kg of CIM at 80 L/day usage rate. This calculation is based on the Langmuir isotherm and instantaneous arsenic removal kinetics. It does not consider the increased efficiency of iron through slow HFO formation. These calculations should be used only as a guide to screen sorbent material. The actual filter efficiency should be measured in the field with real groundwater that are anoxic and may contain high iron, calcium, and other ions.

ARSENIC REMOVAL PROCESSES

There are several methods available for the removal of arsenic from water. The most extensive review on this subject was reported elsewhere (Dinesh and Pittman, 2007). The available technologies can be placed into few broad categories: precipitation–coprecipitation (Gulledge and O'Connor, 1973; Cheng et al., 1994; Edwards, 1994; McNeill and Edwards, 1995; Scott et al., 1995; Karcher et al., 1999; Felds et al., 2000; Guo et al., 2000; Gregor, 2001; Huang and Rong, 2001; Han et al., 2002; Zouboulis and Katsoyiannis, 2002; Altundoğan and Tümen, 2003; Wickramasinghe et al., 2005), membrane filtration (Farahbakhsh et al., 2004; Košutić et al., 2005; Shih, 2005), adsorption, surface-complexation, and ion-exchange processes (Shen, 1973;

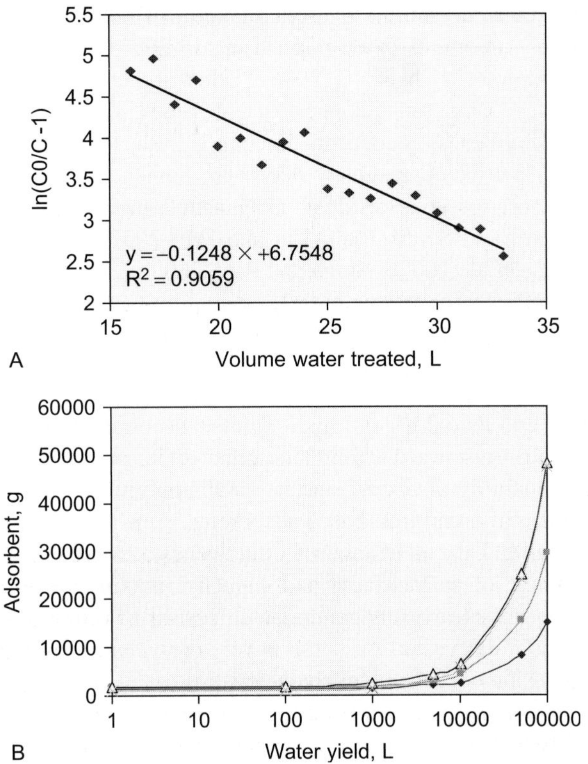

FIGURE 4 (a) Figure shows a typical experimental plot for a laboratory experiment for the sorption of arsenate (1.0 mg/L arsenic) at 5.0 mL/min flow rate on 50.0 g sorbent placed between 10.0 g each of sand layers inside a column (diameter 2 cm). Estimated parameter values: $k = 6.24 \times 10^{-3}$ L/min mg, $q_o = 1.082$ mg/g of sorbent. (b) Scaleup calculation based on 20 L/h flow rate using parameters extracted form above at three different thresholds.

Sorg and Logsdon, 1978; Cheng et al., 1994; Hering et al., 1996, 1997; Joshi and Chaudhury, 1996). Except for ion exchange, the conventionally used sorbents are amorphous iron hydroxide (Pierce and Moore, 1982), HFO (Wilkie and Hering, 1996), granular ferric hydroxide (GFH) (Driehaus et al., 1998), ferrihydrite (Raven et al., 1998), red mud (Altundogan et al., 2002), activated alumina (Rosenblum and Clifford, 1984; Lin and Wu, 2001; Singh et al., 2001), iron oxide–coated polymeric materials (Katsoyiannis and Zouboulis, 2002), iron oxide–coated sand (Thirunavukkarasu et al., 2003a), Fe(III)–Si binary oxide (Zeng, 2004), iron oxide impregnated activated alumina (Kuriakose et al., 2004), blast furnace slug (Kanel et al., 2006), iron–cerium bimetal oxide (Dou et al., 2006), iron-coated sponge (Nguyen et al., 2006), nanoscale zero-valent iron (Kanel et al., 2005; Lien and Wilkin, 2005; Yuan and Lien, 2006), sulfate modified iron oxide–coated sand (Vaishya and Gupta, 2006), HFO incorporated into naturally occurring porous diatomite (Jang et al., 2006), crystalline HFO

(Manna et al., 2003), crystalline hydrous titanium oxide (Manna et al., 2004), granular hydrous zirconium oxide (Ghosh et al., 2006), and iron(III)–tin(IV) binary mixed oxide (Ghosh et al., 2006). Other sorbents include greensand, GFH, iron oxide–coated sand, copper–zinc granules, zeolites, goethite, clay, kaolinites, chitosan beads, coconut husk, coal, fly ash, ferrous iron, zirconium oxide, red mud, petroleum residues, rice husk, human hair, sawdust, manganese greensand, orange juice residues, akaganéite nanocrystals, etc. (Chwirka et al., 2000; Jekel and Seith, 2000; Lin and Wu, 2001; Manna et al., 2003; Thirunavukkarasu et al., 2003b; Singh and Pant, 2004). Clearly, the most widely investigated adsorbents for arsenic mitigation are iron-based.

Arsenic Removal Based on Iron

The use of iron and its oxides and hydroxides is becoming popular because of their chemical affinity toward arsenic and other toxic species naturally present in water, its availability, low cost, and its small footprint on the environment. Table 5 lists some iron and iron-based adsorbents, their arsenic removal capacity, surface area, and the pH range of effectiveness. Iron-based sorbents can have a wide range of surface area and capacity (maximum 134 mg As/g of sorbent) but are not clearly understood with regard to removal mechanisms. The following is a discussion on some novel zero-valent iron and iron-based sorbents and their physicochemical characteristics.

Zero-Valent Iron

Zero-valent iron, Fe(0), was originally developed to mitigate chlorinated hydrocarbon industrial solvents disposed in soil and many other toxic species including arsenic in the environment (Lackovic et al., 2000; Balarama Krishna et al., 2001; Farrell et al., 2001; Su and Puls, 2001a,b, 2003, 2004; Manning et al., 2002a,b; Melitas et al., 2002a,b; Kim et al., 2003; Nikolaidis et al., 2003; Bang et al., 2005a,b; Lien and Wilkin, 2005). In recent years significant improvements were made with iron-based technologies.

Nanoscale (1–120 nm diameter) Fe(0) was found to remove As(III) and As(V) in a first-order process with a rate constant of $k = 0.07$–$1.3 \, \text{min}^{-1}$ (Kanel et al., 2005, 2006). This rate was about 1,000 times faster than that of micrometer-sized iron with As(III) adsorption Freundlich capacity of 3.5 mg As(III)/g of Fe(0). This is a much smaller capacity for high surface area nanoparticles. An inner-sphere surface-complexation mechanism was proposed based on light-scattering electrophoretic mobility measurements of Fe(0)–As(III) complex. Nanoscale iron-containing polymer-supported nanoparticles were formulated for As(III) and As(V) removal where hydrated Fe(III) oxide (HFO) dispersed on a polymeric ion-exchange resin, rendering them magnetically active polymeric particles (Cumbal et al., 2003). The high surface area-to-volume ratios of these nanoscale particles favored both sorption and reaction kinetics. However, extremely high-pressure drops prevented fixed-bed column applications. They also lack

TABLE 5 List of Iron-Based Adsorbents and Their Capacity for Arsenic Removal from Water/Wastewater

Number	Adsorbent	Surface Area (m^2/g)	pH	Capacity (mg/g)		References
				As(III)	As(V)	
1	Iron oxide–coated sand	–	–	0.136	–	Thirunavukkarasu et al. (2005)
2	Iron(III)-loaded chelating resin	–	9 for As(III) 3.5 for As(V)	62.93	54.44	Matsunaga et al. (1996)
3	HFO	200	9	28	7	Lenoble et al. (2002)
4	Gothite	5	103		5	Lakshmipathiraj et al. (2006)
5	FePO4 (amorphous)	53.6	7–9 As(III) 6–6.7 As(V)	21	10	Lenoble et al. (2005)
6	Iron(III) oxide metal slag	196	2.5	30.1	78.5	Zhang and Itoh (2005)
7	Granular ferric hydroxide (GIH)	–	8–9	–	8.5	Driehaus et al. (1998)
8	Iron oxide–coated cement	–	~7		3.39	Kundu and Gupta (2006)
9	Sulfate modified iron oxide–coated sand	–	4–10	–	0.13	Vaishya and Gupta (2004)

(Continued)

TABLE 5 (Continued)

Number	Adsorbent	Surface Area (m²/g)	pH	Capacity (mg/g)		References
				As(III)	As(V)	
10	Modified iron oxide–coated sand	2.9–7.9	4–10	0.14	–	Vaishya and Gupta (2002)
11	Iron hydroxide–coated alumina	95.7	6.62–6.74	7.64	36.64	Hlavay and Polyak (2005)
12	Ferric chloride–impregnated silica gel	–		5.24	5.24	Isao et al. (1976)
13	Hematite	14.40	4.2		0.20	Singh et al. (1996)
14	CIM		6.2 – 7.5	4.25	4.25	Ahamed and Husam (2007)
15	Iron-coated, light expanded clay aggregates	–	–	–	3.12	Nazmul et al. (2008)
16	Iron hydroxide granulates	–	–	–	2.3	Daus et al. (2004)

(–), *not available.*

durability and mechanical strength in flow-through applications. Recently, modified nanosized Fe(0) particles such as NiFe and PdFe were synthesized by borohydride reduction of nickel and palladium salts on Fe(0) particles, and used for arsenate removal (Mondal and Lalvani, 2005). Increasing the temperature caused an increase in arsenate removal, while sorption of phosphate and sulfate inhibited arsenate removal.

As(V) removal in columns packed with iron filings was measured over 1 year of continuous operation. The continuous generation of iron oxide was confirmed, based on As(V) removal on freely corroding vs. cathodically protected iron (Melitas et al., 2002a,b). As(V) diffusion through iron corrosion products determined the rates. Arsenate removal kinetics ranged between zero and first order vs. the aqueous As(V) concentration. The potential use of Fe(0) filings to remove monomethyl arsenate (MMA) and dimethyl arsenate (DMA) from contaminated waters was demonstrated (Cheng et al., 2005). The affinity of MMA for Fe(0) was comparable to that of inorganic arsenate, but lower for arsenite. In contrast, less DMA was retained by Fe(0) filings or their corrosion products.

It was found that oxic conditions can increase the rate of As(III) removal (Bang et al., 2005a, b). High dissolved oxygen content and low solution pH increased the iron corrosion rate and affected As(total) removal. As(V) was reduced to As(III) with Fe(0) under anoxic conditions, but no As(0) was detected in solution after 5 days. X-ray photoelectron spectra showed partial surface reduction of As(III) to As(0).

Plain Fe(0) still has problems associated with low capacity, weak capability of removing As(III) species, and uncontrolled leaching and rusting. The latter can clog the filter media and filter outlets and render the filter useless. Also, the complexity of regenerating and reusing the material in household filtration systems confirms that Fe(0) is not the preferred active material.

Iron Oxides and Iron Oxide–Amended Materials

Although Fe(0) is classified as a separate sorbent, the formation of iron oxide and hydroxides on the surface is the primary substrate where anions, cations and neutral species can adsorb and form complexes. Various forms of iron oxides including amorphous hydrous ferric oxide (HFO: FeOOH), goethite, and hematite ($-Fe_2O_3$) were used to remove both As(III) and As(V) from water (Ferguson and Gavis, 1972; Wilkie and Hering, 1996; USEPA, 1999; Altundogan et al., 2000; Thirunavukkarasu et al., 2003b; Roberts et al., 2004; Saha et al., 2005). Generally, high surface area amorphous HFO appears to have the highest adsorption capacity since it has the highest surface area. The transformation of HFO to form low surface area crystalline iron oxides during preparation, however, can greatly reduce the As removal capacity. Surface area is not the only criterion for high removal capacities; surface-complexation reactions, ion–exchange, and surface precipitation can also play important roles. Some literature data on this sorbent are reviewed here.

The adsorption of As(V) on GFH at concentrations ranging from 100 to 750 µg/L over the pH range of 4–9 was investigated (Saha et al., 2005). The adsorption decreased as the pH of the solution increased, and optimal adsorption was at pH 4, which is far below the circumneutral pH for groundwater. GFH also showed a greater affinity for arsenate adsorption compared to phosphate at the same pH. In another study, GFH demonstrated high treatment capacity of 30,000–40,000 bed volumes until a 10 µg/L breakthrough limit was reached. The sorption capacity was 8.5 mg/g (Driehaus et al., 1998; Swedlund and Webster, 1999). Other studies show that at pH 7 the capacity reduced drastically to 8 µg As/mg dry GFH (Badruzzaman et al., 2004). Surface diffusion was invoked as the primary mass-transfer mechanism. Application of a homogeneous surface diffusion model yields the surface diffusion coefficient value of 2.98×10^{-12} cm^2/s. This value is six orders of magnitude lower than diffusion in aqueous solution and signifies the slow equilibration on GFH surface.

Synthetic HFO was found to have a strong affinity for As(V), which is pH-dependent, whereas As(III) sorption was found to be pH insensitive (Ranjan et al., 2003). It was also found that the columns lose their sorption affinity with further use and can be regenerated with 5 M NaOH. Column regeneration with concentrated NaOH produces highly toxic soluble sodium arsenate and arsenite; therefore a special procedure needs to be implemented for waste disposal. Amorphous HFO immobilized onto a naturally occurring porous diatomite column (Jang et al., 2006) was used for both arsenite and arsenate removal. These were studied only in small scale, using small batch columns.

Studies show that, on ferrihydrite, adsorption of relatively high As concentrations was almost complete in a few hours, and As(III) reacted faster than As(V) species. As(V) adsorption was faster at low As (V) concentrations and at low pH. The high As(III) retention was believed to be because of the formation of a $Fe(AsO_4)$ solid phase and not because of simple adsorption on the surface (Raven et al., 1998). A continuously generated ferric oxyhydroxide in the form of dense granules in fluid-bed technology was developed for the removal of arsenic, where arsenic content of 50 mg As/g or more could be extracted (Stamer and Nielsen, 2000). It was found that 4- and 12-nm-sized Fe_3O_4 nanocrystals can be used to remove arsenic from water under the influence of a magnetic field (Yavuz et al., 2006).The particles appear not to act independently in the separation but rather reversibly aggregated through the resulting high-field gradients present at their surfaces. This is a method whose practice in the field is elusive.

Adsorption and desorption of methylarsonic acid [$CH_3AsO(OH)_2$], methylarsonous acid [$CH_3As(OH)_2$], dimethylarsinic acid [$(CH_3)_2AsO(OH)$], dimethylarsinous acid [$(CH_3)_2AsOH$], arsenate [$AsO(OH)_3$], and arsenite [$As(OH)_3$] on iron oxide minerals (goethite and two-line ferrihydrite) were reported (Lafferty and Loeppert, 2005). Monomethylarsonous acid and monomethylarsonic acid were not appreciably retained by goethite or ferrihydrite. In another study, arsenate and dimethylarsinate (DMA) adsorption kinetics on goethite were described using the Elovich equation (Zhang and Robert, 2005). Replacement of two hydroxyl groups by methyl groups reduces the affinity of arsenic on goethite compared to that of one hydroxyl group. The low affinity of DMA to goethite was due to the formation of outer sphere monodentate surface complexes (Zhang et al., 2007), while As(V) species were known to form bidentate and monodentate corner-sharing complexes (Zhang et al., 2005).

Electrochemical oxidation at steel electrodes in the presence of H_2O_2 is a promising As(III) remediation technology (Arienzo et al., 2002). The mechanism is the adsorption of As(III) on solid HFOs. The removal was fast (3 min) and efficient at pH $<$ 6.5 and at $[H_2O_2]$ = 10 mg/L. The pH control, the availability of H_2O_2, and the need for electricity are the limiting factors for field deployment of this technology.

A fibrous polymeric/inorganic sorbent material was synthesized and used for arsenic remediation (Vatutsina et al., 2007). The sorbent included polymer filaments inside which nanoparticles of hydrated Fe(III) oxides were dispersed. These materials show uniform iron loading and a capacity of about 75–80 mg As(total)/g. In addition, As(III) sorption was not suppressed in the presence of SO_4^{2-}, Cl^-, HPO_4^{2-} at a pH typical for drinking water. Supported HFO particles on anion exchanger offered a high removal capacity for simultaneous removal of both As(V) and As(III). It was found to have $<$10% of arsenic breakthrough after 30,000 bed volumes. Active material regeneration can be done by 2% NaOH and 2% NaCl (Greenleaf et al., 2006). The production of arsenic-concentrated, regenerating effluent solution that has to be decontaminated by the consumer is a drawback for this technology. Similarly, metal-loaded polymers have been used to remove arsenic (Dambies, 2004). The reader can find good descriptions of various resins used for arsenic remediation in the review by Dambies (2004).

It was found that readily available iron turnings (cast iron, low and high carbon steel) can be processed into composite-iron granules (CIGs) by a controlled rusting process to produce mixed oxides of iron and other metals, for example, manganese and then into a CIM (Hussam and Munir, 2007). CIM is different from granular metal oxides in that the active substrate is made from CIGs into a solid, porous matrix through *in situ* processing inside the filter. The active material in the filter removes inorganic arsenic species quantitatively by generating new complexation sites on CIM through iron oxidation and surface chemical reactions, as described in Table 2. The arsenic removal kinetics was found to be a zero-order process. The continuous generation of HFO was confirmed based on As(V) removal on freely corroding vs. cathodically protected iron (Melitas et al., 2002a,b). As(V) diffusion through iron corrosion products determined the rates and the removal kinetics ranged between zero and first orders vs. the aqueous As(V) concentration. Similar conclusions were reached through electrochemical corrosion rate analysis, which showed that arsenate removal was pseudo-first-order at low concentrations and approached zero order in the limit of high arsenate concentrations. It was argued that excess Fe^{2+}, Fe^{3+}, and Ca^{2+} in groundwater could increase the positive charge density of the inner Helmholtz plane of the electrical double layer and enhances binding of anionic arsenates. This is also supported by others (Wilkie and Hering, 1996).

More than 90,000 arsenic filters based on CIM technology are functioning in Bangladesh and Nepal (Hussam et al., 2008).

Groundwater containing Fe(II) also plays a role where inorganic As(III) species are oxidized to As(V) species by the active O_2^-, which is produced by the oxidation of soluble Fe(II) with dissolved oxygen. The presence of manganese (1–2% by wt in most raw iron turnings) can catalyze oxidation of As(III) to As(V). Therefore, the process does not require pretreatment of water with external oxidizing agents such as hypochlorite or potassium permanganate. In addition to arsenic species, =FeOH is also known to remove many other toxic species (Darland and Inskeep, 1997; Alauddin et al., 2001; Munir et al., 2001; Hussam and Munir, 2005). As(V) species ($H_2AsO_4^-$ and $HAsO_4^{2-}$) are removed by surface-complexation reactions on the surface of hydrated iron (=FeOH). New =FeOH is generated *in situ* as more water is filtered.

The removal of arsenic from groundwater can be influenced by the presence of other ions at much higher concentrations than arsenic and also by the presence of organic species known as natural organic matters (NOMs). Presence of competing ions such as phosphate (PO_4^{3-}), silicate (SiO_3^{2-}), and bicarbonate (HCO_3^-) showed negative effects on As removal efficiency (Meng et al., 2000, 2002; Grafe et al., 2001; Holm, 2002). The sorption capacity would also depend on pH (Grafe et al., 2001; Dixit and Hering, 2003; Kuriakose et al., 2004).

NOM showed active redox behavior toward arsenic species on hematite. The NOM may greatly influence redox as well as complexation speciation of arsenic in freshwater environments. Incubation of NOM with arsenic species and hematite dramatically delayed completion of sorption equilibrium and diminished both As(V) and As(III) sorption (Redman et al., 2002). NOM displaced sorbed arsenate and arsenite when NOM and As were introduced sequentially. Furthermore, arsenic species displaced sorbed NOM in significant quantities. These observations are critical for the longevity of iron-based arsenic filters in countries like Bangladesh where NOM could be present in groundwater.

ARSENIC REMOVAL TECHNOLOGIES TESTED IN BANGLADESH, NEPAL, AND INDIA

A number of arsenic removal technologies have been developed by different research organizations as well as business enterprises. Generally, the conventional technologies have been scaled down to meet the requirements of households and communities in the rural environment. Some technologies utilized indigenous materials for arsenic removal. The government of Bangladesh (GOB) has taken some initiatives to test arsenic mitigation technologies through Environmental Technologies Verification Programs for Arsenic Mitigation (ETVAM) (BCSIR, 2003). In this project they have verified and found some suitable technologies for arsenic mitigation and have given temporary approval for installation in the arsenic-contaminated areas of Bangladesh. Figure 5 shows four GOB-approved filters in use in Bangladesh. The recent ETVAM tests of these filters showed approximately 80% of SONO and Alcan, 65% of

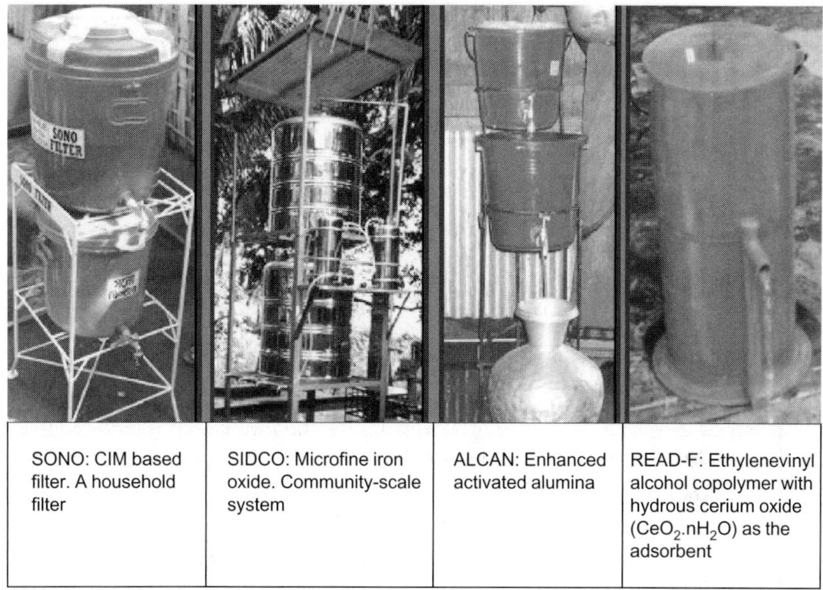

FIGURE 5 Arsenic filters approved by Bangladesh Government through ETVAM testing. Some of these found extensive use in Bangladesh and Nepal for removal of arsenic from groundwater. (Source: BCSIR-OCETA Report on Performance Evaluation and Verification of Arsenic Removal Technologies, 2003.)

Read-F, and 53% of Sidco were still generating filtered water with As(total) below $10\,\mu g/L$ (Technical Performance Monitoring—An Interim Report, 2008). Of these, only SONO removed dissolved manganese from groundwater and produced water that met WHO guidelines for both arsenic and manganese. It was claimed that none of these filters grew, fostered, and/or harbored pathogenic bacteria.

In 1998, West Bengal, India, started installation of household and community-based arsenic removal technologies under government and private initiatives (Hossain et al., 2005). A two-year study covering 18 arsenic removal plants (ARPs) for community use from 11 manufacturers, both local and abroad, were conducted (Hossain et al., 2005), in which 10 of the 18 ARPs failed to remove arsenic below the WHO provisional guideline value $(10\,\mu g/L)$, while six plants could not achieve the Indian standard value $(50\,\mu g/L)$. Only two filters (Oxide India and AIIH & PH) could meet the Indian standard value $(50\,\mu g/L)$ throughout. The AIIH is not based on a solid adsorbent and, therefore, will not be discussed here. During the study almost all the ARPs underwent major or minor modifications to improve performances. Seventy-eight percent of the units were no longer in use. This study points to maintenance and management problems, along with technical issues.

Most of the filters listed in Table 6 did not pass through rigorous ETV procedures and did not find extensive use. The filters that are described outside

TABLE 6 Arsenic Removal Filters and Procedures Tested in Bangladesh, India, and Nepal

Filter	Principle of Separation	Description
Safi Bangladesh	Adsorption of anions on chemically treated active mesoporous material made of kaolinite and iron oxide	The SAFI filter, a household-level candle filter was developed and used in Bangladesh. The candle is made of porous composite materials such as kaolinite and iron oxide on which hydrated ferric oxide is deposited by sequential chemical and heat treatment.
Tetrahedron US	Tetrahedron (United States) is based on anion-exchange resin-based technology to remove arsenate anions.	This filter consists of a stabilizer and an ion exchanger (resin column), with facilities for chlorination using chlorine tablets. The water mixed with chlorine is stored in the stabilizer and subsequently flows through the resin column when the tap is opened for collection of water. Chlorine from the tablet dissolved in the water kills bacteria and oxidizes arsenic and iron.
Shapla Bangladesh	The adsorption medium is iron-coated brick chips manufactured by treating brick chips with a ferrous sulfate solution. It works on the same principle as iron-coated sand.	Shapla arsenic filter, a household-level arsenic removal unit, has been developed and is being promoted by International Development Enterprises (IDE), Bangladesh.
BUET Bangladesh	Adsorption medium is iron-coated sand (Joshi and Chaudhury, 1996).	The BUET iron-coated sand filter was constructed and tested on an experimental basis and found to be very effective in removing arsenic from groundwater. The unit needs pretreatment for the removal of excess iron to avoid clogging of the active filter bed.
Bucket treatment unit US (Meng et al., 2002)	Based on hydrolysis of Fe(III) salt and its coagulation into hydrous ferric oxide and subsequent adsorption and coprecipitation of arsenate and some arsenites	This treatment unit consists of two buckets placed one above the other. Chemicals are mixed manually with arsenic-contaminated water in the upper red bucket by vigorous stirring with a wooden stick and then flocculated by gentle stirring for about 90 s. The mixed water is allowed to settle and then flow into the lower green bucket, and water is collected through a sand filter installed in the lower bucket. The modified bucket treatment unit has been found to be effective in removing iron, manganese, phosphate, and silica along with arsenic.

Apyron Technologies India Pvt. Ltd.	This is on activated hybrid aluminas.	Apyron Technologies India Pvt. Ltd. is the developer and owner of this ARP. The technology is based on the Aqua Bind™-arsenic media, which consists of highly activated hybrid aluminas and alumina composites, which are produced using proprietary technology. These materials are employed to produce particles with enhanced pore and surface properties for cost-effective removal of contaminants. The system has been tested in rural villages in Bangladesh and West Bengal, India.
Anir Engineering India (Kiron and Iftekhar, 2001)	It is based on removal mechanism "adsorption," and slurry/granular ferric hydroxide (S/GFH) is used as filter media.	This arsenic removal plant was designed and developed by International Technology and Product, Germany, in association with Anir Engineering Pvt. Ltd., India. This is a compact ARP consisting of two columns set on a platform.
Pal Trockner (P) Ltd., India	The filter media is GFH (AdsorpAs®), and the arsenic removal mechanism is adsorption.	This ARP was developed by Technical University of Berlin, Germany and marketed in India by Pal-Trockner (P) ltd. The Arsenic Removal Plant (ARP) has been designed to meet the demand of the village people and to suit the realities of rural areas. It is a simple, easy-to-install and compact unit, requiring only $3-4\,ft^2$ space.
Public Health Engineering Department, West Bengal, India	The filter media is based on red hematite (Fe_2O_3) lumps, quartz, and sand-activated alumina based; arsenic species are removed by adsorption.	This ARP was developed and designed by Public Health Engineering Department (PHED, West Bengal India). It has four successive chambers: C_1, C_2, C_3, and C_4. Contaminated water is sprayed in droplets over a hematite lumps (Fe_2O_3) bed before it is led to sedimentation chamber (C_1) at the bottom. The settled water is conveyed through chambers C_2, C_3, and C_4 containing red hematite lumps, quartz, and dual media (sand-activated alumina), respectively. Finally, arsenic-free filtered water comes out from chamber C_4.
Kanchan Nepal (Buzunis, 1995; Ngai et al., 2003, 2002)	It is based on adsorption and surface complexation of arsenic species on hydrous ferric oxide.	This filter was developed by researchers at Massachusetts Institute of Technology (MIT), and is a combination of 3-kolshi filtration system using iron and a slow sand filter. The mechanism is similar to arsenic adsorption on zero-valent iron and arsenic adsorption on hydrous ferric oxides (Lien and Wilkin, 2005). The average arsenic removal efficiency claimed was about 90%. Over 5,000 units are in operation (Ngai et al., 2007). The bio-film on sand is supposed to remove bacteria.

this table passed through several ETVAM or similar technology verification tests and some of them found extensive use.

READ-F Filter Based on Hydrous Cerium Oxide on Polymer Sorbent

The READ-F arsenic filter is a product of Shin Nihon Salt Co. Ltd., Japan. READ-F displays high selectivity for arsenite and arsenates under a broad range of conditions and effectively adsorbs both species. The active material in READ-F is hydrous cerium oxide ($CeO_2 \cdot nH_2O$) deposited on ethylene–vinyl alcohol copolymer. Like hydrous ferric oxide (HFO) the hydrous cerium oxide is a very potent adsorbent and works on the same basic principle of surface-complexation reactions. Breakthrough experiments show that 7,000 pore volume of water containing 1.0 mg/L arsenic can be treated to produce 0.01 mg/L arsenic in the effluent water. One household-treatment unit and one community-treatment unit based on the READ-F adsorbent are being promoted in Bangladesh. The units need iron removal by sand filtration to avoid clogging of the resin bed by iron hydroxide precipitate. In the household unit, both the sand and resin beds are arranged in a container. The community unit has sand and resin beds placed in separate containers. READ-F can be regenerated by adding sodium hydroxide, then sodium hypochlorite, and finally washing with water. The regenerated READ-F needs neutralization by hydrochloric acid and washing with water for reuse. The regeneration is generally done by the manufacturer, or the regenerant sludge has to be disposed in a safe manner.

Alcan Filter Based on Enhanced Activated Alumina

The Alcan filter is based on imported enhanced activated alumina. The primary process relies on the active surface of the media for adsorption of arsenate from water. Although activated alumina is not capable of removing As(III), it does remove As(III) to some extent, and the mechanism of As(III) removal has been discussed elsewhere (Sarkar et al., 2005). The unit is simple in design. Natural Fe(II) present in groundwater can assist in removing arsenite and arsenate through oxidation and complex formation, and can be regarded as a secondary removal step. Other ions such as phosphate present in natural water may compete for active sites on alumina and reduce the arsenic removal capacity of the unit. This filter is prone to clogging by deposition of HFO produced from oxidation and precipitation of natural iron (Fe(II)) in groundwater. The unit can produce more than 3,600 L of arsenic-safe drinking water per day for 100 families. The inactivated alumina at the end of its life requires regeneration or appropriate safe disposal.

SONO Filter Based on Composite Iron Matrix

Some description of SONO has been covered earlier. The SONO filter was developed by a team of researchers from George Mason University, USA, and Manob Sakti Unnayan Kendro (MSUK), a local nongovernmental organization

(NGO) in Kushtia, Bangladesh. The SONO household filter uses CIM, sand, brick chips, and wood charcoal to remove arsenic and other toxic species from groundwater (Khan et al., 2000; Alauddin et al., 2001; Munir et al., 2001; Hussam et al.,2003, 2008; Hussam and Munir, 2007, 2008) through surface complexation and adsorption. CIM is manufactured from CIGs from various iron turnings through a patent-pending process. The CIG (size 0.065–2.00 mm in diameter) is manufactured from iron turnings in a proprietary process to enhance HFO formation. The CIG is turned into CIM inside the filter during the manufacturing of the filter. The CIM has active surface for complexation and immobilization of inorganic arsenic and many toxic metal cations. The filtration system consists of two buckets where the top bucket contains the primary active material CIM, coarse sand, and brick chips. The second bucket contains coarse sand, wood charcoal, fine sand, and brick chips. Tube-well water is poured in the top bucket and filtered water is collected from the bottom bucket. The filter does not require any chemical regeneration and does not produce any regenerant sludge.

About 90,000 SONO filters were deployed in many districts all over Bangladesh, including hundreds of primary schools (Hussam et al., 2008). Many of these filters have been in continuous use for over 5 years without breakthrough. An estimated billion liters of clean drinking water was consumed from these filters and they continue to provide high-quality water for drinking and cooking. Recently, about 1,000 SONO filters were installed in Nepal. The original SONO 3-kolshi filtration system was tested in Nepal by an MIT group with an arsenic removal capacity up to 99% with an effluent arsenic to <10 μg/L (Hurd, 2001; Ngai et al., 2007). The unit SONO filter and its predecessor 3-kolshi filter were also tested by other researchers (Milton et al., 2007). Recently, it was reported that SONO is the only technology that removes arsenic and manganese from groundwater and meets the WHO guidelines (Technical Performance Monitoring—An Interim Report, 2008).

Sidco Filter Based on Granular Ferric Hydroxide (GFH)

This filter was developed by Harbauer GmBH and the Technical University of Berlin. This community filter is based on the sorption and surface complexation of As(III) and As(V) on a high surface area GFH. It does not require any chemical regeneration and does not produce any regenerant sludge. It can filter 50,000–70,000 bed volumes and reduces As to 10 μg/L from 250 μg/L. Groundwater containing high phosphate may affect the performance and reduce the longevity of the filter. A recent test shows that only 53% of the installed filters are working to meet the WHO guideline (Technical Performance Monitoring—An Interim Report, 2008).

Oxide India (P) Based on Activated Alumina

This arsenic removal unit was developed in collaboration with a research group of the Department of Civil and Environmental Engineering, Lehigh University,

USA, and the Bengal Engineering College, India. It is based on adsorption of arsenic species on activated alumina (93.6% by mass), and the average size of the spherical adsorbent particle varied between 0.3–0.6 mm. Activated alumina is manufactured with indigenously available minerals. Since 1997, more than 135 well-head arsenic removal units were installed in remote villages in India. The adsorption column mounted on top of the existing well-head. The unit comprises a cylindrical stainless-steel tank with two compartments. The upper empty chamber of the column contains a splash distributor and atmospheric vent. Here, the oxidation of dissolved iron into insoluble hydrated Fe(III) oxides or HFO particles takes place. There is the fixed-bed activated alumina followed by gravels and water collection chamber below. The flow rate under gravity is 8–10 L/min. The column is backwashed for 10–15 min everyday and the backwash is passed through a coarse sand filter to retain the HFO particulates, which may otherwise clog the filter. This community unit serves approximately 200–300 households and requires about 100 Kg of activated alumina. The units, often exceeding 10,000 bed volumes, are capable of removing arsenates from groundwater and arsenite through a secondary mechanism. The deactivated media, upon exhaustion, is regenerated by backwashing with dilute NaOH (Sarkar et al., 2005). This process produces highly concentrated sodium arsenate that has to be contained through a cementation reaction and disposed appropriately.

Arsenic Filters Tested in the United States

In the United States, the lowering of the arsenic level in drinking water from 50 to 10 μg/L in January 2006 led to a significant increase in cost of supplying water. A consortium under the arsenic water technology partnership (AWTP), in collaboration with national laboratories and professional organizations, was formed to address the problem by developing and testing novel technologies (Siegel et al., 2006). Table 7 shows the list of participating vendors in 2005 and their relative ranking, in parentheses, among 32 technologies tested during 2003–2004. These rankings were based on proponents' presentation and experts' evaluation for possible pilot-scale testing.

SUSTAINABILITY AND MANAGEMENT OF SPENT MATERIAL

The management of spent material and sustainability of a viable filter are interrelated. All filters, irrespective of the nature of physicochemical processes, must have a spent material management protocol. The toxicity of the spent material is generally found by leaching experiments where the sample is subjected to aqueous leaching solutions at different pH and measured for the equilibrium concentration. The U.S. Environmental Protection Agency (USEPA) developed such lab procedures, called toxicity characteristic leaching procedures (TCLPs) to characterize the waste for leaching of known toxic species. The TCLP sets the criteria for disposal of solid waste in sanitary landfill. The USEPA has a

TABLE 7 Arsenic Filters Tested in the United States Under AWTP Technology Verification Protocols in 2005 (Siegel et al., 2006)

Filter (Ranking)	Principle of Separation	Description
Purolite (1) *ArsenXnp; A-530E; A-520E; A-300E; C100E*	Ion-exchange and adsorptive media	Hybrid polymeric media impregnated with iron nanoparticles or iron-impregnated anion resin removes arsenic and uranium. Brine regenerable SBA resins (A-520E and A-300E) will remove arsenic, nitrate, and uranium simultaneously. Media has excellent capacity, flow dynamics, high pressure capacity, no backwashing, no pH adjustment, no waste, and virtually no Operations & Maintenance (O&M). Systems available for Point of Use (POU) as well as small and large municipal applications. Regenerant must be disposed properly.
Graver (3) *Technologies HydroGlobe Division/MetSorb G*	TiO_2 adsorptive media	Non-regenerable titanium dioxide-based media, available in a range of mesh sizes from powder to 16/60 mesh. Claimed to have less sensitivity to common interferences such as silicates, phosphates, pH, and sulfates. It exhibits rapid kinetics, hence low operating cost per thousand gallons of water treated. Disposal of the material is as simple as that for a nonhazardous waste by TCLP and California WET tests.
EaglePicher (16) *Filtration & Minerals/NXT-2; NXT-CF*	Adsorptive media based on $La(OH)_3$ (s)	Lanthanum hydroxide-based media for adsorption and coagulation/filtration arsenic removal, respectively. The lanthanum hydroxide provides pH stability up to pH 10 and removes both As(III) and As(V) without the need for chemical pretreatment. Both media also remove other contaminants such as phosphate, fluoride, selenium, and others.
ResinTech (18) *ASM-10-HP*	Hybrid resin/adsorbent	Iron-based adsorbent dispersed in the gel phase of a strong base anion resin. Arsenic removal first occurs by ion exchange, and then arsenic is adsorbed into the iron. The hybrid resin has very rapid kinetics and robust.

(Continued)

TABLE 7 (Continued)

Filter (Ranking)	Principle of Separation	Description
Brimac carbon (19)	Carbon and bone char adsorptive media	Adsorption media is a granular bone char adsorbent with two components: carbon and hydroxyapatite [$Ca_{10}(PO4)_6(OH)_2$]. The carbon surface adsorbs hydrophobic, lipophilic, and weakly anionic molecules, whereas the hydroxyapatite adsorbs strongly charged molecules together with many inorganic ions (metals).
ADA Tech. (22)	$Fe(OH)_3$ amended silicates adsorptive media	Ferric hydroxide–amended silicate sorbents are based on a process wherein active adsorption sites are distributed onto an inert, inexpensive silicate substrate. The use of the inexpensive silicate substrate allows for efficient distribution of the iron at low cost.
Argonide (23) *Corporation/Alfox GR-3*	Nano iron/alumina adsorptive aedia	Alfox is a granular material consisting of a proprietary nano alumina/ nano iron hydroxide mixture. Laboratory testing shows it has about 2 to 2.5 times the capacity vs. bayoxide E-33.
Inotec (24) *AsTECH*	Chemical/biological	Immobilized functional groups and microorganisms bind and remove arsenic from concentrated and dilute solutions. Arsenic is removed by chemical binding and biological transformation. Pilot-scale tests in mining waters have demonstrated arsenic removal to less than detection levels. Non-regenerable.
Virotec (26) International Ltd. Bauxsol	Adsorption on bauxsol mineral	Ions are trapped through reprecipitation of low solubility minerals, isomorphous substitution, solid-state diffusion, and adsorption. Bauxsol™ has an excellent ability to remove As(V) from water, and field trials show the addition of Bauxsol™ to sulfidic rock reduced the As concentration in leachate from 35 mg/L to less than 0.005 mg/L. Arsenic concentrations have remained below 0.005 mg/L for five years since the treatment, and concentrations of trace metals have remained below regulatory limits.

list of 39 regulated contaminants for which the permissible level is one hundred times the drinking water maximum contaminant level (MCL). For arsenic, this means if the TCLP yields As(total) concentration $>1,000\,\mu g/L$, the spent materials should be regarded as toxic waste and disposed in a sanitary landfill.

Several recent studies have shown that leaching of arsenic can be stimulated or enhanced in a landfill or a hazardous waste site environment under high pH and reducing conditions (Ghosh et al., 2004). A long-term field observation of the Coakley Landfill Superfund Site (NH) showed dissolved arsenic levels increased modestly. These results indicate that reducing environments within organic contaminant plumes may release arsenic (Delemos et al., 2006). Clearly, these conditions must be avoided for all spent materials. It is also reported that iron-bound arsenic has the lowest bioavailability and iron amendment was recommended as a means to remediate arsenic contaminated soil (Jonathan et al., 2007).

In the context of sorbent materials used for arsenic removal, those requiring sorbent regeneration through chemical processing are the most vulnerable. This includes activated alumina, polymer-amended materials, ion-exchange resins, and some GFHs. For example, the spent regenerant solution containing high arsenic could be immobilized by making cement-cinder blocks. This adds to the cost and management problems for household filters; however, it could be successfully managed in a community-based system (Clifford and Ghuyre, 2002). In contrast, the iron-based filters may have inherent advantages in that the spent materials are thermodynamically stable under oxic conditions, and they form self-contained minerals similar to that present in soil. Arsenic species in most iron-based filters are in the oxidized state and similar to a self-contained, naturally occurring compound in Earth's crust. It is like disposing of soil on soil. For example, measurements on SONO household filter's used sand and CIM-Fe by total available leaching protocol (TALP) show that the spent material is completely nontoxic with $<5\,\mu g/L$ As(total) in the effluent, which is 100 times less than the amended EPA limit at $500\,\mu g/L$ (Final Report, 2006). Similar results were also reported by ETVAM using EPA's TCLP methods. Further tests on backwash of filter waste showed SONO produced the lowest concentrations of As(total), 93 mg/kg, in comparison to commercial filters based on microfine iron oxide—2,339 mg/kg, cerium hydroxide based ion-exchange resin—105 mg/kg, and activated alumina—377 mg/kg in solid waste vs. the EPA limit of 500 mg/kg. These numbers, however, should be compared on a total volume basis under similar influent arsenic. Most importantly, the NAE tests of the used CIM of SONO filter was characterized as "nondetectable and nonhazardous (limit 0.50 mg/L)" by the TCLP (Final Report, 2006). It may be noted that USEPA's recommended land disposal limit for arsenic is 2 kg/hectare per year. This corresponds to arsenic from 10 million L of water with $200\,\mu g/L$ of total arsenic (Khan, personal communication, 2007). According to this prescription, $4\,m^2$ of land is sufficient for the disposal of the spent media from a household filter used for 274 years at 100 L/day usage.

FIGURE 6   Example of an integrated arsenic mitigation program in Bangladesh as a sustainable model.

Sustainability of simple arsenic removal technologies requires more than production and distribution of the technology. In most underdeveloped countries, the majority of the production and distribution of humanitarian needs are still achieved through funding from NGOs and financial subsidy from foreign aid organizations. Commercialization through further product development is necessary for a sustainable future. Sustainability of such technologies requires indigenous production using locally available raw materials, recycling, and regeneration of products and active media. Furthermore, merely supplying the filter is not enough to solve the present water crisis. Besides filter manufacturing and quality control, some of the overriding issues (Figure 6) are sanitation, education, training, motivation, medical referral to the arsenicosis patients, and social mobilization (i.e., women empowerment through formation of mother's club, etc.) that must be integrated into a program for a sustainable and progressive solution to the arsenic crisis. These issues are now understood by most NGOs in Bangladesh and elsewhere. Many NGOs have also implemented intensive training and cultural programs to motivate people to drink arsenic-free water.

FILTER EVALUATION THROUGH ENVIRONMENTAL TECHNOLOGY VERIFICATIONS

The evaluation of arsenic water filter is now an integral part of ETV programs (see Table 8). Countries where arsenic in drinking water is a health issue have

TABLE 8 Criteria for Filter Evaluation

Criteria	Description
Performance	The filter must remove arsenic below the MCL ($10\,\mu g/L$), produce potable drinking water as per WHO, and meet local water quality standards (Table 1). The filter must produce enough water for a family of five for drinking and cooking (8–10 L/h). This criterion is essential for Bangladesh, India, and Nepal. The filter's ability to meet the basic technical requirements for As removal and flow rate as specified above, as well as the potential for keeping it pathogen-free is paramount.
Technology	The viability of a technology depends on its performance, availability of raw materials, ease of production, life of the active media, and durability of the hardware parts. Simple technologies using indigenous raw materials and filters that do not require electricity were given priority in NAE–Grainger Challenge. Filters tested under laboratory conditions may not succeed in rural circumstances, so all technologies must be tested in the field. The viability of the technology can also be assessed from the number of working units in the field, manufacturing capability of companies, and distribution network of the products. The viability of a novel technology is also enhanced by its scaleup potential, and the technology must be safe to use.
Cost	Cost to users must be affordable. Capital cost of the device including transport, maintenance, monitoring, and media replacement should be considered. Cost for pretreatment of raw water, such as preoxidation of arsenite to arsenate, and any post-treatment of water should be considered.
Social acceptability	User friendliness is the first criteria for acceptability. The filtered water must be free of unpleasant taste. Space requirements, installation time, skill required for installation of the device at rural level in developing and less developed countries should be considered. Household filters requiring active media regeneration through chemical processes are not viable. Education and training should be a part of filter distribution and dissemination.
Environmental sustainability	All arsenic filters work on the basis of chemical reaction of arsenic species with an active sorbent. At some point, these materials need to be disposed in a manner safe to the environment. This aspect has been described in previous sections. Filters that do not require active material regeneration, produce non-leachable spent materials (based on TCLP, TALP, and modified leaching test procedures), and are environmentally benign should be given priority. Gravity-based pour-and-collect filters using no electricity have no carbon foot-print during use. The recycling and reuse of spent materials without an adverse effect on the environment are indicators for sustainability. Here, iron-based sorbents may have the best potential.

instituted ETVAM projects to screen filters for safe use. This is prompted by the concern that filters may be marketed as arsenic filters without approval and without any waste disposal oversight. In particular, commercial filters should be tested under field conditions and at different places with different water chemistries and varied As(III)/As(V), Fe(II), Mn(II), phosphate, sulfate, silicate and other interfering ions. The evaluation may also include cost and social acceptability. Some of the ETVAM-tested filters have been discussed earlier. All verification projects must start with the knowledge of raw water through physical, chemical, and geochemical characterization of influent water. Each filter must also be tested with a customized protocol based on information from the proponent. The final test should be done in the field with the users, to assess its commercial and environmental potential. Here, we have compiled the common performance criteria from various test protocols (Sandia, ETVAM, and Grainger). Filter evaluation can be divided into five parts: performance, viability of technology, cost, social acceptability, and environmental acceptability as shown in Table 8.

CONCLUSION AND OUTLOOK

The demand for clean potable water in the world is on the rise. The scarcity of drinking water is forcing millions of people in less developed countries to drink water from any source they can find. While the surface water is contaminated with toxic industrial chemicals and pathogenic bacteria of anthropogenic origin, groundwater is contaminated with toxic arsenic from natural origins. This chapter dealt with the present status of arsenic filters based on solid sorbent for the filtration of groundwater. Special emphasis is given to iron-based filters because they appear to be the most successful ones in the field, in terms of cost and performance. The basic surface chemical reactions and computational models, the thermodynamics, kinetics, and scaleup basis for sorbent evaluation have been presented to support reported works. Arsenic filters, particularly those passed through ETVAM tests in Bangladesh, India, Nepal, and the United States are discussed. Finally, we have looked into the issues of sustainability management of spent material and also filter evaluation criteria applied during various ETVAM tests. Although there are some iron-based technologies that hold significant promise and have found extensive use in Bangladesh, India, and Nepal, much work is still needed to develop compact, efficient, affordable, and environment-friendly filters for arsenic and many other toxic species including pathogens. As potable water supplies become increasingly vulnerable to contamination, the development of affordable water-filtration systems is becoming a more attractive option. This is especially true for the majority of the world's population, where water treatment systems and piped water are not readily available.

REFERENCES

Ahamed S., Husam, A., 2007. sorption kinetics onto the composite iron matrix (CIM)—for the removal of arsenic (V) and arsenic (iii), Laboratory Report (Unpublished Data). Department of Chemistry and Biochemistry, George Mason University, VA.

Alauddin, M., Hussam, A., Khan, A.H., Habibuddowla, M., Rasul, S.B., Munir, A.K.M., 2001. Critical evaluation of a simple arsenic removal method for groundwater of Bangladesh. In: Chappell, W.R., Abernathy, C.O., Calderon, R.L. (Eds.) Arsenic Exposure and Health Effects. Elsevier Science, B.V., Amsterdam, the Netherlands. pp. 439–449.

Altundoğan, H., Tümen, F., 2003. As(V) removal from aqueous solutions by coagulation with liquid Phase of Red Mud. Journal of Environmental Science and Health, Part A—Toxic/Hazardous Substances & Environmental Engineering 38, 1247–1258.

Altundogan, H.S., Altundogan, S., Tumen, F., Bildik, M., 2000. Arsenic removal from aqueous solutions by adsorption on red mud. Waste Manage. 20 (8), 761–767.

Altundogan, H.S., Altundogan, S., Tumen, F., Bildik, M., 2002. Arsenic adsorption from aqueous solution by activated red mud. Waste Manage. 22, 357–363.

Arienzo, M., Adamo, P., Chiarenzelli, J., Bianco, M.R., De Martino, A., 2002. Retention of arsenic on hydrous ferric oxides generated by electrochemical peroxidation. Chemosphere 48 (10), 1009–1018.

Badruzzaman, M., Westerhoff, P., Knappe, D.R.U., 2004. Intraparticle diffusion and adsorption of arsenate onto granular ferric hydroxide (GFH). Water Res. 38 (18), 4002–4012.

Balarama Krishna, M.V., Chandrasekaran, K., Karunasagar, D., Arunachalam, J., 2001. A combined treatment approach using Fenton's reagent and zero valent iron for the removal of arsenic from drinking water. J. Hazard. Mater. 84, 229–240.

Bang, S., Korfiatis, G.P., Meng, X., 2005a. Removal of arsenic from water by zero-valent iron. J. Hazard. Mater. 121, 61–67.

Bang, S., Johnson, M.D., Korfiatis, G.P., Meng, X., 2005b. Chemical reactions between arsenic and zero-valent iron in water. Water Res. 39, 763–770.

BCSIR, 2003. Performance Evaluation and Verification of Five Arsenic Removal Technologies: ETVAM Field Testing and Technology Verification Program. Bangladesh Council of Scientific and Industrial Research (BCSIR), Dhaka, Bangladesh.

Bhattacharya, P., Frisbie, S.H., Smith, E., Naidu, R., Jacks, G., Sarkar, B., 2002. Arsenic in the environment: a global perspective. In: Sarkar, B. (Ed.), Handbook of Heavy Metals in the Environment. Marcell Dekker, New York, pp. 147–215.

Bodek, I., Lyman, W.J., Reehl, W.F., Rosenblatt, D.H., 1998. Environmental Inorganic Chemistry: Properties, Processes and Estimation Method. Pergamon Press, USA, New York, NY.

Brookins, D.G., 1988. Eh–pH Diagrams for Geochemistry. Springer-Verlag, Berlin, Germany.

Buzunis, B.J., 1995. Intermittently operated slow sand filtration: a new water treatment process. Master of Engineering thesis. Civil Engineering, University of Calgary, Canada.

Chakraborti, D., Rahman, M.M., Paul, K., Chowdhury, U.K., Sengupta, M.K., Lord, D., et al., 2002. Arsenic calamity in the Indian subcontinent. What lessons have been learned? Talanta 58, 3–22.

Chakraborti, D., Sengupta, M.K., Rahman, M.M., Ahamed, S., Chowdhury, U.K., Hossain, M.A., et al., 2004. Groundwater arsenic contamination and its health effects in the Ganga-Meghna-Brahmaputra plain. J. Environ. Monit. 6, 75N–83N.

Cheng, C.R., Liang, S., Wang, H.C., Beuhler, M.D., 1994. Enhanced coagulation for arsenic removal. J. Am. Water Works Assoc. 86 (9), 79–90.

Cheng, Z., van Geen, A., Louis, R., Nikolaidis, N., Bailey, R., 2005. Removal of methylated arsenic from groundwater with iron filings. Environ. Sci. Technol. 39 (19), 7662–7666.

Chwirka, J.D., Thomson, B.M., Stomp, J.M.I., 2000. Removing arsenic from groundwater. J. Am. Water Works Assoc. 92 (3), 79–88.

Clifford , D.A., Ghuyre, G.L., 2002. Metal-oxide adsorption, ion exchange, and coagulation–microfiltration for arsenic removal from water. In: Frankenberger, Jr., W.T. (Ed.), Enviornmnetal Chemistry of Arsenic. Marcel Dekker, New York, pp. 217–245.

Cumbal, L., Greenleaf, J., Leun, D., SenGupta, A.K., 2003. Polymer supported inorganic nanoparticles: characterization and environmental applications. React. Funct. Polym. 54 (1–3), 167–180.

Dambies, L., 2004. Existing and prospective sorption technologies for the removal of arsenic in water. Sep. Sci. Technol. 39 (3), 603–627.

Darland, J.E., Inskeep, W.P., 1997. Effects of pH and phosphate competition on the transport of arsenate. J. Environ. Qual. 26 (4), 1133–1139.

Daus, B., Wennrich, R., Weiss, H., 2004. Water Res. 38, 2948.

Davis, C.C., Chen, H.-W., Edwards, M., 2002. Modeling silica sorption to iron hydroxide. Environ. Sci. Technol. 36 (4), 582–587.

Delemos, J.L., Bostick, B.C., Renshaw, C.E., Sturup, S., Feng, X., 2006. Landfill-stimulated iron reduction and arsenic release at the Coakley superfund site (New Hampshire). Environ. Sci. Technol. 40, 67–73.

Dinesh, M., Pittman, Charles U., Jr., 2007. Arsenic removal from water/wastewater using adsorbents— a critical review. J. Hazard. Mater. 142, 1–53.

Dixit, S., Hering, J.G., 2003. Comparison of arsenic(V) and arsenic(III) sorption onto iron oxide minerals: implications for arsenic mobility. Environ. Sci. Technol. 37, 4182–4189.

Dou, X., Zhang, Y., Yang, M., Pei, Y., Huang, X., Takayama, T., et al., 2006. Occurrence of arsenic in groundwater in the suburbs of Beijing and its removal using an iron–cerium bimetal oxide adsorbent. Water Qual. Res. J. Can. 41 (2), 140–146.

Driehaus, W., Jekel, M., Hildebrandt, U., 1998. Granular ferric hydroxide—a new adsorbent for the removal of arsenic from natural water. J. Water SRT—Aqua 47 (1), 30–35.

Dzombak, D.A., Morel, F.M.M., 1990. Surface Complexation Modeling: Hydrous Ferric Oxide. Wiley-Interscience, New York.

Edwards, M.A., 1994. Chemistry of arsenic removal during coagulation and Fe–Mn oxidation. J. Am. Water Works Assoc. 86 (9), 64–77.

Elovich, S.J., 1957. The nature of the chemisorption of carbon monoxide on manganese dioxide. In: Schulman, J.H. (Ed.), Proceedings of the Second International Congress of Surface Activity. Butterworths Scientific Publications, London, pp. 252–259.

Farahbakhsh, K., Svrcek, C., Guest, R.K., Smith, D.W., 2004. A review of the impact of chemical pretreatment on low-pressure water treatment membranes. J. Environ. Eng. Sci. 3 (4), 237–253.

Farrell, J., Wang, J., O'Day, P., Conklin, M., 2001. Electrochemical and spectroscopic study of arsenate removal from water using zero-valent iron media. Environ. Sci. Technol. 35, 2026–2032.

Felds, K.A., Chen, A., Wang, L., 2000. Arsenic removal from drinking water by coagulation/ filtration and lime softening plants. EPA report no. EPA/600/R-00/063.

Ferguson, J.F., Gavis, J., 1972. A review of the arsenic cycle in natural waters. Water Res. 6, 1259–1274.

Final Report, 2006. Evaluation of Grainger Challenge Arsenic Treatment Systems- SONO Filter #29. Prepared by Shaw Environmental Inc., under EPA Contract No. EP-C-05-056 and National Academy of Engineering- Shaw PN 118205-03. December.

Freundlich, H.M., 1906. Over the adsorption in solution. J. Phys. Chem. 57, 385–470.

Ghosh, M.M., Yuan, J.R., 1987. Adsorption of inorganic arsenic and organoarsenicals on hydrous oxide. Environ. Prog. 6, 150.

Ghosh, A., Mukiibi, M., Ela, W., 2004. TCLP underestimates leaching of arsenic from solid residuals under landfill conditions. Environ. Sci. Technol. 38, 4677–4682.

Ghosh, U.C., Bandhyapadhyay, D., Manna, B., Mandal, M., 2006. Hydrous iron(III)–tin(IV) binary mixed oxide: arsenic adsorption behaviour from aqueous solution. Water Qual. Res. J. Can. 41 (2), 198–209.

Grafe, M., Eick, M.J., Grossl, P.R., 2001. Adsorption of arsenate(V) and arsenite(III) on goethite in the presence and absence of dissolved organic carbon. Soil Sci. Soc. Am. J. 65 (6), 1680–1687.

Greenleaf, J.E., Lin, J.-C., Sengupta, A.K., 2006. Two novel applications of ion exchange fibers: arsenic removal and chemical-free softening of hard water. Environ. Prog. 25 (4), 300–311.

Greenwood, N.N., Earnshaw, A., 1984. Chemistry of Elements. Pergamon Press, Oxford (Chapter 13).

Gregor, J., 2001. Arsenic removal during conventional aluminium-based drinking-water treatment. Water Res. 35, 1659–1664.

Gulledge, J.H., O'Connor, J.T., 1973. Removal of arsenic(V) from water by adsorption on aluminum and ferric hydroxides. J. Am. Water Works Assoc. 8, 548–552.

Guo, C., Zhang, F., Yang, X., 2000. Treatment of As containing wastewater by lime-polyferric sulfate coagulating process. Gongye Shuichuli 20, 27–29.

Gupta, S.K., Chen, K.Y., 1978. Arsenic removal by adsorption. J. Water Pollut. Control Fed. 50 (3), 493–506.

Han, B., Runnells, T., Zimbron, J., Wickramasinghe, R., 2002. Arsenic removal from drinking water by flocculation and microfiltration. Desalination 145, 293–298.

Hering, J.G., Chen, P.Y., Wilkie, J.A., Elimelech, M., Liang, S., 1996. Arsenic removal by ferric chloride. J. Am. Water Works Assoc. 88 (4), 155–167.

Hering, J.G., Chen, P.Y., Wilkie, J.A., Elimelech, M., 1997. Arsenic removal from drinking water during coagulation. J. Environ. Eng. (American Society of Civil Engineers) 123 (8), 800–808.

Hlavay, J., Polyak, K., 2005. Determination of surface properties of iron hydroxide-coated alumina adsorbent prepared for removal of arsenic from drinking water. J. Colloid Interface Sci. 284 (1), 71–77.

Ho, Y.S., McKay, G., 1998. Kinetic model for lead(II) sorption on to peat. Adsorpt. Sci. Technol. 16 (4), 243–255.

Holm, T.R., 2002. Peer-reviewed-effects of CO_3^{2-}, HCO_3^-, Si, and PO_4^{3-} on arsenic sorption to HFO-common dissolved substances, including bicarbonate, silica, and phosphate, interfere with arsenic removal by iron oxide, but lowering the water pH may help overcome such interference. Am. Water Works Assoc. J. 94 (4), 174–181.

Hossain, M.A., Sengupta, M.K., Ahamed, S., Rahman, M.M., Mondal, D., Lodh, D., et al., 2005. Ineffectiveness and poor reliability of arsenic removal plants in West Bengal, India. Environ. Sci. Technol. 39, 4300–4306.

http://www.nae.edu/nae/grainger.nsf.

Huang, W., Rong, L., 2001. Treatment of sewage with high arsenic by calcium bleach and lime. Mizu Shori Gijutsu 42, 59–60.

Hurd, J.J., 2001. Evaluation of three arsenic removal technologies in Nepal. MS thesis. Civil and Environmental Engineering, Massachusetts Institute of Technology, USA, June. pp. 47–50.

Hussam, A., Munir, A.K.M., 2005. Development and deployment of arsenic filters for groundwater of Bangladesh. In: Green Chemistry and Water Purity. American Chemical Society 37th Middle Atlantic Regional Meeting, Rutgers University, NJ, USA, May 23, 2005. Abstract 252.

Hussam, A., Munir, A.K.M., 2007. A simple and effective arsenic filter based on composite iron matrix: development and deployment studies for groundwater of Bangladesh. J. Environ. Sci. Health A 42, 1869–1878.

Hussam, A., Munir, A.K.M., 2008. Development of a simple arsenic filter for groundwater of Bangladesh on a composite iron matrix. In: Ahuja, S. (Ed.), Arsenic Contamination of Groundwater (Mechanism, Analysis and Remediation). Wiley publications, New Jersey, USA, pp. 287–303.

Hussam, A., Habibuddowla, M., Alauddin, M., Hossain, Z.A., Munir, A.K.M., Khan, A.H., 2003. Chemical fate of arsenic and other metals in groundwater of Bangladesh: experimental measurement and chemical equilibrium model. J. Environ. Sci. Health A Toxic Hazard. Subst. Environ. Eng. 38 (1), 71–86.

Hussam, A., Ahamed, S., Munir, A.K.M., 2008. Arsenic filters for groundwater in Bangladesh: toward a sustaibale solution. The Bridge—Linking Engineering and Society 38 (3), 14–23.

Isao, Y., Hiroshi, K., Keihei, U., 1976. Selective adsorption of arsenic ions on silica gel impregnated with ferric hydroxide. Anal. Lett. 9 (12), 1125–1133.

Jang, M., Min, S.-H., Kim, T.-H., Park, J.K., 2006. Removal of arsenite and arsenate using hydrous ferric oxide incorporated into naturally occurring porous diatomite. Environ. Sci. Technol. 40 (5), 1636–1643.

Jekel, M., Seith, R., 2000. Comparison of conventional and new techniques for the removal of arsenic in a full scale water treatment plant. Water Supply 18 (1), 628–631.

Jonathan, S.I., Mark, B.O., Philip, J.M., Melanie, S.A., 2007. Decreasing arsenic bioaccessibility/ bioavailability in soils with iron amendments. J. Environ. Sci. Health A 42 (9), 1317–1329.

Joshi, A., Chaudhury, M., 1996. Removal of arsenic from groundwater by iron oxide-coated sand. ASCE J. Environ. Eng. 122 (8), 769–771.

Kanel, S.R., Manning, B., Charlet, L., Choi, H., 2005. Removal of arsenic(III) from groundwater by nano-scale zero-valent iron. Environ. Sci. Technol. 39, 1290–1298.

Kanel, S.R., Choi, H., Kim, J.Y., Vigneswaran, S., Shim, W.G., 2006. Removal of As(III) from groundwater using low cost industrial by-products-blast furnace slag. Water Qual. Res. J. Can. 41 (2), 130–139.

Kanel, S.R., Greneche, J.-M., Cgoi, H., 2006. Arsenic(V) removal from groundwater using nano scale zero-valent iron as a colloidal reactive barrier material. Environ. Sci. Technol. 40, 2045–2050.

Karcher, S., Caceres, L., Jekel, M., Contreras, R., 1999. Arsenic removal from water supplies in northern Chile using ferric chloride coagulation. J. Chart. Inst. Water Environ. Manage. 13, 164–169.

Katsoyiannis, I.A., Zouboulis, A.I., 2002. Removal of arsenic from contaminated water sources by sorption onto iron-oxide-coated polymeric materials. Water Res. 36, 5141–5155.

Khan, A.H., Rasul, S.B., Munir, A.K.M., Habibuddowla, M., Alauddin, M., Newaz, S.S., et al., 2000. Appraisal of a simple arsenic removal method for groundwater of Bangladesh. J. Environ. Sci. Health A35 (7), 1021–1041.

Kim, J.Y., Davis, A.P., Kim, K.W., 2003. Stabilization of available arsenic in highly contaminated mine tailings using iron. Environ. Sci. Technol. 37, 189–195.

Kiron, S., Iftekhar, A., 2001. Apyron arsenic treatment unit—reliable technology for arsenic safe water. In: Feroze Ahmed, M., Ashraf Ali, M., Zafar Adeel, (Eds.), Technologies for Arsenic Removal from Drinking Water. Bangladesh University of Engineering Technology, Dhaka and the United Nations University, Tokyo, Japan, pp. 146–157.

Košutić, K., Furač, L., Sipos, L., Kunst, B., 2005. Removal of arsenic and pesticides from drinking water by nanofiltration membranes. Sep. Purif. Technol. 42, 137–144.

Kundu, S., Gupta, A.K., 2006. Adsorptive removal of As(III) from aqueous solution using iron oxide coated cement (IOCC): evaluation of kinetic, equilibrium and thermodynamic model. Sep. Purif. Technol. 52 (2), 165–172.

Kuo, S., Lotse, E.G., 1974. Kinetics of phosphate adsorption and desorption by lake sediments. Soil Sci. Soc. Am. Proc. 38, 50–54.

Kuriakose, S., Singh, T.S., Pant, K.K., 2004. Adsorption of As(III) from aqueous solution onto iron oxide impregnated activated alumina. Water Qual. Res. J. Can. 39 (3), 258–266.

Lackovic, J.A., Nikolaidis, N.P., Dobbs, G.M., 2000. Inorganic arsenic removal by zero-valent iron. Environ. Eng. Sci. 17, 29–39.

Lafferty, B.J., Loeppert, R.H., 2005. Methyl arsenic adsorption and desorption behavior on iron oxides. Environ. Sci. Technol. 39 (7), 2120–2127.

Lagergren, S., 1898. About the theory of so-called adsorption of soluble substances. Kungliga Svenska etenskapsakademiens Handlingar 24, 1–39.

Lakshmipathiraj, P., Narasimhan, B.R.V., Prabhakar, S., Raju, G.B., 2006. Adsorption of arsenate on synthetic goethite from aqueous solution. J. Hazard. Mater. 136 (2), 281–287.

Langmuir, I., 1918. The adsorption of gases on plane surface of glass, mica and platinum. J. Am. Chem. Soc. 40, 1361–1403.

Lenoble, V., Bouras, O., Deluchat, V., Serpaud, B., Bollinger, J.-C., 2002. Arsenic adsorption onto pillared clays and iron oxides. J. Colloid Interface Sci. 255 (1), 52–58.

Lenoble, V., Laclautre, C., Deluchat, V., Serpaud, B., Bollinger, J.-C., 2005. Arsenic removal by adsorption on iron(III) phosphate. J. Hazard. Mater. 123 (1–3), 262–268.

Lien, H., Wilkin, R.T., 2005. High-level arsenate removal from groundwater by zero-valent iron. Chemosphere 59, 377–386.

Lin, T.F., Wu, J.K., 2001. Adsorption of arsenite and arsenate within activated alumina grains: equilibrium and kinetics. Water Res. 35, 2049–2057.

Mandal, B.K., Suzuki, K.T., 2002. Arsenic around the world: a review. Talanta 58, 201–235.

Manna, B.R., Dey, S., Debnath, S., Ghosh, U.C., 2003. Removal of arsenic from groundwater using crystalline hydrous ferric oxide (CHFO). Water Qual. Res. J. Can. 38 (1), 193–210.

Manna, B.R., Debnath, S., Hossain, J., Ghosh, U.C., 2004. Trace arsenic contaminated groundwater upgradation using hydrated zirconium oxide (HZO). J. Ind. Pollut. Control 20, 247–266.

Manning, B.A., Fendorf, S.E., Goldberg, S., 1998. Surface structures and stability of arsenic(III) on goethite: spectroscopic evidence for inner sphere complexes. Environ. Sci. Technol. 32 (1), 2383–2388.

Manning, B.A., Fendorf, S.E., Bostick, B., Suarez, D.L., 2002a. Arsenic(III) oxidation and arsenic(V) adsorption reactions on synthetic birnessite. Environ. Sci. Technol. 36 (5), 976–981.

Manning, B.A., Hunt, M., Amrhein, C., Yarmoff, J.A., 2002b. Arsenic(III) and arsenic(V) reactions with zerovalent iron corrosion products. Environ. Sci. Technol. 36, 5455–5461.

Matsunaga, H., Yokoyama, T., Eldridge, R.J., Bolto, B.A., 1996. Adsorption characteristics of arsenic(III) and arsenic(V) on iron(III)-loaded chelating resin having lysine-N^{α},N^{α}-diacetic acid moiety. React. Funct. Polym. 29, 167–174.

McNeill, L.S., Edwards, M., 1995. Soluble arsenic removal at water treatment plants. J. Am. Water Works Assoc. 87, 105–113.

Melitas, N., Wang, J., Conklin, M., O'Day, P., Farrell, J., 2002a. Understanding soluble arsenate removal kinetics by zerovalent iron media. Environ. Sci. Technol. 36, 2074–2081.

Melitas, N., Conklin, M., Farrell, J., 2002b. Electrochemical study of arsenate and water reduction on iron media used for arsenic removal from potable water. Environ. Sci. Technol. 36, 3188–3193.

Meng, X., Bang, S., Korfiatis, G.P., 2000. Effects of silicate, sulfate, and carbonate on arsenic removal by ferric chloride. Water Res. 34 (4), 1255–1261.

Meng, X., Korfiatis, G.P., Bang, S., Bang, K.W., 2002. Combined effects of anions on arsenic removal by iron hydroxides. Toxicol. Lett. 133 (1), 103–111.

Milton, A.H., Smith, W., Dear, K., Caldwell, B., Sim, M., Ng, J., 2006. Drinking water options in Bangladesh. In: Naidu, R., Smith, E., Owens, G., Bhattacharya, P., adebaum, P. (Eds.), Managing Arsenic in the Environment: From Soil to Human Health. CSIRO publishing, Australia, pp. 355–362.

Milton, A.H., Wayne, S., Keith, D., Ng, J., Sim, M., Rangmuthugala, G., et al., 2007. A randomized intervention trial to assess to arsenic mitigation options in Bangladesh. J. Environ. Sci. Health A 42, 1879–1888.

MINTEQA2 Model System, 2001. Center for Exposure Assessment Modeling, Environmental Protection Agency, 960 College Station Road, Athens, GA.

Mondal, K., Lalvani, S.B., 2005. Arsenate remediation using nanosized modified zerovalent iron particles. Environ. Prog. 24, 289–296.

Mukherjee, A., Sengupta, M.K., Ahamed, S., Hossain, M.A., Das, B., Nayak, B., et al., 2006. Groundwater arsenic contamination: a global perspective with special emphasis to Asian countries. J. Health Popul. Nutr. (Special issue on Arsenic) 24 (2), 142–163.

Munir, A.K.M., Rasul, S.B., Habibuddowla, M., Alauddin, M., Hussam, A., Khan, A.H., 2001. Evaluation of the performance of the SONO3-kolshi filter for arsenic removal from groundwater using zero valent iron through laboratory and field studies. Proceedings International Workshop on Technology for Arsenic Removal from Drinking Water, Bangladesh University of Engineering and Technology and United Nations University, Japan, May 5. pp. 171–189.

Nazmul, H., Gregory, M., Irene, C.-A., Jorge, L.G., 2008. Iron-modified light expanded clay aggregates for the removal of arsenic(V) from groundwater. Microchem. J. 88.

Ngai, T.K.K., Walewijk, S., 2003. The arsenic biosand filter (ABF) project: design of an appropriate household drinking water filter for rural Nepal. Civil and Environmental Engineering, Massachusetts Institute of Technology. http://web.mit.edu/watsan (accessed 10.11.08.).

Ngai, T.K.K., Shrestha, R.R., Bipin, D., Nakhan, M., Murcott, S.E., 2007. Design for sustainable development—household drinking water filter for arsenic and pathogen treatment in Nepal. J. Environ. Sci. Health A 42, 1879–1888.

Ngai, T.K.K., Sen, D., Lukacs, H., 2002. Innovative drinking water technology for Bangladesh, India and Nepal. IDEAS competetion, final application, April 26.

Nguyen, T.V., Vigenswaran, S., Ngo, H.H., Pokhrel, D., Viraraghavan, T., 2006. Iron coated sponge as effective media to remove arsenic from drinking water. Water Qual. Res. J. Can. 41 (2), 164–170.

Nikolaidis, N.P., Dobbs, G.M., Lackovic, J.A., 2003. Arsenic removal by zero-valent iron: field, laboratory and modeling studies. Water Res. 37 (6), 1417–1425.

Noble, R.D., Terry, P.A., 2004. Principles of Chemical Separation with Environmental Applications . Cambridge University Press, Cambridge, UK, pp. 207–212.

Pierce, M.L., Moore, C.B., 1982. Adsorption of arsenite and arsenate on amorphous iron hydroxide from dilute aqueous solution. Water Res. 16, 1247–1253.

Rahman, M.M., Sengupta, M.K., Chowdhury, U.K., Lodh, D., Das, B., Ahamed, S., et al., 2006. Arsenic contamination incidents around the world. In: Naidu, R., Smith, E., Owens, G., Bhattacharya, P., Nadebaum, P. (Eds.) Managing Arsenic in the Environment: From Soil to Human Health. CSIRO publishing, Australia, pp. 3–30.

Ranjan, M.B., Soumen, D., Sushanta, D., Chand De, G.U., 2003. Removal of arsenic from groundwater using crystalline hydrous ferric oxide (CHFO). Water Qual. Res. J. Can. 38 (1), 193–210.

Raven, K.P., Jain, A., Loeppert, R.H., 1998. Arsenite and arsenate adsorption on ferrihydrite: kinetics, equilibrium, and adsorption envelopes. Environ. Sci. Technol. 32, 344–349.

Redman, A.D., Macalady, D.L., Ahmann, D., 2002. Natural organic matter affects arsenic speciation and sorption onto hematite. Environ. Sci. Technol. 36 (13), 2889–2896.

Ringbom, A., 1963. Complexation in Analytical Chemistry. Interscience–Wiley, New York.

Roberts, L.C., Hug, S.J., Ruettimann, T., Khan, A.W., Rahman, M.T., 2004. Arsenic removal with iron(II) and iron(III) in waters with high silicate and phosphate concentrations. Environ. Sci. Technol. 38, 307–315.

Rosenblum, E., Clifford, D., 1984. The equilibrium arsenic capacity of activated alumina. U.S. Environ. Prot. Agency (Rep.) EPA-600, S2-83-107.

Saha, B., Bains, R., Greenwood, F., 2005. Physicochemical characterization of granular ferric hydroxide (GFH) for arsenic(V) sorption from water. Sep. Sci. Technol. 40 (14), 2909–2932.

Sarkar, S., Gupta, A., Biswas, R.K., Deb, A.K., Greenleaf, J.E., SenGupta, A.K., 2005. Well-head arsenic removal units in remote villages of Indian subcontinent: field results and performance evaluation. Water Res. 39 (10), 2196–2206.

Schecher, W.D., McAvoy, D.C., 1998. MINEQL + : A Chemical Equilibrium Program for Personal Computers, User's Manual, Version 4.0. Environmental Research Software, Hallowell, ME.

Scott, K.N., Green, J.F., Do, H.D., McLean, S.J., 1995. Arsenic removal by coagulation. J. Am. Water Works Assoc. 87 (4), 114–126.

Shen, Y.S., 1973. Study of arsenic removal from drinking water. J. Am. Water Works Assoc. 65 (8), 543–548.

Shih, M.C., 2005. An overview of arsenic removal by pressure driven membrane processes. Desalination 172, 85–97.

Siegel, M., McConnel, P., Everett, R., Kirby, C.,2006. Evaluation of innovative arsenic treatment technologies: the arsenic water technology partnership vendors forums summary report. Sandia National Laboratories, Albuquerque, NM, Sandia Report SAND2006-5423.

Singh, T.S., Pant, K.K., 2004. Equilibrium, kinetics and thermodynamic studies for adsorption of As(III) on activated alumina. Sep. Purif. Technol. 36 (2), 139–147.

Singh, D.B., Prasad, G., Rupainwar, D.C., 1996. Adsorption technique for the treatment of As(V)-rich effluents. Colloids Surf. A Physicochem. Eng. Asp. 111 (1–2), 49–56.

Singh, P., Singh, T.S., Pant, K.K., 2001. Removal of arsenic from drinking water using activated alumina. Res. J. Chem. Environ. 5, 25–28.

Smedley, P.I., Kinniburgh, D.G., 2002. A review of the source, behaviour and distribution of arsenic in natural waters. Appl. Geochem. 17, 517–568.

Smedley, P.L., Kinniburgh, D.G., 2005. Sources and behaviour of arsenic in natural water. Chapter 1 in United Nations Synthesis Report on Arsenic in Drinking Water, 2005.

Smedley, P.L., Nicolli, H.B., Macdonald, D.M.J., Barros, A.J., Tullio, J.O., 2002. Hydrogeochemistry of arsenic and other inorganic constituents in groundwaters from La Pampa, Argentina. Appl. Geochem. 17 (3), 259–284.

Sorg, T.J., Logsdon, G.S., 1978. Treatment technology to meet the interim primary drinking water regulations for inorganics: part 2. J. Am. Water Works Assoc. 70 (7), 379–393.

Stamer, C., Nielsen, K.A., 2000. The removal of arsenic: arsenic removal without sludge generation. Water Supply 118 (1), 625–628.

Stephen, J.H., Canonica, L., Wegelin, M., Gechter, D., Gunten, V.U., 2001. Solar oxidation and removal of arsenic at circumneutral pH in iron containing water. Environ. Sci. Technol., es001551s. (ASAP web edition).

Su, C., Puls, R.W., 2001a. Arsenate and arsenite removal by zerovalent iron: kinetics, redox transformation, and implications for in situ groundwater remediation. Environ. Sci. Technol. 35, 1487–1492.

Su, C., Puls, R.W., 2001b. Arsenate and arsenite removal by zerovalent iron: effects of phosphate, silicate, carbonate, borate, sulfate, chromate, molybdate, and nitrate, relative to chloride. Environ. Sci. Technol. 35, 4562–4568.

Su, C., Puls, R.W., 2003. In situ remediation of arsenic in simulated groundwater using zerovalent iron: laboratory column tests on combined effects of phosphate and silicate. Environ. Sci. Technol. 37, 2582–2587.

Su, C., Puls, R.W., 2004. Significance of iron(II,III) hydroxycarbonate green rust in arsenic remediation using zerovalent iron in laboratory column tests. Environ. Sci. Technol. 38, 5224–5231.

Swedlund, P.J., Webster, J.G., 1999. Adsorption and polymerization of silicic acid on ferrihydrite, and its effect on arsenic adsorption. Water Res. 33, 3413–3422.

Technical Performance Monitoring—An Interim Report, 2008. Bangladesh Environmental Technology Verification—support to Arsenic Mitigation (BETV-SAM). Bangladesh Council of Scientific and Industrial Research, Dhaka, Bangladesh. p. 8.

Thirunavukkarasu, O.S., Viraghavan, T., Suramanian, K.S., 2003a. Arsenic removal from drinking water using iron-oxide coated sand. Water Air Soil Pollut. 142, 95–111.

Thirunavukkarasu, O.S., Viraraghavan, T., Subramanian, K.S., 2003b. Arsenic removal from drinking water using granular ferric hydroxide. Water S.A. 29 (2), 161–170.

Thirunavukkarasu, O.S., Viraraghavan, T., Subramanian, K.S., Chaalal, O., Islam, M.R., 2005. Arsenic removal in drinking water—impacts and novel removal technologies. Energy Sources 27, 209–219.

USEPA., 1999. Technologies and costs for removal of arsenic from drinking water, draft report, EPA-815-R-00-012, Washington, D.C.

USEPA, 2000. Arsenic Treatment Technology Evaluation Handbook for Small System EPA 816-R-03-014. USEPA, Washington, D.C.

Vaishya, R.C., Gupta, S.K., 2002. Modeling arsenic(V) removal from water by sulfate modified iron-oxide coated sand (SMIOCS). J. Chem. Technol. Biotechnol. 78, 73–80.

Vaishya, R.C., Gupta, S.K., 2004. Modeling arsenic(III) adsorption from water by sulfate modified iron-oxide coated sand (SMIOCS). Sep. Sci. Technol. 39 (3), 645–666.

Vaishya, R.C., Gupta, S.K., 2006. Arsenic (V) removal by sulfate modified iron oxide-coated sand (SMIOCS) in a fixed bed column. Water Qual. Res. J. Can. 41 (2), 157–163.

Vatutsina, O.M., Soldatov, V.S., Sokolova, V.I., Johann, M.B., Weissenbacher, A., 2007. A new hybrid (polymer/inorganic) fibrous sorbent for arsenic removal from drinking water. React. Funct. Polym. 67, 184–2001.

Wang, S., Mulligan, C.N., 2006. Occurrence of arsenic contamination in Canada: sources, behavior and distribution. Sci. Total Environ. 366, 701–721.

Waychunas, W.A., Rea, B.A., Fuller, C.C., Davis, J.A., 1993. Surface chemistry of ferrihydrite, part I: EXAFS studies of geometry of co precipitated and adsorbed arsenate. Geochim. Cosmochim. Acta. 57, 2251–2270.

Weber, W.J., Morris, J.C., 1963. Kinetics of adsorption on carbon from solution. Proc. Am. Soc. Civ. Eng., J. Sanit. Eng. 89(SA2), 31–59.

Wickramasinghe, S.R., Binbing, H., Zimbron, J., Shen, Z., Karim, M.N., 2005. Arsenic removal by coagulation and filtration: comparison of groundwaters from the United States and Bangladesh. Desalination 169, 231–244.

Wilkie, A.J., Hering, J., 1996. Adsorption of arsenic onto hydrous ferric oxide: effects of adsorbate/adsorbent ratios and co-occurring solutes. Colloids Surf. A Physicochem. Eng. Asp. 107, 97–110.

Yavuz, C.T., Mayo, J.T., Yu, W.W., Prakash, A., Falkner, J.C., Yean, S., et al., 2006. Lowfield magnetic separation of monodisperse Fe_3O_4 nanocrystals. Science 314 (5801), 964–967.

Yuan, C., Lien, H.L., 2006. Removal of arsenate from aqueous solution using nano-scale iron particles. Water Qual. Res. J. Can. 41 (2), 210–215.

Zeng, L., 2004. Arsenic adsorption from aqueous solution on an Fe(III)–Si binary oxide adsorbent. Water Qual. Res. J. Can. 39, 269–277.

Zhang, F.-S., Itoh, H., 2005. Iron oxide-loaded slag for arsenic removal from aqueous system. Chemosphere 60 (3), 319–325.

Zhang, J., Robert, S., 2005. Slow adsorption reaction between arsenic species and goethite (α-FeOOH): diffusion or heterogeneous surface reaction control. Langmuir 21 (7), 2895–2901.

Zhang, N., Blowers, P., Farrell, J., 2005. Evaluation of density functional theory methods for studying chemisorption of arsenite on ferric hydroxides. Environ. Sci. Technol. 39 (13), 4816–4822.

Zhang, J.S., Stanforth, R.S., Pehkonen, S.O., 2007. Effect of replacing a hydroxyl group with a methyl group on arsenic(V) species adsorption on goethite (α-FeOOH). J. Colloid Interf. Sci. 306 (1), 16–21.

Zouboulis, A., Katsoyiannis, I., 2002. Removal of arsenates from contaminated water by coagulation-direct filtration. Sep. Sci. Technol. 37, 2859–2873.

A

Note: The letter "f" following a page number means figure; the letter "t" following a
page number means table.

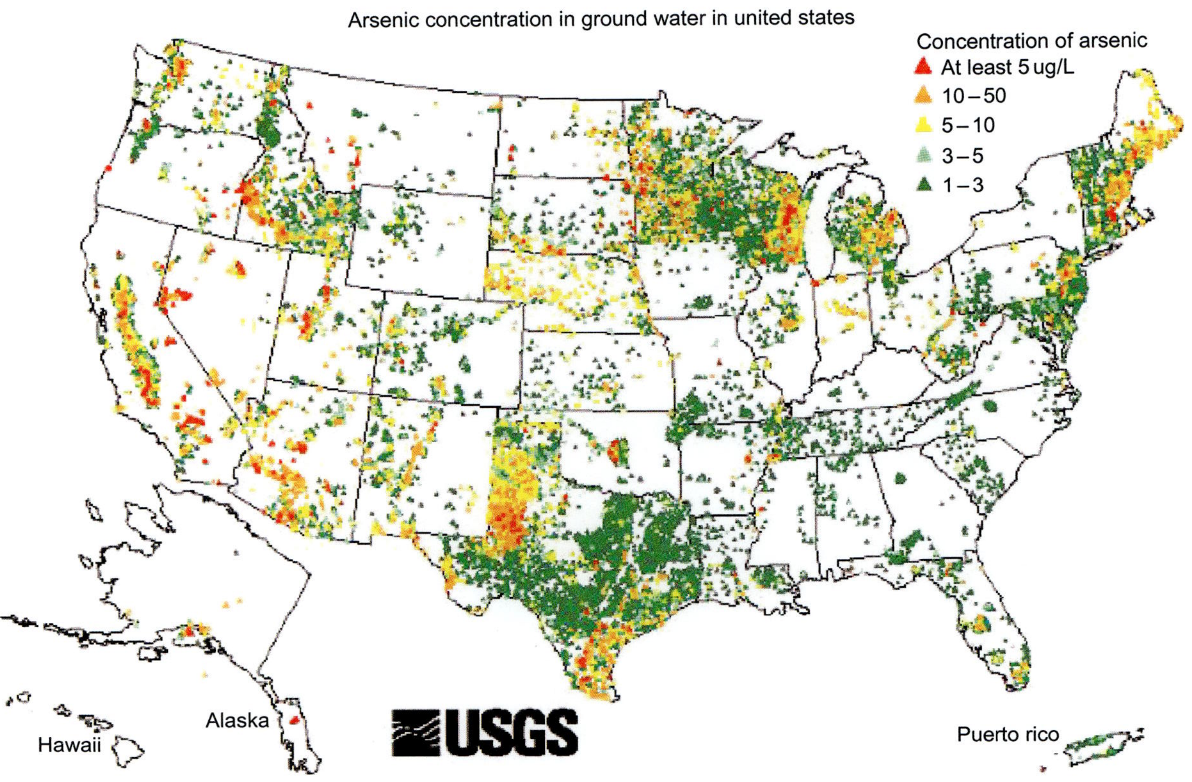

PLATE 1   Arsenic concentration in groundwater in United States. (see chapter 2, figure 1)

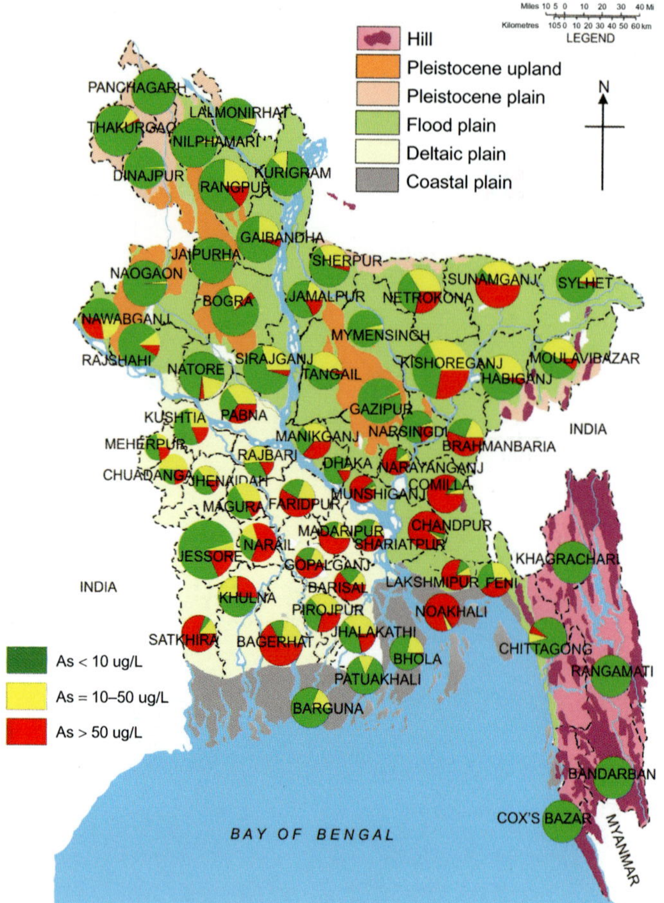

PLATE 2 Four geomorphological regions of Bangladesh and the status of arsenic contamination in groundwater of 64 districts in Bangladesh. (see chapter 5, figure 11)

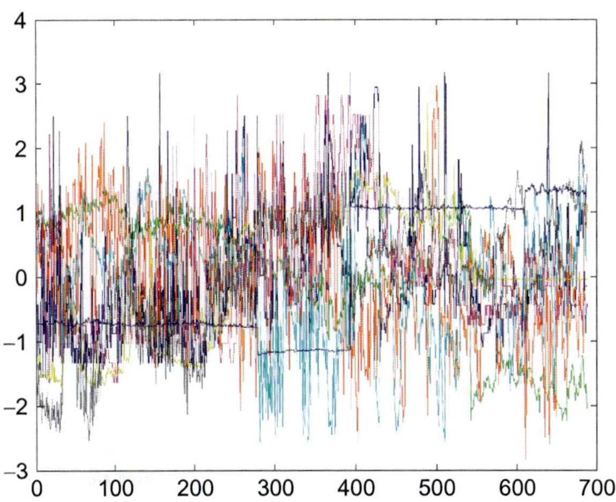

PLATE 3 Real world water quality is highly variable. Graph courtesy of Hach HST. (see chapter 15, figure 1)

PLATE 4 The interior of many distribution system pipes can present a very hostile environment for sensor department. Photos courtesy—US Army Corps of Engineers. (see chapter 15, figure 2)

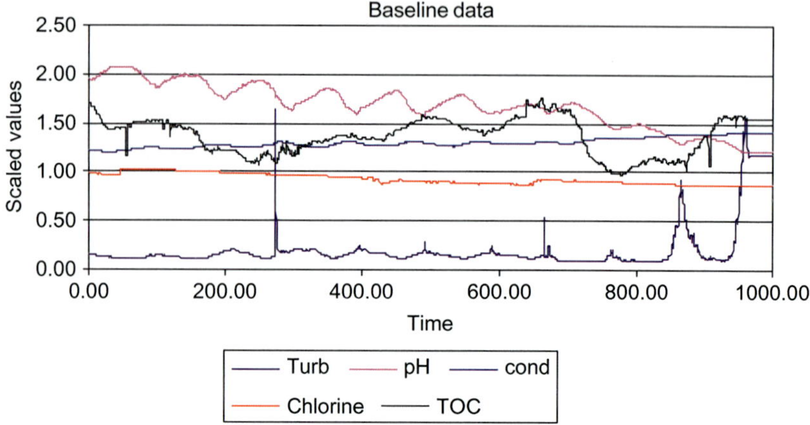

PLATE 5   Actual real-world baseline date. The variability in bulk water parameters that is common in the distribution system requires that any algorithm should contain a workable baseline estimator. Graph courtesy—Hach HST. (see chapter 15, figure 9)

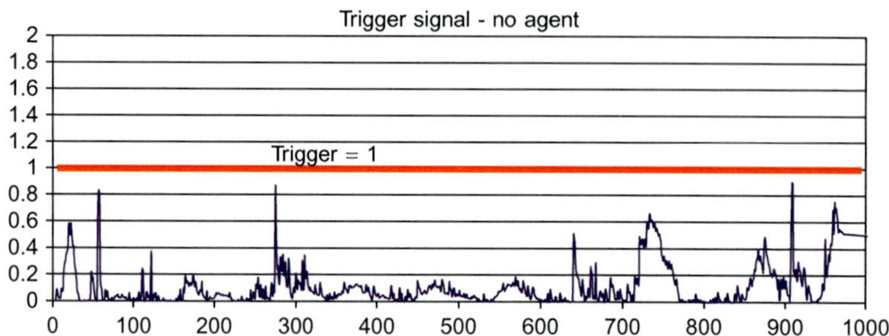

PLATE 6   The noisy data from Figure 9 become easy to interpret when processed through an algorithm. In this care, no significant events above a threshold of one are occurring; therefore, no trigger is initiated. (see chapter 15, figure 10)

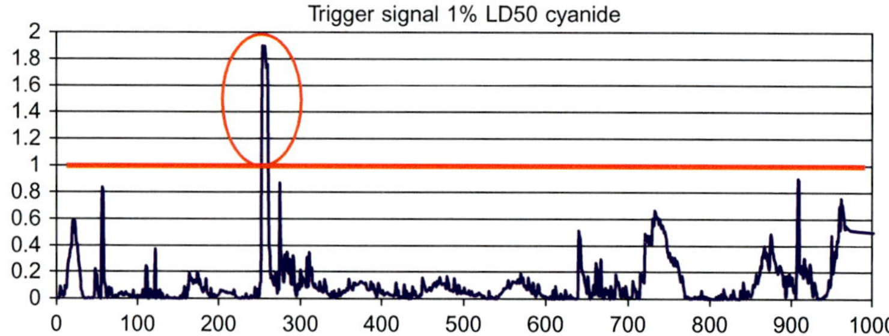

PLATE 7   The algorithm system's ability to differentiate and trigger on low levels of contaminants against a noisy background is shown. (see chapter 15, figure 11)